Comprehensive Security for an Emerging India

Comprehensive Security for an Emerging India

Air Vice Marshal **Kapil Kak** AVSM, VSM (Retd)
Additional Director, Centre for Air Power Studies
Editor

Introduction by
Shri **K. Subrahmanyam**
Former Director
Institute for Defence Studies and Analyses, New Delhi

KW Publishers Pvt Ltd
New Delhi

in association with

Centre for Air Power Studies
New Delhi

The **Centre for Air Power Studies** is an independent, non-profit, academic research institution established in 2002 under a registered Trust to undertake and promote policy-related research, study and discussion on the trends and developments in defence and military issues, especially air power and the aerospace arena, for civil and military purposes. Its publications seek to expand and deepen the understanding of defence, military power, air power and aerospace issues without necessarily reflecting the views of any institution or individuals except those of the authors.

Jasjit Singh
Director
Centre for Air Power Studies
P-284, Arjan Path
Subroto Park
New Delhi 110010

Tele: (91-11) 25699131
E-mail: office@aerospaceindia.org

Published in India by
Kalpana Shukla
KW Publishers Pvt Ltd
NEW DELHI: 4676/21, First Floor, Ansari Road, Daryaganj, New Delhi 110002
T: +91.11.23263498 / 43528107 E: mail@kwpub.in / knowledgeworld@vsnl.net
MUMBAI: 15 Jay Kay Industrial Estate, Linking Road Extn., Santacruz (W), Mumbai 400054
T: +91.22.26614110 / 26613505 E: mumbai@kwpub.in

ISBN 13: 978-93-80502-30-4

Printed at Aegean Offset Printers, Noida

Contents

Contributors

Shri **K. Subrahmanyam**, the country's most prominent strategic affairs analyst, who retired as the Secretary, Defence Production, Government of India, is also a former Director of the New Delhi-based Institute for Defence Studies and Analyses (IDSA). Considered a strong proponent of *realpolitik*, Subrahmanyam is often referred to as the doyen of India's strategic affairs community, and as the premier ideological champion of India's nuclear development.

Air Commodore **Jasjit Singh** AVSM, VrC, VM (Retd), awarded the Padma Bhushan for a lifetime's contribution to national defence and security and former Director of IDSA, is currently Director, Centre for Air Power Studies, New Delhi. He is the author of a number of books, including *Air Power in Modern Warfare, India's Defence Spending, Defence from the Skies, The Icon: Marshal of the Indian Air Force Arjan Singh.*

Dr **Arvind Virmani** was until 2009 senior economic advisor, Ministry of Finance, Government of India. Before joining the government, Dr. Virmani was senior economist in the Research Department of the World Bank, and acting chief of the Public Economics Division. He served as Director of the Punjab National Bank and Allahabad Bank, and as trustee of the Unit Trust of India. Dr. Virmani has written widely on various aspects of economic reforms, legal and institutional reforms, the social sector, agriculture, real estate, and the media. Dr. Virmani holds a Ph.D. in economics from Harvard University, Cambridge, Massachusetts. He is currently India's representative at the International Monetary Fund, Washington.

Dr **Sanjaya Baru** is a member of the Board of the Centre for Policy Research, New Delhi, and of the Committee on Economic Security, Confederation of Indian Industries (CII). He is also a member of the CII-Aspen Strategy Group Dialogue on India-US Relations, and has been a participant in 'Track-Two Dialogues' between India and China, Russia, Japan, European Union, Singapore. He is currently Editor, *Business Standard.*

Gopalaswami Parthasarathy is a career Foreign Service Officer who retired from Service on May 31, 2000. He has served as Ambassador to Myanmar, 1992-95, High Commissioner to Australia, 1995-98, High Commissioner to Pakistan, 1998-2000, and High Commissioner to Cyprus, 1990-92.

Mr. Parthasarathy is presently Visiting Professor in the Centre for Policy Research in New Delhi. He is also a Senior Fellow at the Centre for Strategic and International Studies and a member of the Executive Committee of the Centre for Air Power Studies in New Delhi. His main areas of interest are developments in India's neighbourhood and issues of economic integration, energy and national security and terrorism. A widely read columnist, he writes for a number of newspapers and news agencies in India and abroad on foreign policy and national security issues.

N. S. Sisodia is currently Director General, Institute for Defence Studies and Analyses. He retired from the government in 2005 and has worked as Secretary, Ministries of Defence and Finance. He has also been Vice Chancellor, Mohanlal Sukhadia University, Udaipur, and Member, National Security Advisory Board.

Vinod K. Misra joined the Indian Defence Accounts Service (IDAS) in 1969 and worked with distinction in several key assignments: as Officer on Special Duty with the first ever Committee on Defence Expenditure (1990-91); Principal Controller, Defence Accounts (Pensions), Allahabad, as Additional Controller General of Defence Accounts in the Headquarters office in Delhi from August 2002 to November 2005, where he pioneered a major initiative titled "Mission Excel IT" for comprehensive computerisation of the Defence Accounts Department. He retired as Secretary Defence Finance.

Dr **Manpreet Sethi** is a Senior Fellow at the Centre for Air Power Studies, New Delhi, and has written extensively in national and international journals on nuclear power, proliferation and disarmament. She is the author of *Argentina's Nuclear Policy* and *Nuclear Strategy: India's March Towards Credible Deterrence,* and co-author of *Nuclear Deterrence and Diplomacy*.

General **V. P. Malik** served as Chief of the Army Staff of the Indian Army from October 1997 to September 2000. Concurrently, he was Chairman, Chiefs of Staff Committee, from January 1999 to September 2000. He planned, coordinated and oversaw execution of Operation Vijay to successfully defeat Pakistan's attempted intrusion in the Kargil Sector in 1999. After retirement, he has been a member of the National Security Advisory Board. Currently, he is President of the ORF Institute of Security Studies. General Malik has addressed several military and non-military universities and institutions in Europe, the US and Asia, participated in workshops, seminars and Track II level conferences, and authored several papers on national security, defence planning and military strategy. A regular contributor to Indian newspapers and magazines on these issues, he has authored *Kargil War: From Surprise to Victory*.

Air Marshal **T. M. Asthan**a was commissioned as a fighter pilot in October 1963. He has commanded a fighter squadron in Assam and, has commanded two Air Force stations in the Western sector, one as a Group Captain and one as an Air Commodore. He was nominated to undergo the course at the Royal College of Defence Studies (RCDS) in London for one year in 1995. Assistant Chief of Operations, ACAS (Ops) at Air Headquarters in January 1999, he held this post through the Kargil Ops and was also ACAS (Personnel) at Air Headquarters. Air Officer Commanding-in-Chief (AOC in C) of Southern Air Command during 2002, he was the first Commander-in-Chief of the Strategic Forces Command during 2003-04.

Admiral **Arun Prakash** retired as India's 20th Naval Chief and Chairman, Chiefs of Staff Committee, in end-2006. An aviator by specialisation, during a career spanning over 40 years, he saw wide and varied service in warships and aviation units of the Indian Navy. In flag rank, he commanded the Eastern Fleet, National Defence Academy, Andaman & Nicobar Command, and Western Naval Command. A graduate of the IAF Test Pilots School, the Defence Services Staff College and the US Naval War College, he writes on strategic and defence issues and has published a book of speeches and writings titled *From the Crow's Nest.* Admiral Prakash is currently a member of India's National Security Advisory Board and Chairman of the National Maritime Foundation.

Shri **Bahukutumbi Raman** was Additional Secretary, Cabinet Secretariat, Government of India. He headed the Counter-Terrorism Division of the R&AW from 1988 to 1994. He was a member of the National Security Advisory Board of the Government of India from July 2000 to December 2002 and a member of the Special Task Force for revamping the intelligence apparatus set up by the Government of India in 2000. He is presently Director, Institute For Topical Studies, Chennai and is also associated with the South Asia Analysis Group (SAAG), New Delhi, and the Chennai Centre For China Studies. He is an honorary Editorial Adviser to the *Indian Defence Review* published from New Delhi by the Lancer Publishers. He is the author of four books: *Intelligence: Past,Present and Future , A Terrorist State As A Frontline Ally, The Kaoboys of R&AW—Down Memory Lane* and *Terrorism— Yesterday, Today & Tomorrow.*

Prakash Singh was Director General of the Border Security Force and also Director General of Police (DGP) UP and DGP Assam. He was awarded the Padmashri for his contribution to national security. He is also the architect of police reforms in the country. Books written by him include: *Nagaland, The*

Naxalite Movement in India, Disaster Response in India and *Kohima to Kashmir: On the Terrorist Trail.*

Wajahat Habibullah did his schooling at the Doon School, Dehra Dun, and Masters in History from St Stephens College, University of Delhi. He was a Lecturer in History at Stephens College before joining the Indian Administrative Service (IAS) in 1968. He has held various positions in the Government of India, including Secretary,Ministry of Panchayati Raj, Secretary, Ministry of Textiles, Director, Lal Bahadur Shastri National Academy of Administration, Mussoorie, Vice Chairman and Chief Executive of J&K Lakes & Waterways Development Authority, Srinagar. He has also been Officer on Special Duty, Ministry of Urban Development and Joint Secretary in the Prime Minister's Office where he dealt with the special programmes for poverty alleviation and relief programmes, environmental conservation and with overseeing the Ministries concerned with development.

Shebonti Ray Dadwal is a Research Fellow with the Institute for Defence Studies and Analyses, specialising on energy security and climate change-related issues. She has worked as Senior Editor in *The Financial Express* and has also served as Deputy Secretary at the National Security Council Secretariat. She has presented several papers in national and international conferences and written several peer-reviewed articles and papers on her area of work. Ms Ray Dadwal was awarded the FCO Chevening Fellowship on the Economics of Energy in April 2009. In 2002, she published a book, *Rethinking Energy Security in India* and is now writing her second book on the politics of energy and climate change.

Ashok Parthasarathi is a physicist, electronics engineer and social scientist and involved in S&T policy research. His research degrees in the former areas (specialising in radio astronomy) were taken under Nobel Laureate Martin Ryle at Cambridge University, UK, and in the latter at MIT, USA. He has worked as Special Assistant to the late Vikram Sarabhai in the Department of Atomic Energy and then as Science Advisor to the late Prime Minister Indira Gandhi for several years. He is the only person to have held such a position. He has been Secretary of the Department of Electronics, Non-Conventional Energy and Scientific and Industrial Research. He has published over 100 papers on S&T policy, planning and management and three books in those areas, the most recent of which was *Technology at the Core: S&T with Indira Gandhi.*

M. S. Swaminathan is a Fellow of many of the leading scientific academies of India and the world, including the Royal Society of London and the US National Academy of Sciences. He has received 58 honorary doctorate degrees from universities around the world. He currently holds the UNESCO Chair in Ecotechnology at the M. S. Swaminathan Research Foundation in Chennai (Madras), India.

Chandrashekhar Dasgupta served as Ambassador to China from 1993 to1996 and as Ambassador to the European Community (EC) from 1996 to 2000. His interests include diplomatic history and global environmental issues. He has been involved in international negotiations on climate change, and is currently a member of the Prime Minister's Council on Climate Change. He is a Distinguished Fellow at the Tata Energy and

Resources Institute (TERI) and a member of the UN Committee on Economic, Social and Cultural Rights. He is the author of *War and Diplomacy in Kashmir, 1947-48*.

Ramaswamy R. Iyer, formerly Secretary, Water Resources, in the Government of India, was Research Professor at the Centre for Policy Research (CPR), New Delhi, where he worked on water-related issues, and in particular on cooperation on river waters by India, Nepal and Bangladesh (1990-99). He has been a member of many government committees and commissions, including the National Commission on Integrated Water Resources Development Plan (1997-99), and is the Chairman of a Task Force on Natural Resources, Environment, Land, Water and Agriculture, set up by the Commission on Centre-State Relations. He has published numerous articles and papers, edited /co-edited some books, and authored: *A Grammar of Public Enterprises,* 1991; *WATER: Perspectives, Issues, Concerns,* 2003; *Towards Water Wisdom: Limits, Justice, Harmony* 2007; *Water and the Laws in India*.

Air Vice Marshal **Kapil Kak** AVSM VSM (Retd), a veteran of two wars – 1965 and 1971 – has, post-retirement, served as Deputy Director, Institute for Defence Studies and Analyses, New Delhi and Advisor (Strategic Studies), University of Jammu, apart from representing India in a number of Track II conferences on international security issues. A regular commentator on security issues in the electronic media and a prolific contributor of articles to a variety of journals, he co-edited the book *India and Pakistan: Pathways Ahead.* Air Vice Marshal Kapil Kak is currently Additional Director, Centre for Air Power Studies, New Delhi.

Preface

It would be useful to recall that Jawaharlal Nehru in his many writings and speeches enunciated the idea of comprehensive security for India decades before the Copenhagen School in 1983 argued for the inclusion of human-societal, non-military, economic and environmental dimensions in the security construct. Nehru and other freedom fighters also did not fail to perceive the key linkage between peace, comprehensive security and comprehensive national power. Today, the inextricably linked concept of cooperative security has also acquired far greater traction, as new global threats — quite different from those in the past — can only be resolutely countered through substantial cooperation of the international community. Thus, the definition of national security has not only acquired clarity but also more expansive scope, even as change has become inevitable, uncertain and discontinuous.

Significantly, the international community has not viewed India's emergence with disfavour, unlike past historical parallels when the rise of Germany, Russia, the US, France and Japan generated widespread anxieties. The recent rise of China has also engendered similar apprehensions. Looking back two decades, India's track- record could perhaps have served as a contributory factor. Sustained economic reform and liberalisation, and over 7 per cent average Gross Domestic Product (GDP) growth, addressing nuclear insecurities through Pokhran II, utmost restraint in the face of Pakistan's two-decade long state-sponsored terrorism, especially during Kargil, the terrorist attack on Parliament and the 26/11 Mumbai terrorist assault, have doubtless compelled international attention.

Uncompromising commitment to pluralism, democratic stability and secular ideals — despite the national blot in Gujarat in 2002 — and pursuit of creative and inclusive strategies on poverty alleviation and removal of deep socio-economic inequities have struck a positive chord in the international community. But, as some leading lights of India's strategic community and renowned experts on non-traditional security dimensions have repeatedly highlighted in this book, there have been abysmal failures across a wide spectrum. These have arisen due to systemic, decision-making and organisational inadequacies, and had an impact on the internal, external and non-traditional comprehensive security dimensions. If India has to effectively tackle the many security dangers that loom large in the decades ahead, it must squarely meet the identified shortfalls.

This volume has sought to objectively evaluate the headwinds of the comprehensive security challenges India would confront in its flight path to be a peaceful power of consequence. Issues of economics shaping relations in a strategic environment in which globalisation, economic growth and national security stand intertwined and how India's multi-vector foreign policy fits into such a framework have been addressed insightfully. In the decades ahead, China remains a cause for serious concern, as does Pakistan in the medium term. More importantly, the challenge from the two together has a collusive security dimension. At the same time, an emerging India, the US and China are projected to be the leading players in a new Asia that is taking shape in which simultaneity in competition and cooperation are perceived to be the primary *leitmotif*.

The challenge from Pakistan is unlikely to attenuate over possibly the next two decades. Given its tortuous complexity, this would call for ingenious non-military strategic leaps, in conjunction with leading international players, to reengineer Pakistan's internal DNA and move it onto an entirely new plane. The challenge for India's political leadership would lie in transforming the extant inflamed public opinion in this direction. India must also stay resolute in sustaining its strategic developmental footprint in Afghanistan. Trend-lines suggest other South Asian countries having increasingly perceived the value of economic integration with an emerging India that has a sharply rising economic trajectory.

Contributors to this volume, all eminent personages in their fields of expertise, have incisively analysed multiple facets of security strategy, foreign policy, defence planning and resource management, nuclear strategy and role of the Army, aerospace power and maritime strategy in conventional deterrence as also in pursuit of safeguarding national interests. Internally, India's critical faultlines run along continued socio-economic inequities and near absence of governance that drive internally inspired terrorism and the deeply worrisome left-wing extremism. These aspects have also been comprehensively analysed in the book.

Experts have also drawn attention to the challenges and opportunities in energy security relating to hydrocarbons, nuclear power and renewable energy and in right to information, food security, and climate change and water security. Being key stakeholders, they have put forth insightful recommendations as independent research inputs for public policy on comprehensive security. Due to unavoidable circumstances, two planned contributions on the linkage between technology and security, and the role of education in enhancing India's comprehensive security in the decades ahead remain notable omissions.

When the idea of putting together this book occurred to me two years ago, the first person to be sounded was naturally Air Commodore Jasjit Singh, my Guru, philosopher and senior colleague over a near decade and a half of very close association — earlier at the Institute for Defence Studies and Analyses and since 2002 at the Centre for Air Power Studies. He was instantly receptive in his support for the project and the book design, and offered invaluable advice all along. Mr. K. Subrahmanyam, the doyen of Indian strategic thought, not only agreed to write the introduction but also was, as always, thoughtfully generous with his inspiring support and sagacious advice throughout the time this volume was in preparation.

As it happened, both Mr. K. Subrahmanyam and Air Commodore Jasjit Singh ran into serious health setbacks, which caused a major disruption in the book publication schedule. Mercifully, over time, both regained their health and were able to make their promised contributions.

My profound gratitude goes to them both as also to the other contributors to this volume. It is to the credit of the contributors, renowned experts in their fields, and many of them close friends, who all along showed forbearance and patiently bore the inordinate delay. The views expressed by the contributors as also by the Editor are their own and not of the organisation they represent. Any flaws or mistakes in the preparation of this volume are entirely mine.

It is a pleasure to deeply acknowledge the thoughtfulness and moral support of Air Chief Marshal O.P. Mehra. My gratitude goes to Air Marshal Bharat Kumar for his readily forthcoming assistance in the eventual output of this volume. In the end, my thanks to Ms Kalpana Shukla and Mr Jose Mathew of KW Publishers for their enthusiastic and unstinted assistance in the publication of this book under time pressures.

March 2010 **Kapil Kak**

Introduction

☐ K. Subrahmanyam

It is extremely bold of Air Vice Marshal Kapil Kak to have undertaken this venture at this juncture when there is so much uncertainty about the future global order. A world order which got inaugurated with the Great Depression and the American New Deal, with the hegemonic leadership of the international system shifting from the British Empire to the US, is coming to an end, yielding place to a new order, the dimensions and characteristics of which cannot be foreseen at this stage. To quote Lord Tennyson's words put in the mouth of a dying King Arthur "The old order changeth yielding place to new." The global economic crisis informs us that we are entering a new world order and the only certainty about that new order is likely to be that it will not be a linear extrapolation of the order now passing away.

Though it is difficult to be certain that the present financial crisis will continue to be tackled with a perfectly harmonious consensus displayed among the major economies of the world during the London G-20 Summit, there is no disagreement among them that the crisis can be solved only through global cooperation and not through individual powers adopting "beggar thy neighbour" policies. There is acute consciousness even among the US and UK leaderships that the globalisation of the last two decades was not effectively monitored and regulated in their countries and that triggered off the present crisis, and there is determination not to go back to the pre-globalisation era. The crisis highlighted how closely the nations of the world, especially the industrialised nations, have become interdependent and how much shrinkage in demand in certain nations can throw millions of workers out of employment in other major manufacturing nations. If so much of economic disruption and social turmoil can result only on account of market factors without any violent disruption in international transportation or use of nuclear weapons on population centres, imagine the consequences of a war among major industrial powers, possessing nuclear weapons.

While World Wars I and II were the results of politico-military establishments not being in a position to anticipate the full implications of powerful industrial nations fighting total wars after mobilisation of their entire capacities, there are now no risks of leaderships of nuclear weapon powers initiating wars without the knowledge of the consequences of such wars. It will, therefore, be a reasonable assumption that in any new world order, the nuclear weapons will not play the role they did in the security calculations of nations in the first fifty years of the nuclear era. President Obama, in his Prague speech, has accepted the moral responsibility of the US as the only nation having used nuclear weapons to take the lead to move towards a nuclear weapon free world.

The nature of war and the popular attitude towards war have undergone radical changes. Compare the times of 1914-18 when on one day in the battlefield of Flanders, tens of thousands of fatalities occurred or 1942-45 when one air raid over Germany or Japan inflicted similar order of casualties to the present day when a score or two of casualties in Iraq or Afghanistan is considered unacceptable. There is reluctance among the industrial nations to accept casualties in wars not meant for the immediate defence of their homelands.

During World War II, populations of occupied countries could be terrorised into abject submission. Hitler's Germany could harness most of the European workforce to contribute to the war production of the Wehrmacht. Today, invading countries find populations offering resistance through suicide bombers, and occupation of countries is more costly in men and material than defeating an army and taking over a country. Nazi and Japanese atrocities (war crimes) never became fully public till after the war. Today, in Gaza, Bosnia, Darfur or elsewhere, war crimes come to the notice of the world almost immediately. All these factors make even conventional war no longer as useful an instrument of political policy as it was considered to be in the 19th and first part of the 20th century.

The Cold War clearly demonstrated that political ideologies cannot be enforced or sustained by force. There is a lot of analysis that China is able to sustain its one-party system only on the basis of 8 percent economic growth rate and if that growth rate were to fall below 8 percent, there is likely to be social and political turmoil in China. China has given up Marxist ideology and today the veneer of Marxism is only a justification for a one-party authoritarian state.

There is a widely held view that the 21st century will be a knowledge-based century, and the power, affluence and prestige of nations will be largely determined on the knowledge generated by nations. That will be a function of people engaged in high quality research and investments made on knowledge generation by societies. If this proposition is valid then it is a logical question whether knowledge-based societies can sustain themselves as authoritarian ones.

The populations all over the world are ageing. Some leading nations are at a more advanced stage in age profile than some other nations. As nations age, the proportion of younger population of working age having to support the non-working population living longer becomes smaller, with adverse impact on Gross National Product. The health care costs of nations also go up and a steady decline in population sets in. That, in turn, calls for immigration of alien population into the countries and, in many cases, may lead to heterogenisation of societies. Japan, China, Russia and European Union are very vulnerable to the effects of ageing of populations while in the case of India and the US, the phenomenon will set in after a few decades as their populations are relatively young compared to those of the other four listed .

Given these possibilities, an assessment of the security predicament of nations, particularly large nations like India, should not be on the basis of linear extrapolation of past historical experiences of the 19th and 20th centuries. It is not the contention here that all security issues of the kind faced by nations, especially the large emergent ones in previous centuries, have faded away. No. Many of the old security threats have been replaced by new ones. Terrorism has now become a strategic instrumentality in asymmetric wars against states by faith-inspired extremists, secessionist ethnic groups, minorities with grievances, and organised crime. Failed or failing states often provide convenient safe havens for such terrorist groups and thereby they themselves become a security problem. Sects that interpret religious faith as their manifest destiny to dominate humanity constitute a potent threat to the peace and security of nations. Once terrorism becomes a preferred strategy in asymmetric war, then weapons of mass destruction are likely to become instruments of choice to execute terrorist strategy and, therefore, their spread becomes a potential threat.

If comprehensive security is defined to include avoidance of shortage of basic requirements of a country's population, the provision of clean air, clean water and healthy surroundings, then environmental security becomes a matter of very high priority. So does provision of adequate food, health care, education, energy, employment, old age care and good governance. Such a comprehensive security cannot be aimed at, and sustained by, any single country on its own. It can be done only as part of a cooperative effort by significant sections of the international community.

In earlier periods, nations tried to occupy territories to get additional resources – such as additional food crops, energy resources like oil wells, minerals, etc. That is no longer feasible. Additional resources for expanding populations will have to be generated in future by new research and development. Today, the nations of the world are attempting collective research on fusion energy, on various safe nuclear reactors and clean coal technology. In other words, comprehensive security can be gained only through cooperative international security. The present international financial crisis illustrates this lesson. The US and China are mutual hostages to each other and have to collaborate with each other if both are to emerge from the financial crisis with as minimum damage as possible.

It is now widely recognised in international strategic circles that the emergence of India will be an unprecedented phenomenon and will not be looked upon with fear and apprehension by other nations as happened with the rise of Britain, France, Germany, Russia, Japan and China in the earlier centuries. The US, Russia and the European Union have entered into strategic partnerships with India and hope to build mutually beneficial economic and technological relationships. China and Japan are not averse to the idea, either.

Against this background, the G-20 Summit of nations was held in London to tackle the problems arising out of the financial break- down and recession. The unprecedented nature of this summit should be reflected on. The last time the international community faced a similar situation was in the 1930s and at that stage, no global consensus could be developed. The result was the outbreak of World War II. Without exaggerating the results of the outcome of the G-20 Summit of April 2, 2009, one must admit that it was no mean achievement of the 20 nations responsible for 80 percent of global Gross Domestic Product (GDP) agreeing on a mechanism of global financial stimulus and in principle for a stricter control regime and monitoring of operations of financial markets and steps to eliminate tax havens. Members of the Organisation of Economic Cooperation and Development (OECD), Russia, China, India and other leading emerging economies were able to overcome the irrespective ideological inhibitions to achieve this common security and economic goal. No member advocated going back on the globalisation process or dismantling the market oriented capitalist system. The US President and the British Prime Minister readily conceded that the crisis was largely the result of inadequacies in regulatory mechanisms and had no difficulty in accepting the demand for stricter regulatory regimes.

There were enough indications in the deliberations of the G-20 Summit that the global issues of climate change and efficient energy use are likely to be tackled in the same spirit in which the G-20 results were achieved. US President Obama made a break with the traditional US stand and accepted that it would be unfair to demand of India and China to reduce their per capita carbon footprint without the industrial nations making serious efforts to cut back their carbon emissions. There is now clear recognition that climate change and enhancing energy efficiency cannot be tackled adequately or optimally by individual countries and coordinated international efforts are needed to address this issue.

It is logical to expect that water scarcity, food production, pandemics, narcotics, organised crime, failing states and use of terrorism as an instrument of foreign policy or to spread extremist religious cults need to be tackled in similar international frameworks. In other words, comprehensive security can only be achieved through international cooperative security and not by the efforts of individual nations, however powerful and resource-based they may be.

While it is too early to make a definitive judgement, one can discern an outline of such an approach in the new proposal to establish a contact group to deal with the war on religious extremism in the Pak-Af area. By making the entire North Atlantic Treaty Organisation (NATO), Russia, China, India, Iran, Central Asian Republics and the Gulf states as the contact group, an attempt is made to quarantine, and treat, the extremism-infested states of Pakistan and Afghanistan through a spectrum of measures ranging from military action to development aid to training of armed forces and law enforcers.

While the above are optimistic international trends which India, as a major actor in the international scene, should encourage, it is not possible to overlook that for the next two to three decades, India will face a number of security threats regionally, mostly arising out of obscurantist thinking derived from frozen religious dogmas uninfluenced by renaissance or the evolution of consensual rationalist thought or 19th century ideas of class conflict or an unalloyed Hobbesian vision of the world. Unfortunately, India has a legacy of security problems with its neighbours tracing back to its independence.

Just as the US is an idea and the European Union too is one, so is India though there is a larger reality underlying the vision of India. The idea of a nation-state was only three centuries old when India became independent. The US is one of the first multi-cultural nation-states. The European nation-states were relatively homogenous and, consequently, fashioning a European Union has given rise to a different set of challenges. India emerging out of the freedom struggle as a multi-cultural, multi-lingual, multi-religious and multi-ethnic entity, yet broadly united as a hoary and continuous civilisational identity, was a new concept for many though not to the majority of the Indians who took it for granted. The Marxists talked of Stalin's thesis of multiple nationalities but Stalin's Russia did not survive as an entity. The Pakistani Muslim League thought only Islam would form the basis of unity of Pakistan, while India would break up into multi-linguistic fragments. They still do and, hence, their attempt to bleed India through a thousand cuts. The Chinese pontificated on the Spring Thunder over India, and prairie fire spreading. Sections of Americans did not expect Indian unity to be sustained either.

There are many Indians who blame Nehru for his non-alignment and consequent isolationism, his reluctance to accept the Western market economy and his left orientation. A careful analysis would reveal that Nehru chose non-alignment as a strategy since he, as George Washington advised the Americans in his farewell address, did not want to get entangled into the opposing alliances. With the Communists getting 11 percent of the votes in the 1952 elections, any pro-Western policy would have plunged India into civil strife. Nehru's non-alignment was intended to cultivate Soviet friendship, initially to secure their non-interference in support of local Communists, and, subsequently, against the aggressive Maoist expansionism. At that stage, the US and the West were not conspicuous in supporting industrialisation even of their allies, Pakistan, Iran, Turkey and

Thailand. Up to the Seventies, there was very little Western investment outside Western Europe and Japan.

However, during this period, India's neighbours Pakistan, China, Nepal, Bangladesh after 1971, Myanmar, Sri Lanka, the Southeast Asian and West Asian countries looked at India, partly due to Western propaganda, as a Soviet ally. The neighbours of India tried to take advantage of India's non-cordial relations with the US and China to drive hard bargains with this country. The Chinese went to the extent of arming Pakistan with nuclear weapons and missiles and the US was permissive of it as a price for enlisting Pakistani support for the Mujahideen campaign against the Soviet forces. They also promoted the Wahabi cult propaganda in the Pak-Af region which has nurtured the Al Qaeda and Taliban. History will record that the Carter-Reagan strategy in Afghanistan was one of the worst strategic blunders in all US history.

The unflattering opinion about India among its neighbours was based on its low economic growth rate, the collapse of the Soviet Union, seen as India's ally, and India's poor rating internationally in the 1990s. Added to this was left wing and ethnic violence, rise of caste as a factor in vote bank politics, assertion of the Wahabi cult and its financial support to sections of the Muslim population and the inward looking policy of India in the early Nineties. 1998 marked a turning point in the Indian history: the Shakti tests, the missile and satellite launch prowess, India's proven capability as an Information Technology (IT) power highlighted during the Y2K crisis, the sustained growth rate, beginning with the 1991 reforms. The non-resident Indians' achievements in the US in science and technology and withdrawal of Pakistan in the Kargil War together gave India a new image. During the Clinton years, there were beginnings in the US of a rethink policy towards India. India's growth rate climbing to 9 percent, the country accumulating hundreds of billions of dollars of surplus and Indian companies acquiring foreign assets gave India an image of the second emerging power next to China. During his second term, President George Bush concluded that India's rapid economic growth and its being liberated from the technology denial imposed on it following its nuclear test in 1974 would contribute to a better balance of power in Asia. This view was shared by many other major powers such as France, Russia, the UK, Germany, etc. Thanks to the efforts of President Bush, India was able to get a waiver from the Nuclear Suppliers Group's guidelines on trade and transfer of nuclear materials, equipment and technology.

While the international community's adjusting India in the new international balance of power – economic, technological, political and military – was easily understood all over the world, in India there were severe criticisms, partly on ideological grounds from Marxists and partly on parochial political grounds. The rest of the industrialised world is able to understand and appreciate the role emerging economies like China and India are able to play in the present financial crisis. India will be able to play a significant role on the proposed reformation of the international financial order as a result of the G-20 conclusions.

But in spite of this welcome emergence of India as a global player the politico-military establishments in India's neighbourhood are still to adjust themselves to this reality and are attempting to play the earlier game of exploiting India's perceived differences with other major powers of the world. It will take some more time for them to realise that they cannot do so in the new situation. As India's stake rises in the international community, even China will have to assess the costs and benefits of supporting Pakistan at the cost of alienating India. As India grows rapidly, economically and militarily, and gains further influence and stature, the tipping point will be reached when it will not be cost beneficial for China to continue its anti-Indian pro-Pakistani policies.

Except for China and to a lesser extent Sri Lanka and the Association of Southeast Asian Nations (ASEAN), other neighbours of India are yet to realise that their own economic growth very much depends on the growth hub of India, and their continued hostility towards India will hurt their economic growth more than that of India. After 60 years of the practice of democracy, there are many shortcomings in our democratic structure and practice. It is, therefore, natural that transformation to democracy in our neighbouring countries will not be without stresses and strains. Given the language, ethnicity, religion and kinship our population shares with those across the borders, such turbulence in democratisation in our neighbouring country will involve our own populations. India's efforts in the next two decades should focus on making attractive to its neighbours the prospects of their joining a free trade zone which in due course will become a common market. Though this objective has been spelt out in the South Asian Association for Regional Cooperation (SAARC) declarations, there has been a lot of both overt and covert opposition, especially from Pakistan and Bangladesh. But the international economic climate following the recession and during the recovery period is likely to induce some new thinking in both neighbouring countries.

India has today a tremendous advantage in building up its armed forces. Both the leading arms manufacturers of the world, the US and Russia, are willing to supply sophisticated arms to India. In that respect, India has an advantage over China which has difficulty in getting sophisticated arms from the US. India should fully exploit this factor and build for itself a conventional and nuclear armed force which will be a deterrent for all nations other than the US and Russia. Such an armed force, partnership with the US, Russia and the European Union and implementation of the recommendations of the best Indian brains that AVM Kapil Kak has been able to assemble in the chapters in this book should ensure optimal security for India vis-à-vis its neighbours.

The purpose of this lengthy introduction is to disabuse the notion advocated in certain quarters that India should aim at developing a highly expensive and provocative autarchic defence capability that is bound to raise the apprehensions of our neighbours as well as other major powers of the world. India's motto should be *Vasudeva Kutumbakam*. In its emergence and advancement as a major international player, India has some inherent advantages – our civilisational identity, our population being younger for at least the next three decades, our democracy, our being English-speaking, and our soft power. The most difficult problems for humanity – energy, environment, climate change food and water – are more likely to be solved by cooperative international Research and Development (R&D) efforts rather than by any one power establishing its military dominance over the resources. It is a great advantage that India is seen as a non-threatening partner by most of the industrialised world.

In such circumstances, there are reasons to hope that the Indian civilisational and cultural identity will continue to play its significant role in the world of the 21st and 22nd centuries. The expanding Indian middle class has to pay attention in the next two to three decades to the problems of faith-based terrorism, misgovernance and corruption. If only we can persuade our population to elect better legislators at parliamentary, state legislative and local body levels, nothing will hold back the emergence of India as a major global player befitting its civilisational and cultural heritage.

A Security Strategy for the 21st Century

□ Jasjit Singh

The leadership of the then superpower in the 19th century, based on their own experience and the world they dominated, no doubt cynically believed in the dictum that for a country there are no permanent friends, only permanent interests. But we have found in history that even interests undergo change in many ways, resulting in the world being in an almost perpetual transition, wherein there is both continuity and change, of interests as well as the landscape in which the country must pursue those interests to the extent it can do so. The power of a nation relative to other states, thus, becomes an important factor in deciding the nature, scope and function of its strategy which also is unlikely to remain etched in stone both in time and in space occupied by other states with their own capabilities and perceived interests. It is obvious that the world would keep changing for a variety of reasons and in numerous ways. And with these changes would come a degree of change of interests and the nature of relations with other countries, leading to changes in strategy in dealing with the changes to advantage. Hence, the greatest challenge of our times is to be able to make correct and timely assessments of the changes taking place, and the nature and extent of challenges and opportunities they present.

A World in Transition

It is in this context that we have been witness to a far more rapid pace of change during the past two decades when the collapse of the Berlin Wall symbolically confirmed the end of the Cold War. But, contrary to conventional wisdom, the world during the Cold War was not really bipolar (except in the framework of the Euro-Atlantic-Pacific zone in the northern hemisphere). Two major and large countries—India since its independence and China after the 1950s—stayed out of this bipolar military alliance system; and so did the bulk of the world and its decolonising developing world. The collapse of the limited though powerful bipolarity brought in confusion and chaos in the international system. But very soon it was clear that even during the later decades of the Cold War, the international order was undergoing profound changes toward a polycentric system. The result, only acknowledged recently, has been a shift of global power from the West to the East, largely as a consequence of the rise of China and India, with profound consequences for the world in the future.

But even before the geo-political changes started to impact on the world, the core on which nations historically built their national strategies, that is, military power and its role, had altered. The nature of war, and, hence, the very concept of war (especially a full-fledged conventional war) for those willing to look at the changes taking place, has drastically changed. Historically, territory and its capture (especially for its human and material resources) through employment of military power, had been central to the national strategies of most countries. But, for a variety of reasons that we may not go into here for want of space, territory and its occupation is no longer a viable objective for military power. Saddam Hussein faced this truth after grabbing Kuwait, and the US paid a heavy price for trying to militarily occupy Iraq and Afghanistan, with little prospect of success even after nearly a decade of fighting. And these are exceptions where the force-on-force war of the past took very little time; but the follow-on process of occupation and regime change, which used to be the next set of goals of strategy easily achieved after a victory on the battlefield was confirmed, is far from success. And these wars (1991 Gulf War, 2001 Afghanistan War, and 2003 Iraq War) may well be the last such wars that mankind is likely to experience. On the other hand, the Chinese, as part of their modernisation drive after 1980 and an expansive study of Sun Tzu and the other classics in the context of current trends, came to the conclusion by 1987 that wars in the future are most likely to be "local-border wars."

But territory, even much smaller segments of a nation-state, have assumed increasing importance in a world politically far more conscious than ever before as a consequence of the information-communication revolution. We do see an apparent contradiction of this in the formation of the European Union; but that is the exception that proves the rule since a Frenchman still believes that he belongs to a country called France, and a German who lived for more than four decades separated by the Berlin Wall, rejoices in Germany that is unified. The European states, having lost their colonial empires, impoverished by World War II, felt the need to come together to cope with their weaknesses, and now the process has found great utility in a changing world. The shift in the importance of territory as an object of the military and its strategy has altered the way military power is to be applied in pursuit of national interests. But before we get into that detail, we must take a quick look at the issue of the grand strategy of India so that other aspects can be placed in its context.

India's Grand Strategy?

Going back to the issue of global changes taking place, the national security strategy of India in the future (whether such a document is written down on not) would have to take into account the reality of the ongoing global power shift to (geographical) Asia and the rise to power of key countries of Asia, especially China and India itself. Hence, a number of factors and issues would have to be adequately included in such strategy. India's rise to power in the future, no doubt lagging behind China, poses a set of challenges and opportunities. Historically, rising powers also inherit rivalries from the past as also get involved in new ones. While a lot of people are concerned about the fact that China is well ahead of India, they ignore a historical reality that China was always ahead of India. For example, in the early half of the 18th century, China accounted for nearly

33 per cent of global wealth and manufacturing output, while India was the second largest producer of wealth and manufactured goods, with a share of around 24 per cent. While China did send a sea-faring sortie into the Indian Ocean in the 8th century which made no effort to use force, it did not do so during the following twelve centuries. There was also little contact and less rivalry even when India traded with Southeast Asia, the backyard of China; and nor did India ever interfere with the Silk Route. Hence, in evolving our national grand strategy, we need to look back at our own cultural and historical foundations to build the strategy on the basis of our core values (now clearly enshrined in the Constitution) and also our key interests in future as a rising power.

Like in the earlier centuries, China and India are rising on the foundations of economic productivity though with distinctly different methods and ideology. The factor of ideology would moderate as China itself grows and adopts market strategies more compatible with the globalisation processes in the world. China has set the goals of its grand strategy in terms of matching if not surpassing the United States in power and influence. It argues for a multipolar world to attract states to its side, but in reality seeks a unipolar Asia with China at the core. This can hardly be acceptable to the United States as the sole superpower that knows its relative power is reducing largely because of the rise of China and India. There is also the demographic factor to be taken into account which clearly indicates a reducing population (especially of working age) of US allies, Russia and most of the Muslim countries in the coming decades. China itself will face the problem of ageing population though the impact would be much less. Only India would have an increasing population as well as population in the working age for the coming three decades. This by itself would be a major asset not available to other countries; but a well-educated professionally qualified large population would be a clear advantage in the coming world.

The United States would do everything possible to ensure it continues to be in a leading position through the 21st century at least. And here India becomes an obvious source of assistance because of a market of over a billion people, a rising middle class already a couple of hundred million strong, one of the largest pools of professionally qualified people, a country that has a rule of law (even though in urgent need of repair), a resilient democracy and one of the most professionally capable military forces (even though its weapons and equipment are in need of modernisation at present). Historically, a declining power has sought some sort of alignment with a rising power to retain its influence as far and as long as possible. This is bound to rile the main rising power, China, which had the same choice but apparently is uncomfortable with supporting a rising rival power with which it shares borders in its sensitive regions where it has yet to establish acceptable legitimacy among the people.

However, China in its preoccupation with its own rise and the efforts to woo smaller powers as far away as Africa and Latin America has already made mistakes in its assessments. It had built up a close neo-alliance relationship with Pakistan essentially to increase Pakistan's autonomy from US hegemony and dilute its commitment to its US alliance commitments. Like the proverbial Russian search for warm ports to the south, it chose Pakistan (and now Myanmar, India's eastern neighbour), in the process building a strategy of dismissal of India. In its arrogance of power based on the use of force and the Soviet alliance in the 1950s, it had chosen to violate Indian sovereignty in Ladakh and began to claim territory beyond the McMahon Line along which it had settled its boundary with Burma. It had chosen to ignore India as a poor country beset with problems, at the same time wanting to displace it as the naturally acknowledged leader of the developing world; and Mao was personally deeply jealous of Nehru's popularity and the attention paid to him in the international community as the leader of India that obtained independence through a non-violent struggle rather than a violent upheaval on the basis of which the People's Republic of China (PRC) acquired its rule in the country and was not even recognised by much of the world for a long time. In substance, China has shown no desire to settle the boundary issue and even the agreements of

1993 and 1996 (concluded during a period when China felt vulnerable after Soviet disintegration, with persistent US pressures, especially after the Tiananmen incident, while its own military was a large but non-professional inheritance of the Cultural Revolution) to maintain peace and tranquillity on the borders by measures like demarcation of the Line of Actual Control, have remained unimplemented in spite of numerous meetings.

The flaw in the Chinese grand strategy is not that it seeks to displace the United States in Asia as well as the world leader, but in choosing to raise the level of its discomfort with India's emergence as a major player on the world scene. This is most likely due to the different world views of the Chinese leaders and those of India. It is true that both countries now claim to be "strategic partners" whatever that means. China-India trade has shot up dramatically to touch $50 billion within a few years from a negligible level less than a decade ago, but the nature and composition of that trade creates an imbalance which China could use as a lever any time. China shows no signs of solving the border issues, or taking measures to maintain peace and tranquillity across them. A great deal can be written on China's approach toward India. But, for the purposes of our present study, the bottom line appears to be that China has come to the unfortunate conclusion – like it did before 1962, and was roundly criticised for it by the Soviets and other Communist regimes – that India is moving too close to the United States, and, hence, the need to build leverages against its rising capabilities. Beijing probably feels that pressures on India would make it more cautious about closer relations with the United States without which its growth to power may be stunted. India then would be more vulnerable to Chinese hegemony. The Indo-US nuclear cooperation agreement and Chinese opposition to it even at the Nuclear Suppliers Group (NSG) is but one example.

India's own interests lie in a closer cooperative relationship with China, though the same logic also demands that we base this cooperation in tandem with a hedging strategy in case of any serious reversal of our attempts to cooperate. **The rivalry between the United States, the current superpower, and China, the aspiring superpower, constitutes the greatest strategic challenge in the first half of the 21st century.** Given the nature of developments in the world and the deep economic interdependence between the two countries, this is unlikely to turn into a Cold War of the type we lived through for more than four decades. Also, the earlier Cold War rivalry was geographically located far away in the northern hemisphere although armed conflicts in proxy wars abounded across the globe. But it may be recalled that any time the Cold War came physically close to us, like in Afghanistan during the 1980s and the Indian Ocean earlier, it had a negative impact on our security, some of the effects of which have not dissipated more than two decades later. If anything, it has, on the other hand, intensified as a consequence of the Chinese surrogate state Pakistan's expansionism, as in the Pakistan-Afghanistan region and the *jehadi* terrorism (the main "instrumentality" by which the US fought the Soviet presence in Afghanistan) which has become a Frankenstein threatening the regional civilisations, leave alone states.

India's grand strategy for the 21st century, therefore, must seek to maintain friendly cooperative relations with both the current superpower and the aspiring superpower, the PRC. This does not imply that we ignore the rest of the world with which we must maintain close and friendly relations, not necessarily expecting them to support us in case an adverse situation arises, but to ensure our own growth and building of Comprehensive National Development. In other words, **our grand strategy should be based on the principle of a much maligned term: non-alignment as a national strategy** and not to be confused with the Non-Aligned Movement (NAM) of nations. It may be recalled that non-alignment as the basis of our foreign policy and international relations was not the product of the Cold War, but was adopted a year before World War II broke out and a good eight years before India became independent when the All India Congress Committee (AICC) of the Indian National Congress which led the struggle for independence, adopted the resolution at Haripur that:[1]

> India was resolved to maintain friendly and cooperative relations with all nations and avoid entanglements in military and similar alliances which tend to divide up the world into rival groups and thus endanger world peace.

The principle and strategy of non-alignment was adopted by India as a means of non-involvement in military alliances and staying out of conflicts that were not of direct concern to it. Although grossly misunderstood in the West (and also by Stalin's Soviet Union), it sought to pursue India's own national interests, liberally interpreted as part of the larger human needs. At the broader level, the philosophy of non-alignment also sought a degree of reconciliation between the developed and underdeveloped world where India became a bridge of sorts. It certainly sought what today would be termed "constructive engagement" with both the superpowers and members of their respective alliances. At the same time, the enthusiasm for the UN in India also resulted from the expectation that the UN was the logical and representative instrumentality for a more equitable world order where independent states could freely cooperate for mutual welfare and prosperity.

This is not surprising since leaders like Jawaharlal Nehru had already been arguing against power politics which they held to be particularly menacing. Nehru, on the other hand, hoped to see, with the demise of imperialism, the emergence of an international framework that allowed stabilisation of conflict through maximising cooperation which was the only option in an age in which war had become as disastrous as it was, and national self-respect as universally demanding.[2] His concerns that such developments would have a negative impact on India's freedom and peaceful growth were expressed clearly in the statement, "Only two factors come in the way (of these goals): international developments and external pressures on India, and lack of a common objective within the country."[3]

It may be useful to briefly look back at the early days of non-alignment when there was no NAM. The formal birth of the concept of non-alignment can be traced to the broadcast made by Jawaharlal Nehru (at that time, the vice-chairman of the Viceroy's Executive Council in India) on September 7, 1946, indicating the response of a major nation about to emerge as an independent sovereign nation-state, when he stated: "We propose, as far as possible, to keep away from the power politics of groups, aligned against one another, which have led in the past to world wars and which may again lead to disasters on an even vaster scale. We believe that peace and freedom are indivisible and the denial of freedom anywhere must endanger freedom elsewhere and lead to conflict and war."[4] Nehru also declared, "We shall take part in international conferences as a free nation with our own policy and not as a satellite of another nation."[5]

If there was any hope of reducing the negative impact of the Cold War, especially for the developing countries, it lay in early prevention and a conceptual framework in which to pursue it. This conviction was strengthened by the emergence of Communist China on the international stage. India's early resistance to the great power conflict, in which China rapidly became a part, also provided a model for the newly emerging, but much weaker states of the post-colonial world.

A more specific impetus and imperative of non-alignment for India was the problem of dealing with China which posed a challenge to India in comprehensive terms, besides the specific military terms, especially after it invaded and occupied Tibet by force. Two years after India's independence, Mao Zedong had called for the "brave Communist Party of India" to liberate the country.[6] Implicit in this call also was the concept of export of ideology. The Indian leadership, including Nehru, was conscious of this challenge.[7] After 1954, non-alignment became an even greater imperative since Pakistan became part of the bloc military alliances. In fact, the very logic of non-alignment required enhanced cooperative relations with both superpowers, especially when the Cold War became more intense (as in the 1980s) or it came closer to India.

Military Strategy

The second major element of the grand strategy of a nation-state is its military strategy. Our first thesis is that if the grand strategy is based on the principle and policy of non-alignment, which we strongly advocate regardless of the nomenclature given to it, then the **military strategy would have to be based on self-reliance**. As India expands its capabilities and rises as a power on the basis of economic and technological prowess, it will also increase its vulnerabilities and would have far more interests to defend, beyond its borders, in the future, especially by modern high-technology military power, than was the case in the early years. Ever since the complexities of modern warfare began to increase and the Germans organised their "general staff" to work out the military strategy and synergise it with the operational plan, the need for strategy formulation has always been accepted though not implemented in many cases. Here we must also note that very often we hear that our defence forces plan and fight their own battles at the cost of joint warfare. The three components of the armed forces fight in three distinctly different media, in accordance with their specialisation and capabilities (of weapons, equipment and training of manpower) in that medium. The issue, therefore, is that the success of joint operations rests on whether they complement each other's capabilities and compensate for each other's limitations or not.

Historical evidence points to the reality that while there have been shortcomings in the process of **joint planning**, there is no instance in our history where the armed forces did not actually **fight jointly**. But we must admit that there is a need to move from the "we-they" syndrome (among political, civil bureaucratic and military leaders) to a common sense of ways, means and ends. This change of institutional cultures is one of the key tasks that military strategy must achieve. After all, the purpose of the fighting forces is not merely to fight and lay down lives; it is to win the battle by making the enemy lay down its weapons and lives. However, it is joint planning that is critical for the formulation and implementation of military strategy. Here we must also address the issue of political direction of the use of military power. The military profession is now far too complex and critical to be easily grasped by political leaders who may have no military background and yet the employment of military power must serve political goals and objectives.

Here we must cite the example of the extensive use of the armed forces (mostly the army, though the navy has been vastly employed and the air force has continuously provided combat support) in internal security and counter-terrorism. Jointness must also apply here between the Defence and Home Ministries. It would be logical to deploy the armed forces on such duties where the police organisations do not have the capabilities to deal with the situation. But it is difficult to believe that such capabilities could not be built for more than a quarter century in the police forces which number nearly three times the size of the Indian Army! It is difficult to reject the impression that the government simply finds it easier to deploy the army that is meant to deal with an external threat, for internal security for prolonged periods, with some jawans perhaps having served their full colour service doing what the police should have done to begin with and the army would stand by to tackle serious or sudden contingencies which by definition are rare.

The central reason why countries establish higher defence organisations is to provide an adequate and ongoing interaction between the political and military leadership through discussions of scenarios and policy options and their implications which would provide the logical politico-military direction to joint planning by the three armed forces.

Our second thesis in relation to military strategy is that **you don't win a war by defeating/destroying the enemy's military forces; you win a war by defeating the enemy's strategy**. The study, analysis and assessment of the enemy's grand strategy (to understand its motivations and the limits of its aims) and the understanding of its likely military strategy (which could even undergo changes during operations) are vital to success. Both these interlinked functions require institutions and (strategic) culture to generate strategy options. Unfortunately, in our case, these have comprised a weak area from the time of independence. For example, the two key institutions that would provide the wherewithal for a national military strategy to the apex military leadership—the Chiefs of Staff Committee (COSC)—the Joint Planning Committee (JPC) which would prepare joint operational plans, and the Joint Intelligence Committee (JIC) which would provide the militarily significant intelligence to assess the adversary's likely strategy and help prepare our own military strategy, remained a weak area for decades. In fact, in spite of clear instructions of the Cabinet, no permanent joint staff was ever established for the JPC till the Defence Planning Staff was set up four decades after independence; and which was replaced by the Integrated Defence Staff (IDS) of the COSC six years later.

As regards the JIC, it was detached from the COSC when the COSC was moved out from the Cabinet Secretariat, leaving a big void that has yet to be filled adequately in spite of a Defence Intelligence Agency (DIA) having been established in 2002. To make matters worse, the apex defence decision-making institution at the political level, the Defence Committee of the Cabinet (DCC) established in 1947, began to be eroded by the mid-1950s and was finally abolished after the 1962 defeat in the Sino-Indian War. The issues are now dealt with by the Cabinet Committee on Security which inevitably would spend more time and energy on the vast array of security challenges the country faces, and defence receives little dedicated attention.

As briefly mentioned earlier, territory being the primary objective for the employment of military power in the past is no longer valid. This does not reduce the importance of military power which is becoming more capable due to advances in technology, but to emphasise that there is a clear shift in the nature and concept of war. Occupation of territory will remain an important objective for applying political pressure; and, hence, it may not be cost-effective to occupy the whole territory of another country—even small portions of territory would be important for negotiations in a world where even these become extremely sensitive issues.

But what we want to discuss here is the impact of nuclear weapons on the nature and concept of conventional wars and employment of military power. The reality is that a non-nuclear environment serves India's security interests far better than a nuclearised one which only leads to complexities and uncertainties. But in the context of existing nuclearisation, India has no choice except to maintain a credible nuclear deterrent for its defence against any nuclear threat or use. Hence, our strategy must continue to press for global abolition of nuclear weapons while maintaining a credible nuclear deterrent. Meanwhile, we will need to shape our strategy in such a way that as far as possible, nuclear weapons do not come into play to affect our security calculus. Hence, the need to adopt the doctrine of "no-first-use" which would shift the onus of escalation to a nuclear level onto a potential adversary while we will retain the capability to impose unacceptable punishment with an **assured nuclear counter-strike strategy** in case of such weapons being used against India or its (armed) forces.

But the question remains as to how conventional war, with its risk of escalation to nuclear levels would be fought and won. To begin with, a credible nuclear deterrent should be able to lower the possibility of the enemy reaching out to the nuclear button in a hurry since he must consider its consequences. Empirical scientific studies indicate that the use of even a 15 kiloton yield warhead (of the type dropped on Hiroshima in 1945) in today's increasingly urbanised world would lead to more than half a million dead and another million seriously injured.[8] The enemy's first strike simply cannot wipe out our complete arsenal; and enough would survive to counter-attack and cause unacceptable punishment. Even the loss of half a dozen industrial and economic targets to such counter-strike would destroy the ability of the enemy to function as a state. But prudence requires that we don't push the enemy into a corner where it may feel the survival of the country is at stake and nuclear weapons are the only option left. In fact, our strategic goals and objectives require building our own Comprehensive National Development. Hence, it is obvious that the most efficient military strategy for conventional war would be to undertake a limited war. We must caution here that limited wars would produce limited results.

Nuclear weapons in our security environment obviously have had an impact on the nature and concept of conventional war. At the simplest level, conventional wars have now become less likely and/or being fought at the sub-conventional level. But if and when a conventional war takes place (as indeed happened in the Kargil sector in 1999), these will have to remain limited. Looking back at Kargil, India imposed severe limitations on itself to conducting the war localised to the sector. The Indian Air Force (IAF) was not only not allowed to target Pakistani aggressors across the Line of control (LoC), but the IAF was barred from crossing the LoC even when attacking the Pakistan Army close to it and where the sheer turning radius of a combat aircraft would necessitate use of adequate area. Ultimately, international diplomacy was harnessed to ensure the war did not become a stalemate around the LoC, and Pakistan's defeat and withdrawal was made public. Pakistan also restrained itself from pumping in more troops or expanding the war to other areas of the state/international border. Above all, it had to restrain itself in not employing its air force although its army was getting a severe beating from Indian artillery, infantry and air power.

The point we wish to make, as has been made often in the past, is that **the next conventional war that India may be involved in would have to be a limited war**. It can be stated with certainty that all the wars that we have fought in the past have actually been limited, especially in terms of political aims and employment of military power. And the limitations essentially were imposed on those wars by India. Such limitations in future may or may not be driven by the type of factors that were operative in past wars. But what is clear is that the presence of nuclear weapons with both our potential neighbours would certainly be a constraining factor in war where our own interest would require that, as far as possible, nuclear weapons should not be allowed to come into play. This is not to harbour any doubts on the credibility of our nuclear deterrent, but to ensure that destruction by the use of nuclear weapons should be avoided as far as possible. This implies that the degree to which our armed forces would be free to achieve "decisive military victory" and/or capture large extent of hostile territory and pursuit of military objectives that threaten the "survival" of the adversary would perforce have to be limited. But once the enemy has used nuclear weapons against us, all such restrictions should be immediately waived.

The second major issue that we need to bear in mind is that nuclear weapons must not be allowed to eliminate the choice of a conventional war; all it means is that the way it has to be conducted and the political-military goals set for the war have to take into account the reality that the conflict could escalate to the nuclear level. But a close study would indicate that there would be adequate strategic space below the nuclear level that can be exploited for the successful conduct of a war with conventional military capabilities. Given that our interests require that nuclear weapons should not be allowed to come into play, exploitation of this strategic space also implies that **our conventional military capabilities must be clearly superior to those of the potential adversary**.

For more than a quarter century, Pakistan has been pursuing a strategy of sub-conventional war (especially *jehadi* terrorism across the borders) against India under the belief that its nuclear weapons would deter India from employment of even the conventional forces in a coercive role against it. In principle, this is a flawed assumption even though Indian policy-makers appear to have fallen for it without acknowledging it. In the process, the credibility of our **national deterrent capabilities against sub-conventional war being waged across the borders have been eroded, if not failed.**

The key challenge that comes up in case of a covert sub-conventional war against us under a nuclear umbrella is of how to undertake successful punitive military action to defeat Pakistan's strategy. In the past, when nuclear weapons were not held by Pakistan and it launched covert war with guerrilla forces, backed by the regular military, as in 1947 and 1965, it did not take India long to respond robustly with conventional force even across the borders. But this has not been done even though terrorism levels have been highly provocative and/or terrorist strikes were made on critical Indian targets like the Parliament in December 2001 and more recently in Mumbai on November 26, 2008. Obviously, our response strategy should be based on discrete conventional punitive strikes against selected politico-economic targets (preferably in Pakistan Occupied Kashmir—POK) in a calibrated manner, spread over time, if required. This would best be undertaken in concert by (i) the Indian Army, generally on a strategic defensive posture, but capturing Pakistani territory to shallow depths (possibly up to around 5km each) at a number of (5-7) places along the border to occupy territory for negotiating purposes; (ii) special forces for specific targets close to the border; (iii) the air force combat component to strike military-political and economic targets deeper inside POK; and (iv) the navy applying pressure from the sea as was done during the Kargil War without necessarily any major engagement. The aim of these strikes would be to generate an effect-based outcome. These no doubt would generate military effects, but their real goal should be to create political-economic effects with the aim of influencing change in the policy and strategy pursued by Rawalpindi.

Notes

1. Subimal Dutt, *With Nehru in the Foreign Office* (Calcutta, 1977), p.22.
2. Jawaharlal Nehru, *India's Foreign Policy: Selected Speeches, September 1946-April 1961* (New Delhi, 1961), p.574.
3. Ibid., p.559.
4. Ibid., pp.2-3.
5. Ibid., p.2. See also his speech on December 13, 1946, in *Jawaharlal Nehru's Speeches*, Vol. I, pp.5-16.
6. *New China Agency*, November 21, 1949.
7. For example, see Nehru's statement in 1959 that, "Ever since the Chinese Revolution, we naturally had to think of what China was likely to be. ..... We realised ..... we knew that amount of history --- that a strong China is normally an expansionist China. Throughout history, that has been the case. And we felt that the great push towards industrialisation of that country, plus the amazing pace of its population increase,

would together create a most dangerous situation. Taken also with the fact of China's somewhat inherent tendency to be expansive when it is strong, we realised the danger to India.... As the years have gone by, this fact has become more and more apparent and obvious." See Nehru, n.2, p.369.

8. For example, see O. B. Toon, R. P. Turen et al, "Atmospheric Effects and Societal Consequences of Regional Nuclear Conflicts and Acts of Individual Nuclear Terrorism", published by Copernicus GmbH on behalf of the European Geosciences Union, April 19, 2007, available at www.atmos-chem-phys.net/7/1973/2007/. The study concerns cities across the world. For a more specific study on South Asia, see also Mathew McKenzie, Zia Mian, M.V. Ramana and A.H. Nayyar, "Nuclear War in South Asia", *Foreign Policy in Focus*, June 2002.

Globalisation, Economic Growth and National Security

☐ Arvind Virmani

Introduction

The focus of this paper is on the geo-political system. The world is in the process of moving from an old to a new global system and it is useful to look at some of the projections I made earlier, basically between 2004-06.[1] We then go on to look at the new developments since then— there have been quite amazing and unpredictable developments – and see how well the earlier projections have held up. It is a good test.

There is now a clearer discussion on the transition. Earlier papers had focussed only on the long term. Here, the extra part is the transition where the institutions and the implications for India start coming out. It would, however, be presumptuous to say we have got all the answers. The transition is marked by great uncertainty. Some clarity is possible if we understand where we are coming from and where we are headed. This allows us to focus on the ten years or fifteen years which comprise the period of transition from the old to the new order.

Clearly, globalisation is a major driving force and it is there lurking in the background. The paper will explore the implications of the changing global order. How do we set our objectives and goals for this transition period, which has been assumed to be of the order of ten or fifteen years?

The key is to focus on improving the institutions and rules of global security which include economic, military and other dimensions.

Evolution to A New Global Order

So, what is evolution? Is it an evolution of an old order or is it a new global order? Let me start by quoting from the US National Intelligence Council (NIC) forecast which appeared in December, 2004. This report used the economic forecasts given in a Goldman Sachs paper written in October 2003. The NIC (2004) report indicated that China's Gross Domestic Product (GDP) in US dollars could be the second largest by 2020 and India's GDP in US dollars could be third largest by 2035. Virmani (2004b) which was also published in December 2004, showed that the rise of China and India would be much faster than projected by anyone else till then. The second point that the paper emphasised was that China's GDP at Purchasing Power Parity (PPP) (and this was an outrageous statement at that time), would equal that of the US by 2015 [Virmani (2004b)].

Unipolar to Tripolar+

The power potential, an index (VIPP or VIP^2), was first defined/used in the same paper. The power potential of China, as measured by this index, was projected to reach 70 per cent that of the US by 2035; India's GDP was projected to equal that of the USA at some time after 2035 and the power potential would be about 25 per cent of the USA at that time (in this paper, it was not absolutely clear when that would happen). Virmani (2005a), which came soon after, expounded more clearly on the rise of India, saying that it would be the third largest economy by 2015. It forecast that India's GDP would equal that of the USA by 2040 and power potential would be 80 per cent that of the USA by 2050. Similarly, China's power potential would equal that of the USA by 2030-35 and, therefore, a bipolar world would emerge by 2025. This was the big new message of this paper that contrasted sharply with what most international analysts and scholars were saying. It asserted that we would move first to a bipolar world and then subsequently to a tripolar world, with India coming up by around 2015.

When this paper was presented at seminars, questions were raised about the role of the European Union (EU) [Virmani (2005d)]. It would be recalled that Prof Henry Kissinger (1994) had written about the EU as a potential great power of the future along with the USA, Russia, China and Japan, with India only a distant possibility in sixth place. Subsequently, Virmani (2005e), therefore, defined something called a "virtual state," something like the USSR where decisions regarding external matters, whether economic matters or military matters, were taken in a unified way. The reasoning was that the EU could not be a power unless it became a virtual state.

There were also questions raised about the value of such long-term projections. In response, I had stated that one thing that we can be almost sure of is that any specific futuristic predictions will turn out to be wrong. Projections, no matter how imperfect are essential for any kind of planning. Questions were also raised about the downside risk to the projections, particularly for the Indian and Chinese economies. Does the projection mean that there are no weaknesses, and that these outcomes would happen automatically? The answer is clearly no. The main weaknesses we had identified for China were the mercantilist export Foreign Direct Investment (FDI) oriented policies and I believe that this is what is going to cause it much more problem today in the financial crisis than it is going to cause to India because it will have to fundamentally change its developmental model if it wants to continue to grow as fast or relatively faster. On India's side, we have identified governance, something that everybody is now talking about post-Mumbai. We had specified that the issue of governance was really the issue of the supply of public and quasi-public goods. These are defined as goods which are public in the sense that they cannot be provided in a private market. You cannot sell and buy them. They are collective goods. A good example of that is the rule of law where the police system, the legal system is common. Defence is a classic example. I emphasised the rule of law as it is essential for the existence of competitive markets. This is the fundamental flaw which we see again and again, including in the Naxalite controlled districts and most recently in the Mumbai terrorist attacks [Virmani (2002a b].

New Developments in 2007 and 2008

There have been a number of new developments in the last few years i.e. since our initial projections. Four very significant developments have occurred since then. One, the oil and energy price boom which we have just been through, the roller coaster. Oil prices rose by 130 per cent till July and have since fallen by almost that amount or less than that, but getting there. The important implication of that which I admit freely is that Russia has come back. In my earlier projections, Russia did not have a prominent role, as you will see it has now. Clearly, Russia's comeback has depended on the oil and energy price boom. The fluctuations and trends have been amazing and there are certain implications of that.

The second was a massive change in the PPP estimates by the World Bank/ADB (Asian Development Bank). So, India and China's GDP, according to this new estimate was 36 per cent lower, with one stroke of the pen, it was reduced. That is a huge change. How does it affect my forecasts? The basis of my projections and forecasts is the GDP at PPP which was with one stroke adjusted down by 36 per cent. You would think everything should go haywire.

The third, the end of the nuclear apartheid; and the fourth is the US global financial crisis, its impact on the global economy and big powers. All these factors, with the exception of the ending of nuclear apartheid, a somewhat non-economic factor, have been taken account of in the new projections.

Change in Projections from 2004-05 to 2008

These are summarised in two graphs (Figs 1a and 1b) which depict the evolution of relative economic size of the major powers (vis-à-vis the USA) over the 20-year period from 2008 to 2028. This is the relative economic size which is the size of each country relative to the US, which is the benchmark. Recall that by definition, power is always relative. So, all the numbers, the graphs in this paper, will represent the data for the concerned country divided by the data for the US which is the benchmark. One would say that it is still the sole superpower. Later in the paper, there is a formal definition of superpower which will show the quantitative basis for this statement.

Fig 1a: Relative Economic Size (/USA)- GDP at Purchasing Power Parity

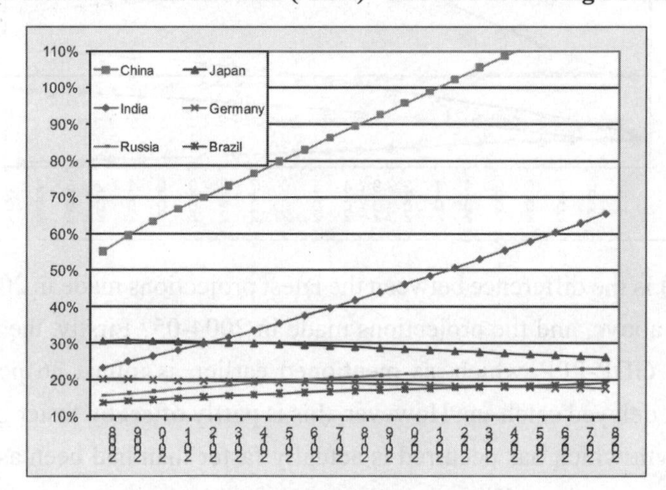

Fig 1a shows that the Chinese economy is now far ahead of the rest of the potential powers; so it is clearly the second largest, and is projected to equal the US economy in size some time after 2020 (crossing 100 per cent, that is, the US). Initially, the Indian economy is smaller than Japan's, but equals and exceeds it somewhere around 2012. Thereafter, it breaks away, with its size increasing rapidly with the "magic of compound interest" or more accurately, the compounding effect of growth asserting itself. This is despite the deceleration of growth that is built into these projections. As long as the growth rate remains higher than the benchmark US economy's growth rate, the relative size keeps rising.

Fig 1b: Relative Economic Size (/USA)- GDP at Purchasing Power Parity

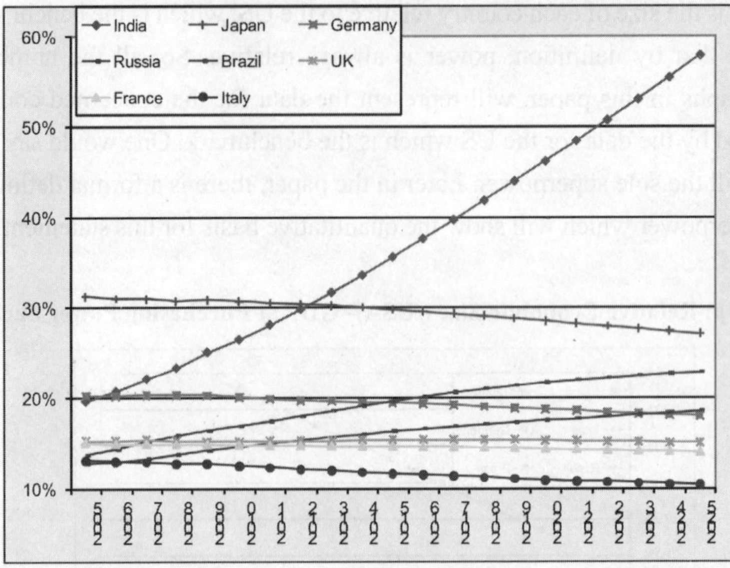

What is the difference between the latest projections made in 2008 and depicted above, and the projections made in 2004-05? Firstly, the impact of lower GDP-PPP, which, as mentioned earlier, is minus 36 per cent, results in delayed catch-up. However, this is partly offset by faster growth. The growth which has occurred is actually faster than had been assumed in earlier projections. What does that imply? China's economy will equal the US economy seven years later than projected in my previous papers —around 2022 instead of the earlier projected year of 2015. There is a similar delay in China equalling the power potential of the USA, which is something different, the graph for which we will just see subsequently. Second, India becomes the third largest country in terms of relative size in 2012, about six years later than in the earlier projections. India's power potential at the middle of the century (2050) would be about 70 per cent rather than 80 per cent. To summarise the current projection is completely within the range of possibilities envisaged earlier; despite these huge shocks, statistically speaking, it is completely within the normal range of errors that characterise virtually all economic variables. This is a mean scenario. The actual trajectory may be higher or lower. This, in turn, means that equivalence occurs some years earlier or later, respectively.

Let us look at the other potential powers such as Germany, Russia and Brazil. As mentioned, the earlier papers had given much less importance to these countries. One major mistake made in the earlier papers, which has been corrected in the current paper, was to downplay the importance of energy demand-supply imbalances and energy price changes. The earlier paper focussed on the effect of energy shortages on countries like China, India and the USA that are net oil importers. It was correctly reasoned that all such importers would be affected to varying degrees and, consequently, their relative position would change only marginally. This has turned out to be, by and large, correct. The mistake was to ignore the effect on oil/commodity exporters like Russia and Brazil. With the experience of the recent oil/energy/commodity price shock, this oversight has now been corrected.

The revision in the commodity and oil price outlook has changed the picture. Recall that Brazil is also a huge commodity producer and exporter. Consequently, towards the end, it became a significant global player (Fig 1a). Fig 1b squeezes the elements and leaves out China so that we can see these countries clearly. This shows that in about 20 years, the Russian and Brazilian economies are likely to be larger in size than Germany's. Brazil overtakes France and the UK by 2012. So, in terms of the current middle powers, if we can call them so, Brazil will be coming up, but only to an extent, substantially due to its natural resource endowment. Russia is somewhere in the middle here. Russia overtakes Germany by 2015. This is new —fifth in size and power potential.

The reason for the rise of Russia which is quite clear in hindsight, is oil and gas endowments and the price of oil rising due to the faster growth of China and increased demand from oil exporters who under-price oil in the domestic market. Another factor is the non-market power of the oil cartel.

This cartelisation has broader implications in terms of power. It is not a normal market. There are a few key differences between oil and certain energy products and the host of normal manufactured and extracted mineral products. All these products are traded in normal broadly competitive markets. The problem in oil/energy is that it is supplied by a cartel of energy producing and exporting countries (which is much worse than a cartel of firms). There are certain other special characteristics of energy, in that it is a universal intermediate. It is not that it is used in one, two or a few industries in relatively small proportions, but is either directly or indirectly an input in virtually every industry and activity. This gives the cartel additional power. This means that power elements beyond pure economics come into the picture here, which Russia is successfully trying to use.

Other Views: US NIC

How does our latest projection compare with other analyses and views? In November 2008, the US NIC (National Intelligence Council) released a new report. This November 2008 report says, "The eight largest economies in 2025 will be in descending order – the US, China, India, Japan, etc." This is consistent with both our earlier projections as well as our revised projections, except that they do not give explicit projections like we have always given. They also say that "by 2025, China will have the world's second largest economy and will be a leading military power." Again, this is consistent with our scenario of a bipolar world. After this, however, NIC (2008) goes on to say something quite different: "A global multipolar system is emerging with the rise of China, India and others which will lead to the development of a globalised economy in which China and India play major roles." So, the US NIC believes that China and India will be important but within what looks to them like a multipolar world. Our perspective on this is somewhat different, as discussed below.

Global System: Long Term

The evolution of the global system will depend not just on the evolution of the different economies, but on the evolution of power. In this section, we start by looking at the evolution of power potential over the long term, where we define the long term as half a century or till 2050. This will help us see where we are coming from and where we are going, and thus put the current period 2025/2028 in perspective. The index of power potential, which has been termed VIP^2, has been defined in a number of earlier papers, such as Virmani (2005e) and Virmani (2005e).[2] Fig 2 shows that currently and for the next five years or so, a number of countries are bunched together, giving the impression of a multipolar power structure. Within a decade, China pulls away from the crowd, and another two decades later, India pulls ahead of the rest. Thus, if you believe these projections, the global power structure would be transformed within three decades.

Global Powers and Superpower(s)

The index of power potential (VIP^2) can also be used to formally define a potential "global power." We define a "global power" as a country with a VIP^2 of more than 25 per cent. As the VIP^2 measures power relative to that of the USA, this means that a country with power potential of more than 25 per cent of that of the USA has the potential to exert its power across the world. So, where does this 25 cut-off come from? This is actually the value of the index (VIP^2) for the USSR at its peak of power — we are able to estimate this using available data on the USSR. It may appear paradoxical that the USSR, with less than 25 per cent of US power potential was able to challenge it and convert the world into a bipolar one. Clearly, there is more to global power and superpower than economic power or power potential.[3] In the globalised world of the 21st century, the relative importance of various elements is changing, as we shall discuss below. So this paper uses a conservative value for defining a global power, a VIP^2 that is higher than the one at which the USSR attained global power.

Towards a Tripolar System

Applying this definition of "global power," China is only a few years away from becoming a global power (2010), while India will take another two decades (2027) to attain this status (Fig 2). Japan has been a global power for a long time and will remain so till 2021. Thus, Japan will cease to have the power potential to be a global power half a decade before India attains this global power potential. We must remember, however, that the index measures power potential. To this has to be added what I call strategic assets or strategic technology, which includes defence assets and technology, to obtain an overall index of actual power [Virmani (2005e)]. We define "superpower", as a country with a VIP^2 of 40-50 per cent, less than double the cut-off for a global power. Based on this cut-off value, China will become a potential superpower around 2020 and India around 2040.

So, what do we have? The past, a unipolar world, with the US as the sole superpower, which peaked in 1999, The future, a bipolar world by 2020, and a tripolar one by 2040 (Fig 2). In between, a situation that the NIC (2008) and others have called a multipolar world. Can anything spoil this simple picture? Some people have talked about the socio-political explosion of "Communist/Socialist" China, others of the implosion of India. Though nothing can be ruled out, I am confident that a tripolar world will emerge by the middle of the 21st century, as predicted in Virmani (2005a) and Virmani(2006a).

Quadripolar: The Role of the EU

The moment we talk about tripolar, the issue of the role and position of the EU comes up. In principle, a quadripolar world is possible, but for it to happen, the EU has to become a virtual state. In other words, the EU will have to develop unified decision-making with respect to external relations – economic, military etc., as for example, prevailed in the USSR. Paradoxically, this ideally has receded over the past few years, since we first addressed this issue, with the EU moving in the opposite direction. A number of country referendums have rejected the EU. The recent global financial crisis saw each member country acting completely independently and focussing on its own national interests, some to the detriment of other member countries. Earlier papers and talks had concluded that the EU had become a virtual state with respect to economics; but the response to the financial crisis raises serious doubts about this conclusion, if not falsifying it. The EU is a virtual state only with respect to trade and World Trade Organisation (WTO) issues. Despite the movement of the EU in a direction opposite to that of a "virtual state", we do not rule out eventual movement in this direction. The degree of uncertainty is so high at present that we cannot make a definite prediction on whether the world will be quadripolar at mid-century.

Fig 2: Evolution of Power Potential (VIP²)

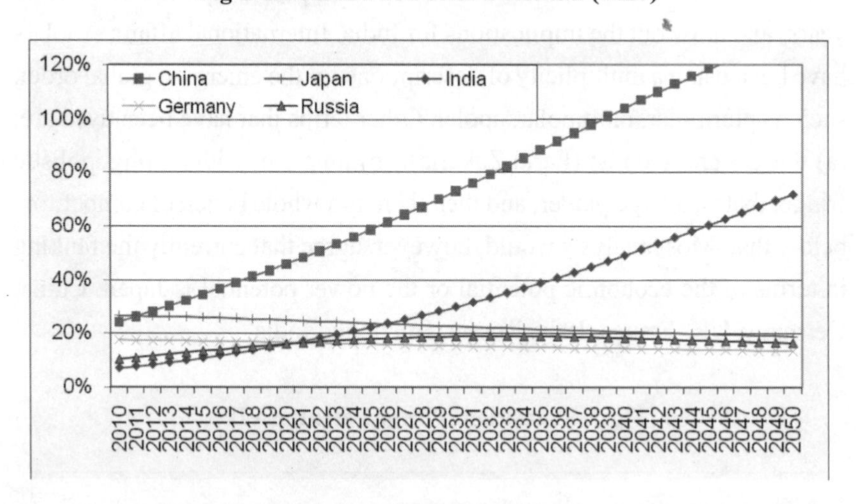

Global System: Transition Phase

With a picture of the possible evolution of power potential over the long term, we can focus our attention on the actions and approaches that India needs to adopt over the medium term or during the transition from a unipolar to a tripolar world. Fig 3 presents the medium term till 2025, with all potentially significant powers represented. China is now about five years from becoming the second most powerful country in terms of power potential measured by VIP² when its power potential will exceed that of Japan. Japan's power potential will, however, remain the third highest during the rest of the transition period. According to this data, India's power potential, currently the lowest among the 10 largest economies, is projected to rise rapidly to equal that of Brazil in two years, Italy in five years, France in seven years, the UK in nine years and Germany in twelve years. The rising power potential of Russia will remain above India's till they both exceed Germany in 2021, when India will simultaneously overtake Russia. According to this data, Brazil's rising power potential will intersect Italy's declining one, to exceed it by the end of the period, but will remain below that of the UK and France, which are likely to retain their relative power vis-à-vis Germany, the most powerful European country.

We are now ready to analyse this transition phase, a period of 15 to 20 years, and draw out the implications for India. International affairs scholars have been using a multiplicity of descriptions for the emerging global order, such as pluripolar, multipolar, apolar. Other terms that have been used are: (a) the US and the rest (Farid Zakaria); (b) unipolar, with an oligopolistic fringe, that is, a large player; and then there is a whole bunch of competitors below that. Most analysts would, however, agree that currently the ranking in terms of the economic potential or the power potential is Japan, China, Germany, UK, France, Italy, Russia, Brazil and India.

Fig 3: Transition: Power Potential VIP[2]

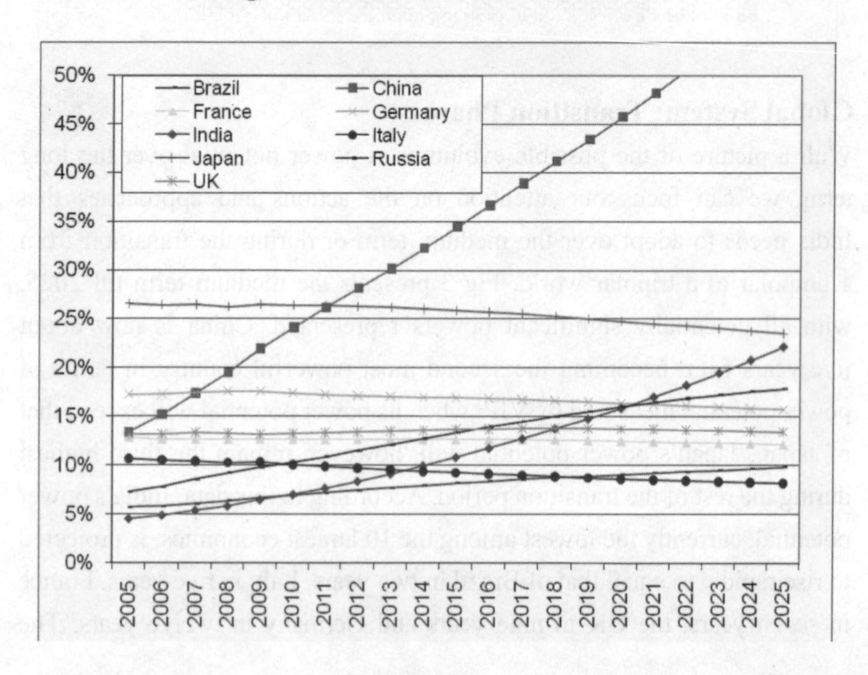

This is a good point at which to digress briefly and say a little bit about actual power. The index of power potential (VIP^2) is based on economic theory and is an economic concept, but at some point, one has to bring in strategic assets, including a country's strategic military assets, the strategic technology (defence, nuclear, space) into the picture. Virmani (2005e) formally defines an index of power, termed VIP, which is composite of VIP^2 and the strategic assets of the country, including strategic military technology, assets and capability. Unfortunately, it is very difficult to get data on strategic assets and, therefore, we have not been able to construct this index VIP for all the large powers.[4]

There are, however, a few important implications that one can draw even without calculating VIP. One is that Russia, because of its legacy of strategic assets, which it developed as the USSR, is actually more powerful than this ordering would suggest. Similarly, China has very determinedly, from the days of Mao's nuclear and space deals with the USSR, built up its strategic capability. In contrast, since its defeat in World War II, Japan has turned inwards, shunned the acquisition of strategic capability and consequently comes lower in the power ranking than its rank on VIP^2.

Multipolar: Concentration Index

The first issue that we have to address is the issue of multipolarity raised by many scholars. If we ignore the US which is at 100 per cent and is not shown in Fig 2, the world seems to have some elements of multipolarity. Virmani (2006b) had argued that though the world economy could be described as multipolar, because of the near "virtual state" nature of the EU in terms of trade-WTO and some other issues, the global power structure looked distinctly unipolar with an oligopolistic fringe. In the current paper, we apply certain established economic concepts to this issue with the hope of getting more definitive conclusions. The Harfendahl Concentration Index (HCI) is commonly used to measure the concentration of market share among supplier firms in a given industry.

We use this to measure the concentration of economic size and power potential.

This can be done, stretching back to the start of the millennium, using Angus Madison's data and the VIP[2], which had earlier been calculated and presented in Virmani (2005e). Here, we use the data back to 1870-90 to find out what has been happening to concentration of global power potential, using the VIP[2]. When we measure the concentration of power among all the countries for which the data is available, it turns out that the highest concentration of power occurred in 1950, immediately after World War II. Thereafter, it declined rapidly till 1982, but it has been rising slowly since then. Though the absolute value of the HCI is not important, we can get an idea of the order of magnitude, by noting that it fell dramatically in 30 years from about 1.0 in 1950 to about 0.07 in 1982. Therefore, the concentration of power potential has been increasing since 1980, but it remains well below that in the post-war years, according to the Madison GDP data used in the HCI.

The data we have used earlier in the paper is, however, the World Bank GDP-PPP data, which is available only from 1980 onwards. We get a slightly different but not too different a picture when we use this data and look only at the top ten countries. We first calculate the concentration among the ten largest economies i.e. the concentration in terms of economic size. As shown in Fig 3, among the ten largest countries, economic concentration peaked in 1999 and has been declining since then. Projected forward, the concentration reaches a trough in 2014 after which the concentration increases progressively to exceed the 1999 peak by 2030. Thus, the growth of the Chinese and Indian economies initially serves to disperse power among the large powers as China catches up with the USA, the largest economy (Fig 1a) and India catches up with Japan, the third (earlier second) largest economy(Fig 1b). Subsequently, economic concentration is projected to rise as China's economy rivals and exceeds that of the USA and India's economy does the same vis-à-vis Japan's economy

(Fig 3). What appears to be an unprecedent rise in the concentration of economic activity among the large powers by mid-century, is however, a return to a pre-industrial revolution era, when China and India were the two largest economies by a big margin.[5]

Virmani (2005d) and other papers argued that post-war globalisation is leading to a convergence of Asian per capita income to developed country levels. As China and India converge to world average income levels, the population size of these two countries will assert itself. From the start of the millennium (zero) to around 1800, these two countries were both the richest and largest countries in the world. So, the projected future would just be a return to a situation which existed for centuries in the pre-industrial era, before the industrial revolution transformed Europe and the USA into high income countries with economies larger than those of India and China.[6]

Fig 4: Economic (GDP PPP) Concentration Among the 10 Largest Countries (HCI)

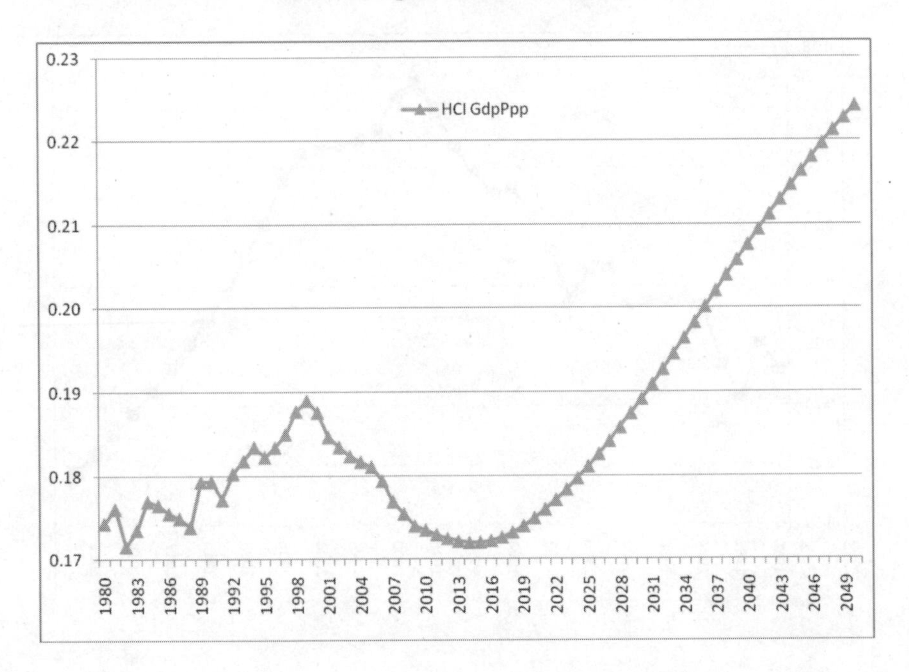

The measurement and analysis of concentration in terms of VIP², is slightly more complicated. The concentration of power potential (HCI using the VIP²) among the same set of ten countries (ten largest in terms of size) is shown in Fig 4. As in the case of economic concentration, the concentration of power potential is on a rising trend from 1980 to 1999. After peaking in 1999, it trends downwards much more sharply. This pattern suggests that the "unipolar moment" occurred in 1999. As conventional wisdom classifies the post-war years till 1980-85 as bipolar, the fact that the HCI is still larger than in any year between 1980 and 1990, is consistent with the view that the world is still unipolar. The HCI is, however, projected to go below the 1980s' trough around 2014. This can be interpreted as being consistent with emerging multipolarity. However, it is also consistent with emerging bipolarity or tripolarity.

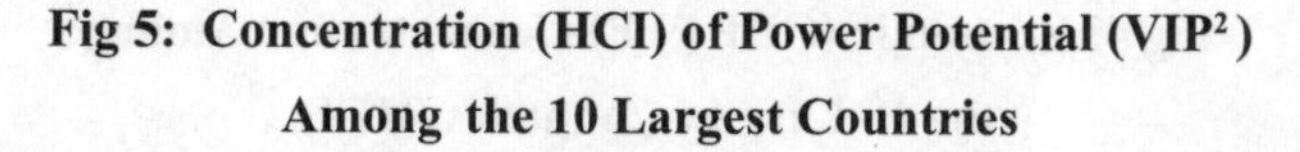

Fig 5: Concentration (HCI) of Power Potential (VIP²)
Among the 10 Largest Countries

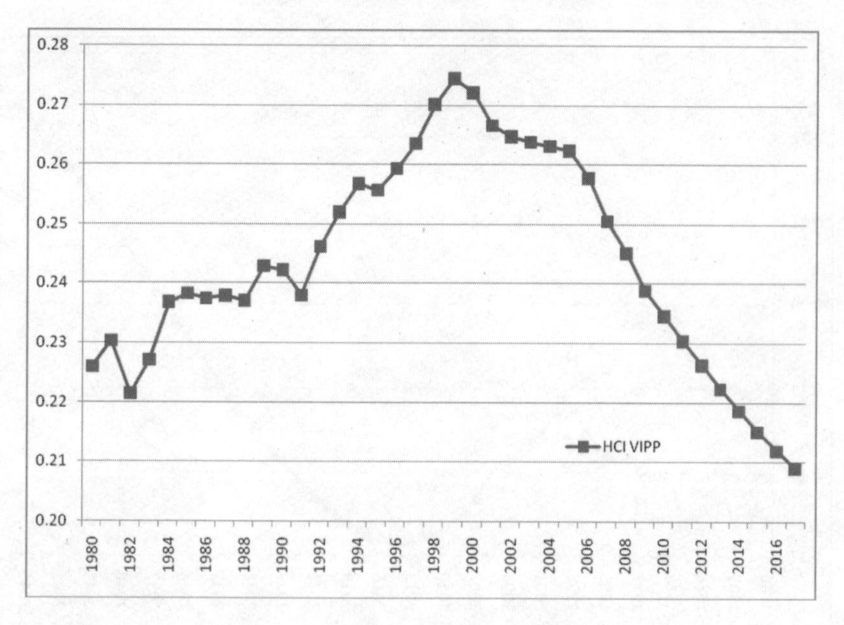

Global Institutions and Rules

This brings us to the most important part of this paper: the implications of the evolving global order for India's national security and international relations. In this section, we explore the implications of this milieu of powers at this point and the rise of India from our perspective. The assertion of power depends on relative power and balance of power, which students of international relations are quite familiar with. But the assertion of power also depends on the constraints imposed by international and national institutions. According to Alfred North, who got a Nobel Prize for his work on institutional economics, "institution" does not mean the physical structure like the UN, but the rules, both formal and informal, under which individuals and organisations operate. By analogy, global institutions are the formal and informal rules under which countries operate in the world. For simplicity, we will also encompass the global organisations such as the UN under this overall rubric of global institutions. From this perspective, a key distinction between countries that can be classified as global powers and those that cannot be so classified is that the former help shape and/or participate in the formulation of global rules. Similarly, the distinction between a superpower and a global power is that the former makes the rules along with any other superpower, with or without the participation of global powers. The main message of this paper is that India as an emerging global power must aim to influence the formulation and modification of these global rules under which countries operate on the world stage. The global powers help shape the rules, and we must participate in changing the rules of this global game. That is the main point and I will try to elaborate it. The second point is that globalisation is changing the factors which underline power, the economic, military and social elements of power. Globalisation is leading to a change in the relative importance of these factors. This is the underlying theme of this section.

Economic Institutions and Rules

Since the classic 19th century balance of power systems of Europe, which have been widely analysed, and depicted in history, there has been a lot of evolution in terms of both economic and social elements. What are these? Let us start with economic elements. As emphasised strongly in Virmani (2004b), economic size and strength is the foundation of national power. Globalisation of the world economy has resulted in an increase in interdependence since the end of the Cold War. This is the first time that every country, except Cuba, North Korea and Myanmar, wants to be part of this global system of trade and financial interdependence. This is a complete change from the so-called bipolar world of the USA-USSR. The USSR and many of its block partners chose to remain completely or partially outside this system. Currently, every potential global power, including China and Russia, wants to be a part of the global economic system. So, the nature of the rules and institutions of global economic governance are changing. In trade, we now have a WTO, and everybody wants to be a member, including Russia, which is still not a member, but has applied for it — so everybody wants to subject themselves to, and be part, of this economic system. Members voluntarily accept these rules of trade and competition. As North had said, the rules are not always formal. Even in the UN, WTO, etc, the formal rules are not necessarily always the most important. These institutions have informal rules that have evolved over time, and members voluntarily accept these rules in the expectation that they are in their own long-term interests.

The only exception, which was mentioned in an earlier section, is trade in oil and gas or energy. The oil/gas cartel continues to operate outside the rules of the world trade system with apparently complete impunity. It is an amazing thing. Nobody bothers to haul them up in the WTO. This probably requires some thinking. We are probably not strong enough to do that, but we should certainly be thinking about it. Similarly, the World Bank — again, there is a voluntary membership which deals with aid and social factors. The International Monetary Fund (IMF) deals with short-term BOP (Balance of Payment) loans, and though it is called the International Monetary and Financial Organisation, it rarely deals with other financial issues. The recent US-EU-global financial crisis exposed the fact that there are big holes in the global financial system. The IMF really does not perform the functions which are equivalent to the WTO trade. There is no global organisation which deals with capital flows in a manner anywhere comparable to the WTO in terms of trade and competition. These are some of the things that we are actually discussing in various fora.

For instance, these issues are currently before the G-20 meeting or the P-20. From our perspective, it is more important to participate in some way in the formulation of rules of the global financial system. One reason for being considered a power, or to become a great power is, therefore, to be a part of the club which frames the rules. Thus, for instance, if a new club emerges to deal with financial issues, we must endeavour to be a part of it. One possibility is a G-4 consisting of the US, EU, China and India, a club that would clearly be in our interest to bring about. We have to all think of how India can start creating, and becoming part of, the clubs which shape the rules of the game, so to say, the economic game.

Military Power and Rules

We move next to military power and its rules. Earlier sections have given the approximate dates at which China and India are likely to meet the benchmark for becoming a potential great power and superpower. Superpowers clearly define the rules of the game among themselves. Global powers can influence them if they play their cards well. Military power is an important instrument for enforcement of global rules, whatever they are. Of course, we have two recent glaring examples of failure of military power — Iraq and Afghanistan. However, even these examples show that the failure was less that of military power *per se* and more of other broader elements of power and the context and competence of its use. Military power also creates a space that allows a global power to act outside the rules. If you are a superpower, you have the liberty of acting outside the rules. You just invent your rules. It is not always possible to detail it now, but we will discuss how that happens. But other military powers can also create the space and have done so in the past to violate the rules of global military action. Pakistan's aggressive militarisation, way beyond its economic size and power potential, is a good example.

Two important new elements have come in and become a significant part of the canvas of military power and its use. Most people are familiar with the fact that nuclear weapons are game changers. As Mr. Subrahmanyam has emphasised, full-fledged war between nuclear weapon states is unthinkable; may be, that is too strong a word, but clearly it is a game changer. The emergence of alternative instruments of warfare, terrorism, guerrilla warfare, whatever you want to call it, is also familiar. Virmani (2006c) gives an indication of Pakistan's strategy of combining nuclear weapons and terrorism (along with incredible diplomacy) to violate global rules and conventions for 25 years. Pakistan's use of these elements is a case study worthy of inclusion in a graduate textbook.

Social-Political Power

The third element, which is new and an important element of post-war globalisation, relates to the issues of human rights. This is an element of power in which there is the greatest philosophical divide. Democratic societies put much more emphasis on individual rights, "The right to life, liberty and the pursuit of happiness,..." as the first democratic Constitution said. In contrast, Communist-Socialist ones such as China's emphasise collective rights. But leaving this aside, the issue of human and social rights is beginning to shape global rules in a way that has never happened historically. This change is consistent with our national, social and human development objectives. We must, therefore, incorporate it into our global objectives and strategy. We must develop a world vision or vision for the world which is consistent with our own ethos, our interest in, and commitment to, democracy. This would intelligently combine our national interests with a wider spectrum of peoples and countries.

Soft power has been widely talked about since it was first proposed by Joseph Nye. Instead of putting this as one of three elements of power, this paper treats it as a sub-element of socio-political power in a global environment. This, because globalisation is increasing the connect between national elements (economy, security and social) and international elements (economy, military and socio-political), I have to call it social, political elements of power as the important issue here is the perceived consistency in the narrative of power. When a country takes a global action, it is consistent with the vision that it has been propagating. Thus, Iraq is a failure because there was no consistency between professed vision and the actions taken after the initial military victory. From the perspective of global public opinion clearly, there was what psychologists call dissonance.

If India is to become a global power, we have to start thinking of this for ourselves. The perceived consistency of the narrative and the credibility and the legitimacy of the use of power. If we are going to use power, for instance, to fight the terrorists in either neighbouring country, we have to develop and focus much more on establishing this credibility. If I may say so, and this is not meant to be a criticism of the Foreign Office, I think there is too little of this propagation. At least, I do not see it; may be, they do it somewhere, but I have not seen enough of it.

In overall power, the will to power (i.e. the will to acquire and use power) is very important. For example, when the nuclear deal was hanging in the balance here, one wondered whether we really have the will to power. That has been partly but not wholly corrected.

Globalisation of National Security

Virmani (2006b) argued, "The changing role of India in the global environment requires a change in the institutional arrangements and approaches that we have for interacting with the current and potential global powers as well as with other countries and regional organisations. Over the next decade or so, we have to change our approach from one in which we react to the actions of other more active and hyperactive countries to one in which we try to influence the policies of others. We need to bring about a greater conceptual clarity to our foreign policy and revamp our international intelligence capabilities. We also need to ensure that the rise in the 'actual power' of India is commensurate with the rise in 'power potential,' by developing a comprehensive approach to Strategic Technology."

Since then, the Indo-US nuclear deal has been signed and the Nuclear Suppliers' Group (NSG) waiver obtained. These and other new issues have arisen to which we return after looking briefly at the issues raised in the above referenced EPW paper:

Foundation of National Security: Internal Security

"The most fundamental role of the State or the Government that claims to govern it is to provide security of life, limb and property to all its citizens. The State is defined by its monopoly over violence. In return, it must provide physical security to all. Only the latter can legitimise the former. The rule of law is the basic foundation of a modern market economy. It is a prerequisite for development and equitable economic growth. A country that cannot ensure rule of law and public security within its borders cannot become a global power. The writ of the government must be established over the Naxalite dominated interiors of the country and lawlessness (kidnapping and murder industry) eliminated from states/regions of the country where it has become prevalent, because of the abdication of responsibility by the political rulers of these states and the gross failure of governance. Given the limited capabilities of the governments in these states / regions, they must focus on universal provision of the basic public goods (and quasi-public goods) such as police, courts, roads…" [Virmani(2004a)].

The foundation of national security has to be internal security. Post-Mumbai terror attack, the issue of an anti-terrorism organisation has returned to centre-stage. In my view, having an anti-terrorism organisation(s) is necessary but not sufficient. Unless we correct our basic system of law and order so that we can protect the innocent and punish the guilty, special efforts can only be sustained for a few years, after which they invariably fizzle out or revert to the mean/average quality. I have been collecting and narrating horrendous stories about law and order for 20 years. The first time such a story came up about two decades ago, I said, well, Bihar has arrived in Delhi. The most recent story is of an average middle class person arrested and kept in jail for six months and released without charges. Though these are mere examples, there are very serious issues underlying them and we cannot have a sustained and sustainable anti-terrorism strategy unless we do something about them. The solutions are well known: the Supreme Court order on police reforms needs to be accepted and acted upon – fine-tuning and correction of anomalies can always follow.

Foreign Policy Constructs and Narrative

"If India is to play the 'Balance of Power' game to its advantage the Perspective Planning functions, intellectual effort/output, information/knowledge base and training systems for operational personnel must be reoriented and strengthened. Similarly, the conceptual basis of our foreign policy approach to our region (South Asia, East and Southeast Asia and West and Central Asia) must be widened and developed with greater sophistication" [Virmani(2006b)].

Foreign policy constructs have to be weaved into a foreign policy narrative that can be understood by engaged citizens and the media. For example, one narrative could be the development of South Asia for all South Asians, an essential element of which is the declaration/creation of South Asia as a Terrorism Free Zone (TFZ). Ideas have power when they can be understood by all. We have to have the words, sentences and paragraphs that express these ideas in simple terms that resonate with the intelligentsia and media of all relevant countries. We have a problem of terrorism. We have to globalise it in some way to our advantage. We have to think of ways in which to do that, how to involve the people, convince the people of other countries and get them to help push these things.

Global Intelligence Capability

"The Subrahmanyam committee long ago recommended the revamping of institutional structures for intelligence collection, integration, analysis and assessment and others have reiterated these recommendations since then. This becomes even more imperative in the changing context. Modern technology must be introduced to upgrade and widen the intelligence network (satellite, internet, radio waves) while strengthening traditional methods of human intelligence (e.g. foreign language training). We must develop an extensive capability for covert action to address asymmetric threats from non-state actors, fundamentalist terrorists, their handlers/ controllers/ motivators/ financiers" [Virmani(2006b].

This issue is well known, but perhaps needs some reorientation from domestic to cross-border or global threats.

Strategic Assets and Technology

"Strategic technology and assets [major defence platforms (submarines, fighter aircraft)], nuclear and aerospace technology, robotics, communication (interception/eavesdropping), internet (pattern recognition), radar (stealth, ECM) are critical to India's becoming a global power. This requires better planning of procurement and development of strategic technology and an integrated view across different organisations and departments currently engaged in them. Issues like the nature and amount of offset purchases and the trade-offs between acquisition and development would have to be addressed. Attention has to be given to specialised education and training of scientists and technicians (perhaps in-house) in nuclear, space, oceanic and other technologies. The National Security Council should have a special wing that carries out the planning and monitoring of the development of strategic technology and skills" [Virmani(2006b)].

Post-nuclear apartheid, a comprehensive and non-ideological view has to be taken to accelerate the flow of nuclear technology into India. For instance, the flow of private Foreign Direct Investment (FDI) into the nuclear and other strategic sectors must be facilitated, for the same reasons that we generally prefer FDI over Foreign Institutional Investment (FII) or debt flows. The former brings with it new ideas, skills and managerial capabilities that benefit the whole economy beyond the specific firms.

We must also encourage and incentivise the acquisition, adaptation and development of Communication-Interception-Security Technology (CIST) and remotely piloted aircraft (Predator).

Defence: Flexible Response

"The nature of the challenge facing India has been evolving over the years. The nature of constraints under which defence and foreign policy operates will also change – some will loosen, others will tighten. As open violent wars between major powers are becoming increasingly unlikely, those who wish us ill have developed indirect means of undermining our security. We need to increase the range of options available to us to attack the source or fountainhead as well as to pay back in the same coin that they choose to use or which are more cost-effective from our perspective and reduce the potential adversary's benefit-cost ratio. This requires a sharp increase in capability for unconventional warfare and unconventional means of defense.

News reports suggest that former Prime Minister, Shri I. K. Gujral dismantled the limited capacity we had for covert action against non-state actors who have been waging covert warfare against us for decades. This capability must be rebuilt and enhanced.

Economic Interdependence

"In this age of globalization the size of the market a country offers for other countries' exports is a measure of the (actual or potential) influence it has over them. It is easier to switch most imports from one country to another than to switch exports from an established buyer/country to others."[7] Thus, paradoxically, the size of a country's imports is a better measure of its economic power over others than the size of its exports. In some cases, the latter can even be a source of dependence and weakness in terms of power relations."[8] "FDI inflows into an economy also measure the relative attractiveness of an economy."[9] Though the stock of FDI, equity and debt owned by foreigners in an economy denote a mutual dependence of the host and the source country, once invested and to the extent they are difficult to disinvest, the balance of advantage shifts to the host country" [Virmani (2006b)].

An important issue for the future is the role of oil and gas and other energy resources. As already discussed, the cartelisation of petroleum/mineral oil supply by national oil companies/nation-states and the fact that on the demand side it is essential to the functioning of every country, makes it unique. Unlike all other goods and services, it is capable of being used as an instrument of power. Energy security (oil, gas, uranium) must, therefore, be an element of our overall national security policy.

International System Objectives

The question to be addressed is what should be our objectives in the international arena in terms of international security and national security? This has to be done keeping in mind our earlier analysis of the need to change global institutions and rules and the cross-border developments over the past few years, including the recent terrorist attacks. The paper suggests three short-term and two medium-term objectives. These are by no means exhaustive and readers will hopefully come up with many more suggestions.

Permanent Membership of UNSC

It is the thesis of this paper that national and international security are interlinked. Globalisation has, and will continue, to result in increasing these linkages. We, therefore, start with the most important institution of global security: the UN Security Council (UNSC). Virmani (2004b) suggested that the veto should be based on a voting share that depends on relative power, with a definition of what decisions can be taken with a majority and which require super-majority. Till such reforms come about, we must aim, as a globally recognised nuclear weapon power, to have veto power as a permanent member of the United Nations Security Council (UNSC).

The UNSC is much more important than some people assert, based on its demonstrated ineffectiveness in many situations of international security. People say it is a dying club, how does it matter whether you are a permanent member or whether you have a veto or not? What is overlooked by these observers is its informal role as a club where bargaining among permanent members takes place. There are also informal bargains made between permanent and non-permanent members. The formal rules determine whether you are a member or not and the informal rules determine how much of your power you can use to obtain bargains. Thus, it is important for India to aim to become a permanent veto bearing member of the UNSC, if it wants to be able to assert its emerging global power.

Global Anti-Terror Strategy

We need to develop a national anti-terror strategy that is part and parcel of a global anti-terror strategy and vice-versa. Viewed from another angle, we must help develop a global anti-terror consensus that views our anti-terror strategy as an essential part of fighting global terror.

The anti-terror strategy would include the following elements.

UN Consensus

India had introduced, several years ago, a resolution on terrorism. This needs to be pursued in the UN as well as in other forums. We also need to pursue the development of a globally accepted definition of terror that is consistent with the terrorist problems that we face and the domestic constraints under which the war on terror will have to be fought. The development of a narrative with which leaders across the world can empathise, could be helpful in attaining our objective.

As part of this consensus, we could build global public opinion to impose a travel ban and freeze the global financial assets of current and retired members of state intelligence agencies (such as the Inter-Services Intelligence ISI)—associated with terrorists or their activities.

Redefine Aggression and Self-Defence

The UNSC definition of aggression and self-defence needs to be modified in the light of the increasing importance of non-state actors and the use of failing states or parts of states as a base for cross-border terrorism. We must aim to change the UN definitions of aggression to include irregular guerrilla warfare, terrorism, etc and the definition of self-defence to include attacks against them on the territory of such failing states, regions. This requires an identification of conditions under which such counter-attack would be acceptable and the development of markers for identifying cross-border terrorism.

Elimination of Terrorist Networks

Elimination of terror networks was a key objective for us even before the Mumbai attack, as we have been the targets for close to a quarter century. One has often asked the question, whether there is any other case in history of a large country that was so pacifist as to not take any military action against a neighbouring country, that despite being,

- 13 per cent to 19 per cent of its size, trained, financed and supported terrorists to attack the larger country, and simultaneously convinced intellectuals and ayatollahs of the great powers that it was constantly under threat? [10]
- Allows its territory to be used by terrorists of different varieties and persuasion against a neighbour?

It is an amazing situation. There is something fundamentally wrong in India's conceptualisation and approach and we cannot afford this as a great power.

Access to Strategic Technology

Access to all possible areas of strategic technology must be an important goal. An important area of technology, where a minor apartheid regime still prevails is missile and space technology, which is subject to the Missile

Technology Control Regime (MTCR). We must aim to become a member of this club. At a more operational level, we must try and acquire pilotless attack aircraft, communication intercept equipment, etc. from countries that possess such technology. [11]

Objectives: Medium Term

Over the medium term, there are two objectives worth pursuing: the setting up of a new institution for global governance and a new strategic technology club of which we become founder members along with the potential great powers of the 21st century.

Global Governance Institution

Virmani (2004b) suggested a reform of the United Nations to take account of both current economic and power balance and evolving global socio-political principles such as democratic/human equality [also in Virmani (2006a)].

India has traditionally been hesitant to get involved in global democratic conclaves. This was partly justified when there was clear dissonance between the words and actions of those who proposed these issues. With this dissonance likely to be greatly muted, India being the largest democracy in terms of population and (soon) the third largest in terms of economic size, our approach needs to change. Democratic principles are of both kinds — globally acceptable and in our own favour. So, I do not think we should shy away from them. We should conceptualise a system of global governance which is based on some kind of democratic principles. The basic idea is to divide global governance into two components, which have fundamentally different management challenges and, therefore, require different distribution of global representation. These are global socio-economic-development issues and global security (national) issues, corresponding approximately to the current UN General Assembly (and the UN social agencies —WHO, etc) and the UNSC. Voting power in the first must be distributed according to democratic principles (all humans are equal and must, therefore, have an equal vote), while voting power in the second must be based on relative power, as the key issue is enforcement of rules through exercise of power.

There are two ways to implement the democratic principle. One is through direct election of representatives to the global assembly through individual votes or to let country governments have a vote in proportion to their population. In both cases, there is asymmetry between relatively free democratic countries and relatively controlled dictatorial or oligarchic systems. We, therefore, need to devise a system that adjusts the vote share for the degree of democracy. For this, we need a system in which an independent and credible rating organisation rates the degree of democracy – not only must such an agency be independent and unbiased, it must be seen to be so across the world. As existing US rating organisations may not be credible, we need a global organisation in which all democratic countries are shareholders. The rating (0 to 1 with 1 as perfect democracy) would be used to discount the individual vote or representation in the body that manages global socio-economic-developmental objectives.

The other component is global security. The Security Council is the instrument of military power and the use or threat of use of force to enforce the rules. By definition, the veto means that even the weakest member of the veto bearing powers can delegitimise any action even if the other members support it unanimously. How can countries with completely different amounts of power accept such a rule? A superpower or superpowers will never feel bound by such a ridiculous principle. Thus, the new system must have some kind of proportional vote based on a measure of power, if it is to be both fair and realistic and, therefore, credible and sustainable. Thus, it may be reasonable to expect a superpower with less than 50 per cent of the global power to convince a set of countries with more than 50 per cent of the global power that it should be authorised to take certain actions. Conversely, a global power that can convince 25 percent /50 per cent (whether or not it requires a super-majority respectively) of the power votes to oppose any action against it would provide acceptable safeguards to the powerful countries to subject themselves to the same global rules. As enforcement of rules through use of military power (or the threat of its use) is an important element of international security decisions, a powerful country will not surrender (on a long-term basis) its freedom of action, unless its vote is proportional to its power.

Therefore, the international security part of the new global governance organisation must provide for voting shares in proportion to global power (measured, for instance, by the VIP[2] or VIP). It must define situations which require majority decisions and those that require a super majority (66 per cent or 75 per cent). It must also provide for periodic reestimation and readjustment of voting proportions according to shifts in global power.

Strategic Technology Club

The second one is a new security technology club to supplant the NPT, NSG, MTCR, etc. In the existing structure, we are like beggars, trying to get in in each case. Why do we not float an alternative idea which is based on some kind of a global rationale and a global system? But one has to take account of the views of those who are powerful today, for instance, the issue of Weapons of Mass Destruction (WMDs), that is on the global table, would have to be addressed.

Conclusion

When we first projected (in 2004-05), the development of the tripolar global power structure by the middle of the 21st century, leading US foreign affairs journals refused to even consider this possibility seriously. Despite major upheavals in the last few years, including the reduction of India's per capita GDP and GDP at purchasing power parity by 36 per cent virtually at the stroke of the pen (i.e without a clear and transparent explanation and comparative data on the components of this unprecedented change), we have shown that our projections have changed only marginally. As we had anticipated, adverse (favourable) changes may delay (accelerate) the achievement of the bipolar and tripolar global power structure but not derail it. Thus, the dates for these two events have been pushed back by about 7 years.

The latest National Intelligence Council (2008) report recognises the eventual emergence of China and India as powers, but still asserts that they will be operating in a multipolar world. In this paper, we have tried to show that what has been termed a multipolar structure is merely the phase of high uncertainty during the transition between the unipolar structure at the end of the 20th century and the bipolar/tripolar structure that is likely to emerge during the second quarter of the 21st century. However, as Keynes said, "In the long run we are all dead."

The paper, therefore, focusses on the transitional phase spanning the first quarter of this century, which can be characterised as multipolar, but also by various other terms that have been used by other scholars. After analysing the nature of this transition phase it tries to draw out the implications for India, and what we must do to enhance national and international security. Besides traditional measures to enhance national security, it emphasises the need to modify/change the institutions and rules of the global economic, military and socio-political game and to work to set up new institutions that will better serve our national and global security interests.

Notes

1. Virmani (2004), Virmani (2005a b c d e f), Virmani (2006a b).
2. VIPP or VIP² or Virmani Index of Power Potential measures the overall economic capability of an economy, including capital, skills and technology. This is then combined with an index of strategic assets and defence capabilities to derive an overall index of power.
3. See Virmani (2005e) for an analysis of the other critical element of power, strategic technology.
4. In India, VIP is a common abbreviation for Very Important Person. However, countries are not persons, so to keep an allusion to the common usage of this abbreviation, the Index was christened Virmani Index of Power.
5. Virmani (2006d), Virmani (2007) and Virmani (2008).
6. Virmani (2006d), Virmani (2007) and Virmani (2008).
7. Perhaps the only exception is oil imports, where there is a monopoly/oligopoly

(OPEC cartel) with a small competitive fringe. It is highly misleading to treat the highly volatile spot and futures markets where most of the these small competitive suppliers operate as the market for oil.

8. This will become clear when the impact of the US-EU financial crises and the consequent global slowdown and trade is clear. I have asserted over the past year that export-led economies such as China will suffer a sharper decline in growth than those with deficits on merchandise trade account such as India. Others in the former category are Japan and Germany.
9. Luxembourg has shown a surge of investment as many European MNCs have shifted their headquarters to it in recent years.
10. The ratio which was 17 per cent in 1980, rose to a high of 19 per cent in 1992 and has since fallen to 13 per cent.
11. My thanks to Arundhati Ghosh for suggesting this in response to a specific question.

References

1. Kissinger, Henry, *Diplomacy* (New York: Simon & Schuster, 1994), ISBN 0671510991
2. National Intelligence Council (2004), Mapping the Global Future, Report of the NIC 2020 Project, Washington DC, December 2004.
3. National Intelligence Council (2008), Global Trends 2025, A Transformed World, Report of the NIC 2025 Project, Washington DC, November 2008.
4. Virmani (2002a), Arvind, "A New Development Paradigm: Employment, Entitlement and Empowerment," *Economic and Political Weekly*, Vol. XXXVII No. 22, June 1-7, 2002, pp. 2145-2154.
5. Virmani (2002b), Arvind, "A New Development Paradigm: Employment, Entitlement and Empowerment," *Global Business Review*, International Management Institute, SAGE Publications, Vol. 3, No. 2, July-December, 2002, pp. 222-45.
6. Virmani (2004a), Arvind, "Accelerating Growth and Poverty Reduction – A Policy Framework for India's Development," Academic Foundation, New Delhi, January 2004.
7. Virmani (2004b), Arvind "Economic Performance, Power Potential and Global Governance: Towards a New International Order", Working Paper No. 150, ICRIER, December 2004(www.icrier.org/wp150.pdf).
8. Virmani (2005a), Arvind, "A Tripolar Century: USA, China and India," Working Paper No. 160, ICRIER, March, 2005 (www.icrier.org/wp160.pdf).
9. Virmani (2005b), Arvind, "China's Socialist Market Economy: Lessons Of Success", ICRIER Occasional Policy Paper, April, 2005 (http://www.icrier.org/China05_policy7.pdf).
10. Virmani (2005c), Arvind, "VIP^2: A Simple Measure of Nations (Natural) Global Power," ICRIER Occasional Paper, July, 2005 (http://www.icrier.org/vipp4.pdf).
11. Virmani (2005d), "A Tripolar World, India, China & US," Lecture delivered at India

Habitat Centre on May 18, 2005 http://www.icrier.org/public/TripolarWrld_IHC5.pdf).

12. Virmani (2005e), Arvind, "Global Power From the 18th to 21st Century: Power Potential (VIP[2]), Strategic Assets & Actual Power (VIP), Working Paper No. 175, ICRIER, November 2005. http://www.icrier.org/WP175VIPP8.pdf
13. Virmani (2005f), Arvind, "China's Socialist Market Economy: Lessons Of Success," Working Paper No. 178, ICRIER, December 2005.
14. Virmani (2006a), Arvind, *Propelling India from Socialist Stagnation to Global Power: Growth Process*, Vol. I (Policy Reform, Vol. II), Academic Foundation, New Delhi, 2006.
15. Virmani (2006b), Arvind, "World Economy, Geopolitics and Global Strategy: Indo-US Relations in the 21st Century," *Economic and Political Weekly*, Vol. XLI No. 43-44, November 4-10, 2006, pp. 4601-4612 (http://www.epw.org.in/showArticles.php?root=2006&leaf=11&filename=10717&filetype=pdf).
16. Virmani (2006c), Arvind, "Trilateral Nuclear Proliferation: Pakistan's Euro-Chinese Bomb," *IDSA Monograph Series No. 1*, Institute of Defence Studies and Analysis, New Delhi, December 2006.
17. Virmani(2006d), Arvind, "Evolution of World economy and Global Power: From Unipolar to Tripolar World in Economic Diplomacy," Editor I. P. Khosla, Konarak Publishers, 2006.
18. Virmani (2007), Arvind, "Future Economic Contours of the Asia-Pacific and South Asian Region and their Impact on Global Security Architecture," USI National Security Lecture, 2006, in National Security, Global Strategic Architecture and Information Security, National Security Series, 2006, KW Publishers Pvt Ltd, New Delhi, 2007.
19. Virmani (2008), Arvind, "The Economic Foundations of National Power: From Multipolar to Tripolar World?" in V. R. Raghavan (ed.), *Economic Growth & National Security, Centre for Security Analysis*, Chennai, 2008.

India in a Multipolar World : How Economics is Shaping Strategic Relations

☐ Sanjaya Baru

The world is in the midst of a global power shift. The financial crisis and the economic recession, bordering on a depression, in North America and Europe, with varying global ramifications, is likely to reduce the size of the trans-Atlantic economies relative to that of East Asian economies. While the United States remains the world's pre-eminent military and technological power, the locus of economic power is shifting eastwards.[1] A multipolar world order is in the making, shaped as much by high politics as low economics!

India's strategic world view will increasingly be shaped by the requirements of what it will take for India to emerge as one of the "poles" of a multipolar world. Beginning with the new turn in India's economic policies in the early 1990s and followed by India's emergence as a nuclear weapon power, India has slowly been acquiring the necessary attributes of a regional power. If India can sustain its economic growth process and stabilise its neighbourhood and resolve its internal security conflicts, it will emerge as a regional and global power. These issues have been discussed in my book, *The Strategic Consequences of India's Economic Performance*.[2]

As a developing economy that is home to the largest concentration of poor and illiterate people, India has some distance to travel in terms of its economic development before it can acquire a higher global profile.

Moreover, how India responds to the current global crisis and slowdown and how it is impacted by it, will also shape the new dynamics of multipolarity. If China emerges even stronger, relative to India and the West, it will alter the balance of power in Asia and globally. India's economic strength and resilience will matter even more in the coming years in shaping its strategic profile.

India's strategic world view derives from this ground reality. This perspective was best elaborated by Prime Minister Manmohan Singh when he addressed the Combined Commander's Conference in Delhi in October 2005:

> Our strategy has to be based on three broad pillars. First, to strengthen ourselves economically and technologically; second, to acquire adequate defence capability to counter and rebut threats to our security; and third, to seek partnerships both on the strategic front and on the economic and technological front to widen our policy and developmental options.

Widening India's development options lies at the core of Indian strategic policy. India's relations with major powers, neighbours and with other powers, especially in Asia, are shaped by this single-minded focus on improving the well-being and livelihood security of its people and in creating an external environment conducive to India's sustained long-term development. Further, India's investment in human development and the building of a globally competitive knowledge-based economy will determine to what extent India is able to face the challenge posed by the rise of China.

The Primacy of Economics

The Indian economy is on an upward growth trajectory. Compared to virtual stagnation in growth in the first half of the 20th century, the first three post-Independence decades saw the Indian economy growing at an average rate of 3.5 per cent per annum. Between 1980 and 2005, the economy grew at close to 6.0 per cent and in the past five years, the growth rate has been closer to 9.0 per cent. This acceleration of growth is based on an increase in the savings and investment rate. These are likely to rise further and remain high, given the demographic trends, the spread of banking and of urbanisation. The Indian economy can be expected to grow anywhere between 6.0 per cent per cent and 8.0 per cent in the next two decades. The challenge for India is to ensure upwards of 8.0 per cent growth and to be able to sustain this over a long enough period of time. This requires investment in infrastructure, urban and rural, in education and skill building and in manufacturing and agricultural development. This also requires economic reform, especially in land and labour markets and public services delivery.

India's biggest strength on the economic front is the growth and spread of private enterprise. Indian business and professional classes are able to increasingly compete in the global market. However, India has to increase the productivity of all its factors of production, namely, land, labour and capital. To sustain such historically high rates of growth over the medium term, India will also have to address the challenge of adequacy of natural resources. On a per capita basis, India is not well endowed in terms of availability of water, energy, food and other natural resources.

Hence, food and energy security emerge as two vital areas of strategic importance. India requires a Second Green Revolution, with a wider ambit covering non-food crops, to help address this challenge in the context of rising incomes and changing consumption habits. India's energy needs will increase exponentially with growth, modernisation and urbanisation.

The strategic importance of energy security has acquired a new profile in the context of global warming and climate change. India has vast reserves of coal and a hydropower potential, but environmental concerns may limit the ability to tap into these resources. Hence, developing nuclear energy and non-conventional energy sources is a strategic priority. India's search for nuclear energy has shaped its diplomatic relations as well, in the context of the existing nuclear order.

It is, therefore, understandable that Prime Minister Manmohan Singh took the initiative to secure the approval of the Nuclear Suppliers Group (NSG) for an India-specific International Atomic Energy Agency (IAEA) safeguards agreement that enables India to access nuclear fuel and develop its energy programme. Prime Minister Singh told Parliament last year:

> It is my solid conviction that mass poverty can be removed only if we have a fast expanding economy. ……. If India has to grow at the rate of 8 % to 10 % and, maybe more, India needs rising amounts of energy. ……… In this context, we must never forget the primary motivation for India's nuclear programme was the production of energy, defence came much later…… all development is about widening human choices. And, when it comes to energy security, widening our choices means that we should be able to make effective use of nuclear power.

India will also have to import other natural resources and technologies to sustain its growth process. This means India will remain a trade dependent nation and will have a strategic stake in the nature of the global trading and maritime order. A rule-based, transparent multilateral trading system is in India's strategic interest. As a founder-member of the World Trade Organisation (WTO), India would like to see this organisation strengthened. India would like it to function in a manner that will be supportive of the developmental aspirations of the developing countries.

Given the importance of external trade, and the dependence on external sources of energy, ensuring the safety and security of the entire Indian Ocean region is vital to India's national security. These maritime interests will increasingly shape India's defence strategy, placing more importance on India's naval capabilities.

The Multilateral Dimension

In the transition from a bipolar to a multipolar world, with the 'unipolar' interregnum coming to an end, multilateralism will gain ground. The

United Nations system, related multilateral institutions and plurilateral associations like the G-20 and so on, will gain salience. India will be an active member of all such forums. The revitalisation of multilateral institutions will have to be based on a restructuring reflective of the new global order.

India has a strategic stake in strengthening multilateral economic institutions like the WTO, International Monetary Fund (IMF) and World Bank (WB), and in seeking a greater voice in their management and their policies. The past decade has witnessed the emergence of new groupings of countries based on changing economic and strategic equations, at both the global and regional levels. India will remain actively engaged in as many of these groups as necessary.

India's overdue membership of the UN Security Council will enable India to participate more actively in the management of multipolarity. However, how effectively India is able to play this role depends largely on domestic politics and the ability of the government of the day to carry public opinion at home with it in the articulation of Indian interests abroad.

Given India's energy needs, the outcome of multilateral negotiations on climate and carbon emissions will be of vital strategic importance to India. India cannot allow environmental security to undermine its economic security. Hence, the former will have to be incorporated into any development strategy.

The Regional Dimension

India's strategic neighbourhood spans a wide area from the Persian Gulf and East Africa to the Malacca Strait and the eastern reaches of the Indian Ocean. India will be actively engaged with all the littoral states of the Indian Ocean region. With the industrial economies of East Asia, on one side, and the markets of Europe, West and Central Asia and Africa, on the other, India is placed at the strategic crossroads of the emerging global economy. Hence, it will have to be actively engaged with this entire region, both economically and strategically.

Active participation in the East Asian Summit, and the creation of an East Asian Community, has strategic importance for India, for both political and economic reasons. The growth of East Asian economies and India's increased economic interaction with them has raised the profile of this community. Equally, India will remain engaged in the South Asian Association for Regional Cooperation (SAARC), strengthening regional cooperation in South Asia, and, at the same time, seek to keep other regional groups such as Bay of Bengal Initiative for Multi-Sectoral Technical and Economic Cooperation (BIMSTEC) and Indian Ocean Rim Association for Regional Cooperation (IORARC) active and relevant.

India's full participation in an "East Asian Community" has become the main focus of India's "Look East Policy" in recent years. Towards this end, India has vigorously pursued an India-ASEAN (Association of Southeast Asian Nations) Free Trade Agreement (FTA), in the face of considerable opposition at home, from business interests that feared competition, and political resistance within ASEAN, initially from Malaysia and subsequently from Indonesia.

Prime Minister Singh told the Third India-ASEAN Business Summit in New Delhi in October 2004

> A decade ago, we unveiled our "Look East" policy. This is more than a mere political slogan, or a foreign policy orientation. It has a strong economic rationale and commercial content. We wish to "Look East" because of the centuries of interactions between us……. We envision an *Asian Economic Community*, which encompasses ASEAN, China, Japan, Korea and India. Such a community would release enormous creative energies of our people. One cannot but be captivated by the vision of an integrated market, spanning the distance from the Himalayas to the Pacific Ocean, linked by efficient road, rail, air and shipping services. This community of nations would constitute an "arc of advantage", across which there would be large-scale movement of people, capital, ideas, and creativity.

Drawing attention to the strategic dimension to the ASEAN FTA, Dr Singh said in a letter addressed to Congress President Sonia Gandhi:

> Our approach to regional trade agreements in general, and FTAs in particular, has been evolved after careful consideration of our geo-political as well as economic interests. Although India has a large domestic market, our experience with earlier relatively insular policies, as also the global experience in this regard, clearly bring out the growth potential of trade and economic cooperation with the global economy. [3]

India has also launched a new "Look West Policy" in the neighbourhood by strengthening relations with the Gulf states. The historic visit to India of the King of Saudi Arabia and the initiative to seek closer economic relations with member countries of the Gulf Cooperation Council has shifted the emphasis away from politics to economics in India's relations with Arab nations.

For reasons of energy security and given the large dollar remittances sent home by the Indian community in the Persian Gulf region, as well as the employment potential of the region, relations with the Gulf region have great relevance for India's economic security.

As Prime Minister Singh told a session of the All India Congress Committee in November 2007:

> We have had a historic and long-standing relationship with the countries of the Middle East and the Persian Gulf. Over fifty lakh citizens are working there for their livelihood. We have always sought peace in this region – be it Iraq, be it Iran or be it any other country. The bulk of our petroleum and energy requirements come from this region and our energy security is critically dependent on the conditions there. It has been — and will be — our effort to reduce tensions there and promote peace and harmony.

In short, India will 'walk on two legs' of multilateralism and regionalism, making use of both to sustain a global environment conducive to its growth aspirations.

Economics also defines India's renewed engagement of Africa. Energy security and the search for natural resources and markets lends to Africa a new strategic dimension. India will invest in Africa and support African economies. At the First India- Africa Summit held in New Delhi, India extended a duty free Tariff Preference Scheme for Least Developed Countries (LDCs), 34 out of 50 LDCs being African countries. Under the scheme, India agreed to provide preferential market access for exports from all 50 LDCs, covering 94 per cent of India's total tariff lines, and 92.5 per cent of global exports of all LDCs.

India has allowed the corporate sector to be more actively engaged in Africa, rather than pursue government-based initiatives. Indian private enterprise has been far-sighted in employing local managers and workers, providing employment to the locals and training them in the process.

Challenges Before India

In dealing with this new world order while promoting economic and social development at home, India will face four challenges in four directions.

North. India will have to secure the support of the countries of the North to influence the policies of multilateral institutions and the outcomes from multilateral negotiations. The countries of the North will have to take a benign view of India's rise and accommodate India in the new managerial structures of the UN, WTO, IMF, WB and so on. Equally, in negotiations on climate change and in dealing with terrorism, India would like a more cooperative stance to be adopted by the countries of the North.

As a democracy, India is a natural partner of this group of countries. The North has a strategic stake in India's success as a "free market democracy", especially when non-democratic societies and monotheistic societies are challenging pluralism and liberalism across the world. As a plural and multi-cultural democracy, India has a strategic stake in the strengthening of democratic and plural nations around the world, including in its neighbourhood.

West. India's west, and, to an extent, its east, is mainly Islamic. India will need friends in the Islamic world to deal with the challenge of religious fundamentalism and *jehadi* terrorism. India will also need the economic support of the Arab world and the security of energy supplies from this region. Capital and labour flows bind the two regions together and create a win-win equation. Closer economic interaction with the Gulf region will benefit India, and vice-versa. At the same time, the Arab world must be made to appreciate India's strategic concerns, especially with religious extremism and *jehadi* terrorism.

East. India seeks closer integration with the economies of the East Asian region. In this region, it faces China's growing power and influence as well as the challenge of Islamic radicalism. How India and China manage their bilateral relations will influence to a great extent the new equilibrium of the emerging multipolar world. India and Japan must work together, including in the field of maritime security. Japan has been slow to invest in India. Ensuring closer relations between the two is of strategic importance for both.

South. The countries of the South are India's natural constituency. However, they will expect a rising India to be more supportive and helpful. South-South solidarity will come with a price tag. Does India have the deep pockets of a China, not to mention the West, to win friends in the developing world, and keep them? India can play the role of a "bridge" power between the countries of the South and North. India's vision of "inclusive globalisation" and its policies of "asymmetric liberalisation" should help consolidate this equation, especially with LDCs.[4] India's triangular relationship with Brazil and South Africa can also help strengthen this dimension of India's external engagement.

Global Economic Crisis and India

The ongoing financial and economic crisis will have implications for India's economic rise. It will alter the global economic and strategic environment. How this will impact on India's growth prospects remains to be seen. The crisis may have the effect of weakening US power and enhancing China's. Where it will leave Russia, Japan and Western Europe is still not clear.

China's massive "bailout offer" of close to US$600 billion, twice the size of India's total foreign exchange reserves, is a pointer to change underway. China's response to the trans-Atlantic financial crisis is similar to its response to the Asian financial crisis. It would like to be seen as being part of the 'solution' by the rest of the world, not as part of the problem. Just as the Asian financial crisis enhanced China's profile in Asia, its response to the present crisis will enhance its global profile. [5]

While it is easy enough to forecast a decline in US power and an increase in China's, one cannot rush to the conclusion that America is in terminal decline. The United States has shown remarkable ability to renew itself as an economy and as a technological and military power. Its openness and pluralism have been great assets. Under Barack Obama's leadership, America can renew itself and reemerge as a global power with global influence. In doing so, the US may well reach out to other economies and peoples, especially in Europe, Asia and Latin America. This can only enhance US power and profile, not diminish it. If China has to compete with the US for global influence, it will have to imitate the US and become a more open economy and open society. It remains to be seen if China can make that transition.

Therefore, to appreciate the geo-political implications of the current financial crisis for India, one must wait and see how the respective political leaderships of the US, Europe and Asia respond to the challenge at hand, and how India responds in turn. What is, however, clear is that reform of global governance is long overdue, and India will want its voice heard. The International Monetary Fund has failed in disciplining the developed economies when they falter on fiscal and financial policies. The global regulatory and supervisory regime requires improvement, with countries like India and China given a role commensurate with their size, experience and record in economic management. India's recent record in sound management of its financial sector and its record over the past two decades in external economic liberalisation and management deserve to be commended.

In recent years, India has benefited from a benign global environment and the robust performance of the domestic economy, with savings and investment rates reaching record highs of over 35 per cent of Gross Domestic Product (GDP), and ever burgeoning foreign exchange reserves. This has also enabled India to adopt a more generous profile within its own neighbourhood. The "asymmetric trade liberalisation" that Manmohan Singh offered all LDCs was one manifestation of this. India's willingness to walk an extra mile with the Southeast Asian nations and conclude an FTA with ASEAN was another manifestation.

However, if the Indian economy comes under renewed pressure, if the US is unable or unwilling to be as supportive as it was during the Bush II regime, and if China becomes more assertive, especially with India, while being more accommodative of the US and its other neighbours, the global environment for India could easily deteriorate. Such a deterioration of the external environment could encourage elements in India to adopt more assertive and less accommodative postures on national security issues. This would complicate relations with the major powers and neighbours, slowing down the process of change in the subcontinent, with attendant consequences.

Hence, to prevent any negative geo-political fallout of the present crisis, India must remain focussed on the economy and ensure sustained economic growth in the medium term. If the Indian economy can sustain an average of 8.0 per cent growth over the next decade, it would have finally succeeded in altering geo-political equations within the region. China may continue to outperform India, but strong economic performance by India can enhance its strategic autonomy and enable it to deal with immediate challenges closer home. India must remain actively engaged with all its immediate and wider Asian neighbours, in South, Southeast and West

Asia. India must also remain engaged with all major powers, especially the United States.

India must work on the assumption that China will continue to perform well and will use its economic prowess to strategic advantage. China's influence is on the rise within Asia and globally. If the US and Europe remain preoccupied with domestic economic issues and the management of the global economy, China will increase its global influence both by continuing to perform well on the economic front and by working with the major powers in the management of the global economic and political order.

India's options stand out: continued focus on the economy; making the growth process inclusive and more efficient; increased public investment in infrastructure, in education and skill-building; financial stability and prudent fiscal management; exchange rate stability and sustainability of the external account. In short, sustaining its growth process is the only insurance India can take against any adverse geo-political fallout of the global financial crisis.

Notes

1. See Paul Kennedy, *The Rise and Fall of the Great Powers, Economic Change and Military Conflict from 1500- 2000* (Random House, 1987); Kishore Mahbubani, The *New Asian Hemisphere - The Irresistible Shift of Global Power to the East* (New York: Public Affairs 2008).
2. Sanjaya Baru, *Strategic Consequences of India's Economic Performance* (New Delhi: Academic Foundation, 2006).
3. Prime Minister's letter to UPA Chairperson, April 20, 2006. See "Manmohan Allays Sonia's FTA concerns", *Indian Express*, May 8, 2006.
   For an elaboration of Manmohan Singh's "strategic world view," see Sanjaya Baru, "India and the World – Economics and Politics of the Manmohan Singh Doctrine in Foreign Policy", *ISAS Working Paper*, November 2008, www.isas.nus.edu.sg
4. On this, see Baru, Ibid.
5. The Western powers would have approved of China's decision not to bail out Pakistan but encouraged it to go to the IMF. China will seek to increase its vote share in the IMF and a role in its management. China has jealously held on to its management positions in the Asian Development Bank.

India's Foreign Policy Concerns and Strategy

□ G. Parthasarathy

India's foreign policy strategy in the 21st century will be guided by a number of diverse considerations. Now regarded an "emerging power" in the councils of the world, the predominant focus of attention of the people in India is going to be on devising an environment, both external and internal, which will help the country proceed on the path of around double-digit economic progress, with economic growth being as inclusive as possible. In a diverse and pluralistic country like India, the very process of economic growth will inevitably generate social, ethnic, linguistic and sectarian tensions. While corruption and criminalisation of politics have severely strained our body politic, adversely affected economic growth and added to public cynicism, there is growing confidence, domestically and internationally, that India has the strength and resilience to overcome these challenges.

The foreign policy challenges having national security dimensions that India will have to deal with arise from a number of issues and factors. Terrorism sponsored by radical Wahabi oriented Islamic groups, whose ideological leanings influence sections of minority opinion within India and motivate groups like the Students Islamic Movement of India (SIMI), is going to remain an important challenge for at least the next decade or two. Developments within Pakistan, where the army establishment has for long used radical Islamic groups to promote its strategic goals, has become a major challenge for regional security, especially in the aftermath of the ouster of the Taliban in Afghanistan. The "War on Terror" has dispersed, but not destroyed the terrorist threat emerging from India's neighbourhood. Two other factors will continue to play a crucial role in determining the contours of Indian foreign policy. With its demand for energy resources rising rapidly, India will have to focus increasing attention on the Persian Gulf, where over two-thirds of the world's resources of oil and gas are located. Finally, the rise of Russia under the leadership of Vladimir Putin appears set to bring in new rivalries between Russia and the Western world. China's role in this emerging situation will be an important factor in shaping the emerging balance of global power.

In these circumstances, one has to analyse how global power equations are going to take shape. A crucial question for India is how it figures in the emerging equations of global power. Few analysts would disagree with some of the findings of the US National Intelligence Council (NIC) in its report about emerging global power equations till the year 2020, entitled "Mapping the Global Future." The principal finding in the NIC report is that even in the year 2020, the US will remain the most powerful actor in the world economically, militarily and technologically. US pre-eminence will, however, not be undisputed and its position cannot be sustained if it loses its present technological edge. Studies by a large cross-section of economists across the world broadly tally with this assessment.

The NIC report recognises that the emergence of India and China, as well as other new global players, will transform the geo-political landscape in this century, with impacts similar to the rise of a United Germany in the 19th century and the United States in the 20th century. Given the rapid rates of economic growth in Asia, the balance of economic power will inevitably shift from Europe to Asia, which will become the world's manufacturing hub, in the coming decades. The main challenges that countries in Asia will face to sustain their growth rates will arise from instability in the oil rich Persian Gulf region, which could jeopardise security of energy supplies and from terrorism and the proliferation of nuclear weapons. The NIC report notes that facilitated by global communications, radical Islamic ideology will spur terrorism globally in the coming years. While this may be true, the US will hopefully recognise that the use of force for effecting "regime change", or selectively targeting countries for non-proliferation goals can only exacerbate an already volatile situation in the Persian Gulf. Thus, we are going to see greater trends towards multipolarity in the global order, though the US appears likely to remain the pre-eminent global power.

In this emerging scenario, we need to analyse how the US and China will view other players in the coming years. While the Bush Administration followed a dual policy of engagement and containment in dealing with both Russia and China, will such a policy continue under a new Democratic Party dispensation? Senator Hillary Clinton (formerly a frontrunner in the US Presidential elections of 2008) and now Secretary of State, stated: "Our relationship with China will be the most important relationship in the world in this century." While acknowledging differences with China on issues like human rights, religious freedoms and Tibet, Senator Clinton noted: "There is much that the US and China can and must accomplish together." Hillary Clinton, however, warned the Russians against "regional interference", evidently alluding to the ongoing rivalry between the US and Russia over the oil and gas resources of the Central Asian and Caspian regions and in the former Soviet Republics. Many Clinton aides subsequently hitched their bandwagon to Senator Obama and the Democratic Party President may well adopt China-centric policies, akin to those the Clinton Administration adopted in the 1990s. The McCain campaign was also marked by strident rhetoric about relations with Russia.

One of the major premises of American foreign policy has been that with the disintegration of the Soviet Union, a policy of "containment" of a weakened, impoverished and dispirited Russia could succeed by an aggressive expansion of the North Atlantic Treaty Organisation (NATO) alliance to Russia's borders. This was to be accompanied by domination of the oil and gas resources not only of Russia, but also of the former Soviet

Republics in Central Asia and the Caucasian regions, by American and Western oil companies. The aim was to integrate the Caucasian Republics like Georgia, Armenia and Azerbaijan into NATO and construct pipelines bypassing Russia, to carry oil and gas from countries like Kazakhstan and Azerbaijan to ports like Ceyhan in Turkey, for onward shipment to America's NATO partners in Europe.

The American strategy for access to Caucasian energy resources was spelt out by Ariel Cohen a leading analyst of the Washington- based neo-conservative think-tank, the "Heritage Foundation" in 1999. Cohen then proclaimed: "US interests in the Caucasus boil down to providing guarantees of greater independence to Georgia, Armenia and Azerbaijan; controlling Iran; ensuring access to energy resources, and precluding the possible revival of Russian imperial ambitions in the region." To achieve these objectives, Cohen urged more political support for an oil pipeline project bypassing the Russian pipeline networks, from Azerbaijan to the Turkish port of Ceyhan. He argued that if this was not done, Russia and Iran would control access to, and investment in, a major part of the Caucasian energy resources, making the West dependent on Russia and Iran. As the pipeline was to be constructed through Georgia, Cohen urged the promotion of "security collaboration with Georgia" and expanding ties with Azerbaijan and Armenia, as a "signal" to Moscow that its support for separatism in South Caucasus would lead to an end of US economic assistance. Worse still, Cohen urged that for the US to achieve its strategic objectives, it should open talks with leaders of North Caucasian ethnic groups— a euphemism for promoting Muslim separatism in Russia's Chechen and Dagestan regions.

What Cohen and American policy-makers failed to anticipate was that under the leadership of Vladimir Putin, Russia would stage a remarkable economic recovery. In less than a decade, Russia has emerged as a global player, shrewdly using its position as holding the world's largest resources of natural gas, the second largest resources of coal and as the world's second biggest producer of oil, to effectively make America's European allies look to it with respect and realism. Under Putin's leadership, Russia's economy has grown at over 7 per cent annually since the year 2000. Russia has wielded diplomatic clout as a Permanent Member of the UN Security Council and its participation in groupings like the G-8, the Middle East Quartet, the Six-Power initiative on North Korea's nuclear programme, Asia-Pacific Economic Cooperation (APEC), Organisation for Security and Cooperation in Europe (OSCE) and in the Russia-NATO Council. Russia has also expanded its power potential in crucial areas like its defence and space industries. Moreover, with significant Russian minorities in the former Soviet Republics like Kazakhstan and Ukraine, Moscow has signalled that it will not remain unconcerned about how ethnic Russians are treated in these countries.

Unmindful of these changes in power equations, the Americans have attempted to virtually encircle Russia by proposing missile defences in former Warsaw Pact members and by encouraging Russia's neighbours like Ukraine and Georgia to join NATO. On the eve of the last NATO Summit in Bucharest in April 2008, President Bush commended the "bold decision" of Ukrainian President Viktor Yuschenko to apply for NATO membership and for dispatching Ukrainian troops to Iraq and Afghanistan. Bush added: "In Bucharest this week, I will continue to make our position clear about our support for MAP (NATO membership) for Ukraine and Georgia." The Kremlin strongly opposed NATO expansion and warned that it could lead to Moscow's recognition of the breakaway regions of Abkhazia and South Ossetia in Georgia, which had asserted their independence and were protected by Russian peace-keepers. While the US and the European Union (EU) were opposed to independence for these regions, Russia's Parliament proclaimed that if the Western powers could recognise the independence of Kosovo after military intervention, there was no reason Russia could not do likewise in Georgia.

Georgia's US educated President, Mikhail Sakashvili, gave the Russians the opening they seized when, bolstered by arms supplies from Ukraine and expectations of full scale American support, he mounted a military operation to establish control over South Ossetia on August 8. Within days, the Georgians were humiliatingly defeated by the Russians and forced to accept EU mediation by French President Nicholas Sarkozy. The proposals agreed to between Sarkozy and Russian President Dmitri Medvedev include a provision for "international talks on the future status of Abkhazia and South Ossetia and ways to provide for their security." Russia views this as EU acceptance of the impossibility of return to the pre-war status quo. With Chancellor Merkel of Germany and French Prime Minister Francois Fallon having opposed NATO membership for Ukraine and Georgia, the US now finds that apart from support from the ever loyal British, its other major partners in NATO like Germany, France and Italy, which are increasingly dependent on Russian oil and natural gas, have no desire to embark on another Cold War against Russia.

These developments are going to have profound implications on global politics in the coming years. The Americans are not going to give up their attempts to encircle Russia. The Russians, in turn, could make American diplomacy on issues like the nuclear programmes of North Korea and Iran very difficult, should the Americans become confrontational. Former Soviet Republics like Kazakhstan, which have huge energy resources, will now become more cautious in their dealings with the US out of fear of Russian reactions. Given the growing instability and the growth of Taliban influence along the Pakistan-Afghanistan border region, the US is looking for alternate supply routes through Central Asia to Afghanistan. This can only be achieved if there is a measure of strategic understanding between the US and Russia even though the Russians agreed to provide transit facilities for non-military supplies to Afghanistan during the Bucharest NATO Summit in April 2008. It remains to be seen how the Americans reconcile their traditional policies of containment of Russia with their need for Russian understanding on a wide range of issues.

In the face of rivalry from Russia, the Americans will now seek closer ties with Beijing—a development of considerable importance for India and the balance of power in Asia. Like in the Nixon and Clinton years, China will seek to prove that it is a useful and "responsible" partner of the US and endeavour to isolate India diplomatically. But intrinsic differences between the US and China on issues like China's support for undemocratic governments in countries like Sudan and Myanmar and China's relentless drive for securing control of energy and raw material resources in Africa, Latin America and Central Asia will remain sources of rivalry and differences between China and the USA. The dynamics of emerging relationships in the US, Russia, China triangle are now set to play a far more important role than in the days immediately following the end of the Cold War.

India has traditionally maintained close relations with Russia. Even today, Russian supply of enriched uranium keeps the Tarapur nuclear power plant functioning in the face of an American embargo. Moreover, cooperation in crucial areas of our defence requirements, like the acquisition of cruise missiles and futuristic fifth generation fighter aircraft are based on joint collaboration and development with the Russians. New Delhi would be well advised to ensure that on energy related issues like the proposed pipelines with Iran and Turkmenistan and developments in Central Asia, it pays greater attention to Russian policies and imperatives. Moreover, Indian diplomacy should seek to promote a dialogue between the US, on the one hand and Russia and Iran, on the other, on issues like the resurgence of the Taliban in Afghanistan, where US policy has been to exclude these countries, as far as possible. Even though the Americans were inclined to show accommodation of Chechen separatism earlier, they now appear to have a better understanding of Russian imperatives, after the terrorist strikes of 9/11.

The NIC report notes that with relatively young populations and work forces, expanding educational facilities and the benefits of globalisation, Asia, with 60 per cent of the world's population, will become the manufacturing hub of the world in the coming years. China and India will alone provide 1.1 billion of the labour force of 1.7 billion in the

Asia-Pacific region. In the next half century, as the developed world, and especially Europe, ages, a younger and better educated work force in Asia will become the driving force for global manufacturing and growth. The balance of power will shift — particularly from Europe to Asia. It is in this context that India has to develop a comprehensive policy in promoting widespread engagement not only with the fast growing Asia-Pacific region to its east but also with the oil rich Persian Gulf region to its west, where nearly four million Indian nationals now live and work, remitting around $ 25 billion to India's economy every year. This has to complement a drive to deal with challenges that India faces in its South Asian neighbourhood, in the conduct of relations with its South Asian Association for Regional Cooperation (SAARC) neighbours.

India faces an uncertain political situation in virtually every one of its South Asian neighbours, which are members of SAARC. Maldives is facing a crisis of domestic credibility of its leadership, with periodic acts of terrorism perpetrated by radical elements motivated by religious-political ideologies prevalent among, and advocated by, politico-religious organisations in Pakistan and Saudi Arabia. Moreover, the fairness and impartiality of elections held by President Gayoom have been called into doubt. The ethnic conflict continues in Sri Lanka, with President Rajapakse's government appearing determined to seek a military solution — a policy that can only complicate the search for a durable political solution. Bangladesh also faces an uncertain political future, with no sign yet about precisely how the political process is going to be put back on track and when the army will return to the barracks. Past experience during the rule of President Ershad suggests that given the political awareness of its people, military rule over a prolonged period is neither desirable nor feasible in Bangladesh. Moreover, radical Islamic groups operating from Bangladesh are now getting increasingly involved in acts of terrorism across India. It is heartening that the army rulers recognised these realities and conducted free and fair elections in the country — elections that resulted in a rejection of fundamentalist parties by the people in Bangladesh. In Nepal, a Maoist led government has assumed office with the primary task of drafting a new Constitution for the country, after years of a Maoist insurrection and a difficult transition from a monarchy. A quiet but major shift has now occurred in India's approach to developments in its immediate neighbours in South Asia. Whereas in the past, India was averse to the involvement of external powers in shaping developments in these countries, India now works together with the US, EU, Japan and even the UN in dealing with internal developments in countries in South Asia, on the basis of shared interests.

There has been some movement forward in India-Pakistan relations since the SAARC Summit in Islamabad in January 2004. The statement issued at the end of this meeting on January 6, 2004, contained the following salient points: (1) "Prime Minister Vajpayee said that in order to take forward and sustain the dialogue process, violence, hostility and terrorism must be prevented;" (2) "President Musharraf assured Prime Minister Vajpayee that he will not permit territory under Pakistan's control to be used to support terrorism in any form"; and (3) "To carry forward the dialogue process, the President of Pakistan and the Prime Minister of India agreed to commence the process of the composite dialogue in February 2004. The two leaders are confident that the resumption of the composite dialogue will lead to peaceful settlement of all bilateral issues, including Jammu and Kashmir, to the satisfaction of both sides."

Coinciding with the commencement of the "composite dialogue" process was a statement by Pakistan's Minister for Railways and former Chief of the Inter-Services Intelligence, Lt Gen Javed Ashraf Qazi in Pakistan's National Assembly on March 10, 2004. Gen Qazi stated: "We must not be afraid of admitting that the Jaish-e-Mohammed was involved in the deaths of thousands of innocent Kashmiris, in the bombing of the Indian Parliament, in Daniel Pearl's murder and in attempts on President Pervez Musharraf's life." Till this admission, the Pakistan government had steadfastly claimed that the Jaish-e-Mohammed, which was declared an international terrorist organisation in the aftermath of the terrorist strikes in

New York and Washington DC, had nothing to do with the terrorist attack on India's Parliament. The appointment of two negotiators, Mr. Tariq Aziz from Pakistan and Ambassador Satinder Lambah from India, after a new government assumed office in India following general elections in 2004, for "back channel" discussions on a wide range of issues, with particular focus on Jammu and Kashmir (J&K) signalled that both sides were keen to seek new and innovative ways to address even complex issues.

The Composite Dialogue Process is made up of discussions on a range of issues at the senior official level. Under this dialogue process, the two countries discuss issues pertaining to peace and security, confidence-building measures, economic and commercial cooperation, terrorism and drug smuggling, cultural ties, people-to-people contacts and also seek to address issues like J&k, Siachen, the Sir Creek issue and the construction of the Wullar Barrage in J&K. Perhaps the most important agreement that has contributed to relaxation of tensions between the two countries is the respect shown by both sides to the ceasefire along the Line of Control (LoC) and the International Border (IB) in J&K since November 2003. Reinforcing this has been the establishment of more secure communication links between the Directors General of Military Operations (DGMOs) of the two countries and a direct hotline between the Foreign Secretaries of India and Pakistan. Similar communication links have been established between the Coast Guards of the two countries, while an agreement to avoid incidents at sea between the Navies of the two countries is under discussion. President Musharraf visited India in April 2005 for meetings with Prime Minister Manmohan Singh. The two heads of government met again on September 14, 2005, in New York and September 17, 2006 in Havana. The India-Pakistan Joint Commission headed by their Foreign Ministers which was established in 1983 and has languished for the last several years has been revived, providing the framework for regular ministerial level interaction, to give added impetus to the official level composite dialogue.

The continuing dialogue has also resulted in certain important Confidence-Building Measures (CBMs) on nuclear related issues with the signing of agreements on "Reducing the Risk from Accidents Relating to Nuclear Weapons" and the "Pre-notification of Flight Testing of Ballistic Missiles." Pakistan has not formally enunciated a nuclear doctrine, evidently because influential cross-sections of its establishment felt that "a policy of ambiguity would appear best for Pakistan's security." Subsequently, however, in 2002, Lt Gen Khalid Kidwai, head of the Strategic Plans Division that manages Pakistan's nuclear arsenal and operations, made it clear that Pakistan's nuclear deterrent was entirely India specific. Lt Gen Kidwai spelt out four distinct thresholds for use of nuclear weapons: loss of large parts of territory (space threshold); destruction of large parts of land and air forces (military threshold); economic strangulation (economic threshold) and political destabilisation or internal subversion (political threshold). This enunciation, while not constituting a formal doctrine, nevertheless confirms what most Indian strategists have long held—that sections of the Pakistan establishment have wrongly believed that India and the international community would be unduly influenced by rhetoric, indicating that Pakistan had a low and unpredictable nuclear threshold.

The Composite Dialogue Process has also led to some progress in resolving the issue of demarcating the boundary in the Sir Creek area, though differences on resolving the Siachen issue still remain. But the last three years have seen significant progress in furthering people-to-people contacts between the two countries, with the opening of new rail and road links. There has been some progress in expanding trade ties, though Pakistan has stated that it will not reciprocate the Indian grant of Most Favoured Nation (MFN) status to imports from Pakistan. This has inevitably resulted in tardy implementation of the agreement reached by countries in SAARC on the establishment of a South Asian Free Trade Area (SAFTA). The implementation of SAFTA, which requires member states to provide transit facilities to each other for intra-SAARC trade will also be adversely affected till Pakistan provides India transit facilities for its exports to Afghanistan, which was admitted to SAARC only earlier this year.

While one can derive some satisfaction from the dialogue process with Pakistan in the past few years, there are signs that following the recent elections in Pakistan, the exit of Gen Musharraf and escalating tensions across the Afghanistan-Pakistan border, there are growing differences within the Pakistan establishment over the country's approach to relations with India and Afghanistan. The Pakistan Army establishment, led by Gen Ashfaq Pervez Kiyani, now seems to be asserting its preeminence, in regard to traditional policies of "bleeding India" in J&K and elsewhere and seeking "strategic depth" in Afghanistan. This is clearly at odds with what has been articulated by President Zardari on relations with Pakistan's northern and eastern neighbours. With the army evidently determined to reassert its traditional role in the country's national life, Pakistan appears to be headed for greater radicalisation internally, with the state machinery maintaining tenuous control over the Northwest Frontier Province (NWFP) and uneasy relationships, with the potential for escalation of tensions, with both India and Afghanistan. The terrorist attacks on our embassy in Kabul and in Mumbai clearly signalled that the Pakistan Army establishment is determined to use terrorism as an instrument of state policy.

While India has been commended universally for showing "restraint" in the aftermath of the 26/11 terrorist attack on Mumbai, questions are now being raised about whether this "restraint" is not a reflection of Indian weakness, indecision and a lack of a coherent policy in addressing Pakistan sponsored terrorism on its soil and on its interests in the neighbourhood. The Mumbai attack has also internationally exposed the weaknesses in India's internal security mechanisms, with Chinese commentators even suggesting that the attacks have caused a setback to India's claims to being an emerging power. No foreign policy can succeed if India is perceived as lacking basic strengths in its internal security and lacking the will to respond swiftly and appropriately to grave provocations against its territorial integrity. It remains to be seen how India develops the mechanisms, covert and overt, to deal with such provocations.

While there has been a marked improvement in the climate of Sino-Indian relations in recent years, the relationship between India and China is still clouded by mistrust. While China views improved US-Indian relations with suspicion, India retains memories of close Sino-US cooperation detrimental to its interests, during the Nixon and Clinton Administrations. There is concern in India about what is perceived as China's policy of "containment" of India, marked by growing Chinese interest in maritime facilities in countries like Myanmar, Sri Lanka, Maldives and Pakistan. China's supply of weapons to the beleaguered regime of King Gyanendra at a time when the international community was endeavouring to assist in a process of democratic change in the Himalayan Kingdom and its continuing cooperation with Pakistan in nuclear and missile development, have only accentuated Indian misgivings. China's growing assertiveness in its territorial claims on Arunachal Pradesh, its efforts to undermine India's efforts for regional influence by opposing India's participation in forums like the East Asia Summit and the summit level Asia Europe Meetings (ASEM), its undermining of India's candidature for permanent membership of the UN Security Council and its crude attempts at the Nuclear Suppliers Group (NSG) to torpedo efforts to end global nuclear sanctions on India, indicate that dealing with China is going to be major challenge for India in the coming years.

Despite these differences and challenges, bilateral trade and economic relations between Beijing and New Delhi are booming and the two countries have embarked on a series of measures to enhance mutual confidence. Moreover, on multilateral issues like global warming and in the Doha Round of the World Trade Organisation (WTO), common and shared interests and perceptions have led China and India to cooperate with each other. Till now, the Indian response to Chinese policies of "containment" and "strategic encirclement" has been largely defensive and indeed almost apologetic. But, as India's economic and military potential grow and the country's soft power expands in an electronic age, India would need to deal with Chinese policies with more proactive measures in its relations with countries like Japan and Vietnam, by developing a larger footprint in our relations with the Association of Southeast Asian Nations (ASEAN) and a more imaginative economic engagement with Taiwan.

There are several major challenges that India is going to face in the conduct of its relations in the coming years. As US-Russian relations and rivalries grow, India will have to closely observe China's moves in the context of such rivalry, while ensuring that it does not get drawn into the vortex of the tensions and rivalries between the US and its NATO allies, on the one hand, and Russia and its Collective Security Treaty Organisation (CSTO) allies on the other. We should also be clear about the limitations of our power and avoid getting too involved in conflicts like the Arab-Israeli issues in which we have little leverage to influence events. China's role in the Middle East is a striking example of its recognition about avoiding getting directly involved in disputes it has little leverage to influence constructively and in keeping with its own strategic interests. But, this should not prevent our going ahead with wider engagement economic, diplomatic and strategic, with the Arab countries in the neighbouring Gulf region. We should move ahead expeditiously in finalising a Comprehensive Economic Cooperation Agreement with members of the Arab Gulf Cooperation Council (GCC), with provisions for free trade in goods, investments and services. Closer consultations with Iran on issues like developments in Afghanistan are also essential in our diplomacy in the Persian Gulf region.

India's "Look East" policies have visibly increased India's diplomatic profile in its eastern neighbourhood over the past two decades. As a full Dialogue Partner of ASEAN and as a member of the ASEAN Regional Forum (ARF), India is now on the verge of concluding a Free Trade Agreement (FTA) with ASEAN after concluding bilateral FTA with two ASEAN members — Thailand and Singapore. Within South Asia, the SAFTA, though limited to trade in goods and services, should be treated as only the first step in moves towards free trade in investments and services, with the goal of progressively moving towards a Customs Union and Economic Union in South Asia. Pakistan will resist moves towards such economic integration in SAARC, unless there is a radical change in policies that it has deemed to be in its national interests for over three decades now. But, this should not deter us in seeking to isolate Pakistan within SAARC, if it chooses to remain negative. Pakistan will find itself further isolated, if we move purposefully towards greater economic integration within the Bay of Bengal Initiative for Multi-Sectoral Technical and Economic Cooperation (BIMSTEC) — an organisation that can act as a bridge between South and Southeast Asia. With India now a member of the East Asia Summit and of the ASEM, efforts should now be made to seek a more meaningful role in G-8 Summits, to act as a bridge in reconciling differences between the developed and leading developing countries. We will have to recognise that groupings like our association with Brazil and South Africa have far greater relevance in today's world than anachronistic forums like the Non-Aligned Movement.

After the Pokhran nuclear tests of 1998, India was relatively indifferent towards reiterating its commitment to nuclear disarmament, or in undertaking imaginative initiatives on issues of nuclear disarmament. This approach has, however, changed in recent years and we are once again renewing our commitment to the goal of universal nuclear disarmament. In this effort, we have been encouraged by the fact that even strong proponents of nuclear deterrence like Henry Kissinger, Senator Sam Nunn, William Perry and George Schultz have now realised the importance of seeking to work towards the goal of a nuclear weapon free world. But, we will have to realise that with the passage of time, pressures on us to sign the Comprehensive Test Ban Treaty (CTBT) and cooperate in finalising a Fissile Material Cut-off Treaty (FMCT) will grow. While we are committed to not standing in the way of the CTBT coming into force, we will have to stand firm on insisting that for us to accede to any FMCT, the treaty will have to be non-discriminatory and verifiable. There should be no compromise on this score.

The 21st century is now widely alluded to as "Asia's Century." As the largest democracy in Asia, India will have to sustain a near double-digit economic growth, with far better standards of governance than we now have, if we are to play a role commensurate with our size and potential in the councils of the world.

What Ails Defence Planning in India?

☐ N.S. Sisodia

Defence concerns the very existence of a nation-state. It is a public good and an essential part of a state's sovereign functions. Unlike some other sectors of national planning, defence cannot be left to market forces and, therefore, must be carefully planned for. Yet, for a variety of reasons, defence planning has not received the same attention and exalted status in India as development planning.

India's Five-Year Defence Plans have almost never been finalised in time. In the earlier years of India's independence, defence expenditure remained lower than even 2 per cent of Gross Domestic Product (GDP) and till India's 1962 War with China, there was no serious defence planning effort.

The first Defence Plan was prepared for the period 1964-69, the second for 1969-74, which was later converted into a 'roll-on' plan for the period 1970-75. The roll-on plan was reformulated covering the period 1979-84. Thereafter, Defence Plans were made co-terminus with national plans.

The sixth and the seventh Defence Plans were approved after a delay of 2 and 3 years, respectively. The eighth and the ninth Defence Plans were never formally approved. The tenth Defence Plan could not be finalised and two years into the Five-Year Plan period, the eleventh Defence Plan continues to remain under "active consideration" of the government. Parliament's Standing Committee on Defence, in its 29th Report has been constrained to observe: "The Committee regrets to note that even after a lapse of two years after commencement of the 11th Defence Five-Year Plan, the Plan proposals have not been finalised yet."[1]

This state of affairs has adverse implications for defence preparedness, one of them being low fulfillment of capital acquisition plans. Based on a performance audit of the Indian Army and Ordnance Factories, the Comptroller and Auditor General of India has observed that due to abnormal delay in approval of both long and medium-term plans, percentage fulfillment of the last three medium-term plans has "varied from 5 to 60 per cent in respect of various Arms and services of the Army."[2]

What then are the principal problems and issues besetting defence planning in India? And in the context of emerging challenges, what should be the way forward? This essay is an attempt to address some of these issues and suggest a few ideas to reform the system.

Planning Sans National Security Strategy

An important issue raised time and again in India is that the country has no formal documents systematically articulating national security objectives, national security strategy and defence policy. This has been acknowledged by the government in the Parliament and its Standing Committee on Defence.[3]

Ideally, long and medium-term Defence Plans should follow from a consensus-based and coherent articulation of national security objectives and strategy. These would be based on an assessment of India's security environment, threats and technological developments. Determination of national security objectives and strategy would also depend on the country's overall national interests, objectives and ethos.

As regards security environment assessments, some examples to consider are the American National Intelligence Council's assessment titled "Mapping the Global Future." The more recent assessment has been brought out in the document titled "Global Trends 2025: A Transformed World." The formulation of these long-term assessments is an outcome of an intensive collaborative effort among strategic experts and intelligence professionals in the government, think-tanks and universities. The Development, Doctrine and Concept Centre (DCDC) of the UK Ministry of Defence (MoD) has undertaken a similar exercise and published a report called "The DCDC Global Strategic Trends Programme 2007-2036." For more specific guidance in defence planning, the UK MoD had brought out a "Strategic Defence Review" in 1998. In the United States, the US Department of Defence publishes the Quadrennial Defence Reviews. The former was a one-off exercise which triggered some major reforms, while the latter is periodic and more incremental in its scope.

Periodic assessments of the international regional and national security environments are being carried out in India by the intelligence agencies, defence forces and Joint Intelligence Committee/National Security Council Secretariat. India's National Security Advisory Board (a group of strategic experts) has also undertaken similar assessments and it has in fact, produced comprehensive reports from time to time. The findings and recommendations of these reports now need to be synthesised at the level of National Security Adviser to produce a coherent document on India's national objectives and strategy. These could then constitute the basis of the nation's defence policy and objectives as well as capability development plans. If such exercises could follow a linear process, which may not happen in actual practice, the model would look something like Fig 1 below.

Fig 1: Defence Planning Stages

National and International Security Environment Assessment

↓

Threat Scenarios

↓

National Security Objectives

↓

National Security Strategy

↓

Strategic Defence Review

↓

Defence Policy

↓

Military Strategy/Objectives

↓

Military Doctrine

↓

Capabilities-based Integrated Defence Plan

In actual practice, the above process may not be linear, but iterative and complex. Nevertheless, the chart lays down the essential preparatory steps needed for sound defence planning. Absence of coherent national security objectives and strategy handicaps defence planners in India as they have to often resort to their own interpretations of a variety of statements and speeches which do not have the same sanction as formally approved documents would. A former Chairman, Chief of Staff Committee has identified the "total absence of a central focus and direction, as far as the articulation/formulation of national interests and national strategy are concerned," as a major challenge for India's defence planners.[4]

Scope of Strategic Defence Review

If the Ministry of Defence were to undertake a Strategic Defence Review, what should be its scope, so that it can facilitate the task of defence planners and managers? Some of the questions which such a review would need to answer should include:

- What is the short, medium and long-term assessment regarding India's security environment? The assessment should look at both the traditional and non-traditional threats. It should particularly focus on the newer forms of warfare, surprises and unanticipated contingencies.
- What are India's strategic objectives and priorities?
- What are the missions and tasks that the country's armed forces are expected to undertake? In India's context, these would include the full-spectrum of warfare; yet realistic plans would have to take into account the probability of each form of conflict.
- What should be the shape of the forces that will carry out these missions and tasks and what changes are, therefore, required in the existing forces structure and numbers? In particular, India's defence forces need to examine whether the existing structures created for conventional wars would continue to be useful in the future.
- What strategy is sought to be adopted to attract, train, support and retain the right persons for the defence forces?

- What capabilities will be needed for the missions and tasks expected, and what changes does this involve, taking into account the development and potential impact of new technologies that may be available to the defence forces as well as to India's friends and adversaries?
- What are the best means to acquire and maintain those capabilities and what changes should be made to existing methods?
- Finally, what is the feasible and realistic level of expenditure to put the answers to all the preceding questions into practice, and what are the best ways to finance this expenditure?[5]

Many of these issues are addressed by different agencies even at present but for informed planning decisions, a coordinated approach would be far more effective.

How Much is Enough for Defence?

How much money is enough for a country's defence? This question, the title of the classic work of Alain C. Enthoven and K. Wayne Smith on defence plans and budgets in the US during 1961-1969, is a perennial dilemma confronting planners and decision-makers. There can, however, be no perfect answer to this question, as it would depend on a country's goals, resources and the judgment of its political leaders.

In the aftermath of India's independence, defence received a rather low priority. Having waged a successful non-violent struggle against the world's most powerful imperial power and determined to live in peace with all its neighbours, the government of the day saw no need to lay particular stress on the armed forces. However, India's debacle in its war with China in 1962 jolted the leadership into a more realistic assessment of the threats facing the country. Thereafter, defence outlays were stepped up significantly. In the aftermath of the Cold War and during the early Nineties when the country's economy was under stress, the outlays as a share of GDP once again started declining. (See Fig 2 for defence outlays as a share of GDP). However, in terms of constant prices, defence expenditure has grown quite substantially since the early 1990s. (See Fig 3 below for defence expenditure at constant prices and its annual growth.)

Fig 2: Share of Defence in GDP (%)

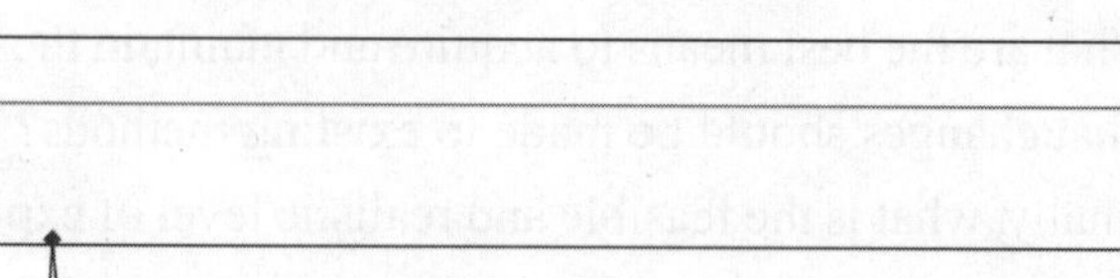

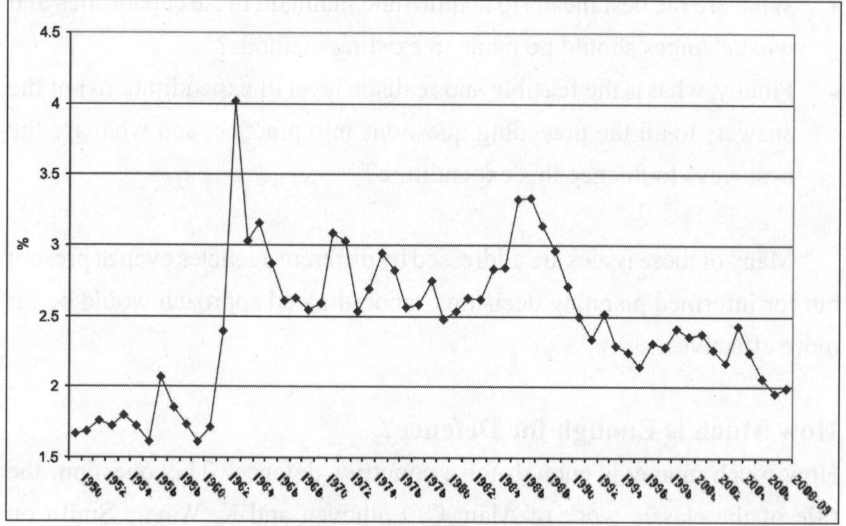

Fig 3: India's Defence Spending and its Annual Growth (%) at Constant Prices

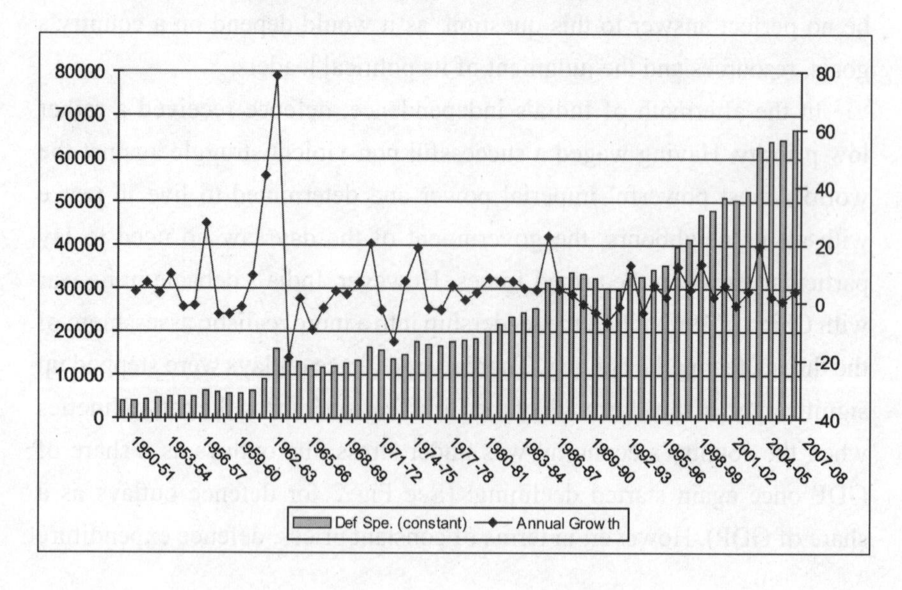

Allocation of resources for defence always poses a particularly difficult choice for the political leadership in democratic countries. Every rupee spent on defence cannot be spent elsewhere for other priorities like poverty alleviation, health, education and infrastructure. Neglect of such vital needs in a developing nation creates internal fault lines, which are prone to exploitation by hostile actors outside. Recent history provides examples of cases where military strength alone has failed to guarantee security. The collapse of Soviet Russia is a case in point. A nation's security also depends on its comprehensive national strength, which includes its economic and technological prowess.

The Remainder Method

In the past, alternative criteria have been adopted to determine defence outlays. American political scientists Glenn Snyder and Samuel Huntington in their classic work on defence policy- making under President Harry Truman and Eisenhowever have pointed that in calculating defence expenditure, they followed the "remainder method." Thus, they estimated tax revenues, "subtracted domestic spending, and gave whatever was left over to defence."[6] On the other hand, President Kennedy imposed a budget ceiling on defence. Unable to say that more money was unavailable, civilian mangers now had to reject programmes proposed by the military by stating that they were unnecessary. The process thus "forced them to substitute their judgments for those of the military professionals and turned budget discussions into a test of Civil-Military Relations."[7]

Defence Plans Outlays as Share of GDP

It has sometimes been suggested that defence outlays should be fixed as a percentage of GDP over an extended period of time to ensure adequate resources for defence. The Standing Committee on Defence (1995-96) in its Sixth Report proposed a long-term commitment of 4 per cent of GDP.[8] The proponents of this school, however, seem to forget that fixing a percentage of GDP is irrational, as it does not take into account the size of the country's economy. This standard is "just as disconnected from a net assessment of enemy threats as was the remainder method."[9]

In India's own case, it has been possible for the government to provide the required resources for modernisation, given its impressive economic growth. A higher share of GDP becomes meaningless if the economy itself is doing poorly. At less than one per cent of GDP, Japan's defence expenditure is nearly twice that of India. While a share of GDP is a good way to make comparisons, finally a country's defence outlays must be determined in the light of a realistic assessment of the threats it faces and the capabilities it may actually need. More effective planning exercises will greatly ease the task of rational allocation of resources.

A Defence Reserve Fund for Modernisation?

Another related issue concerning resources is the suggestion that a non-lapsable Defence Reserve Fund should be created outside the annual budgets to ensure regular flow of funds for modernisation. There is insufficient justification for such a fund as substantial budget provisions meant for modernisation are being returned unspent year after year and the government has to resort to deficit financing. The knowledge that unspent money can be retained in a non-lapsable fund adversely impacts speed of decision-making, and generates complacency. There is sufficient flexibility in budget-making to ensure uninterrupted flow of funds to vital projects, like a proper assessment of obligatory liabilities and earmarks. Apart from this, under India's constitutional provisions and parliamentary financial discipline procedures, the propriety of maintaining such a reserve fund is doubtful. Thus, the remedy to this situation lies in more realistic budgetary exercises which must provide for all essential defence needs and efficient procurement procedures.

Long-term Financial Commitments

The government's failure to commit itself to a series of long-term modernisation plans for the armed forces and uncertainty about long-term availability of funds for such plans has been cited as a major handicap for defence planners.[10] Given the long gestation of defence modernisation plans and the risks attached to underfunding and delays, the expectation of the Services about long-term commitments is fully justified. However, the problem arises when indications given by the Ministry of Finance about availability of funds for future periods are unacceptable to the MoD/Services. Decisions on resource allocation will have to be finally made by political leaders. What can improve the quality of these decisions is the rigour of the planning process, which should also propose alternatives to match different levels of defence outlays.

Synergy with Other Sectors

Since defence plans are a part of the union government's non-Plan expenditure, there is at present no satisfactory mechanism to ensure inter-sectoral coordination with other sectors of national plans. Currently, efforts in this direction are mostly piecemeal and ad hoc. A more systematic attempt to coordinate Defence Plans with other related sectors like surface transport, ports and shipping, civil aviation, information and broadcasting, telecommunications, industry, science and technology will maximise the value of money spent on defence and ensure more effective defence preparedness.

Capability-based Planning

In the past, defence planning both in India and elsewhere has tended to be equipment or platform-centric rather than capability-based. Thus, modernisation has been traditionally viewed in terms of acquisition of aircraft, fighting ships, submarines, artillery regiments, etc. Such Defence Plans were essentially an aggregation of the Services' projected requirements of weapons and equipment. This approach has led to acquisition of equipment, but not necessarily the relevant capabilities for a diverse range of threats. How is capabilities-based planning different from traditional planning? Paul K. Davis defines the process as follows:

> Capabilities-based planning is planning, under uncertainty, to provide capabilities suitable for a wide-range of modern-day challenges and circumstances while working within an economic framework that necessitates choices.[11]

Capabilities-based planning adopts an integrated approach, looking at complete capabilities, rather than discrete equipment; and makes choices among alternative capabilities, for meeting the same threat or a set of threats. It focusses on outputs and outcomes instead of inputs. It considers costs and benefits of each alternative and selects what in the judgment of planners would be the most cost-effective alternative meeting the same objective. While doing so, it should look at the total cost of owning and maintaining a capability. This assessment would have to take into account the life-cycle costs of the capability involved. Capabilities-based planning would also involve abandoning over time, capabilities no longer required. The process calls for a certain degree of ruthlessness and entáils rigorous intra-Service and inter-Service prioritisation.

Given the fact that in the future military forces may be needed less for conventional inter-state wars and more for entirely unanticipated surprises and shocks, there is a need for flexible and broad-based capabilities. These may have to be inter-Services and inter-agency structures. While some effort is reported to have been made in India to move towards capabilities-based planning, the success so far has been limited. The former Chairman, Chiefs of Staff Committee (COSC), Admiral Arun Prakash (Retd) has stated that while a beginning has been made in "capabilities-based planning (especially in the Navy), cost-benefit analyses and life-cycle costing are still at an embryonic stage."[12] Integrated capabilities-based planning can be truly effective once there is much greater integration amongst all the defence forces. This leads to the next important question of structuring our higher defence organisation for proper planning.

Are the Organisational Structures Effective?

In India, defence planning really began after the 1962 War with China and the first plan was formulated for the period, 1964-69. A planning cell was set up in 1965 in the Ministry of Defence to focus on "the wider aspects of defence planning." However, this cell proved unequal to its task. Thereafter, the Minister for Planning, in 1974, suggested that long-term defence programmes would be more cost-effective and to integrate defence planning with development plans, the two should be made co-terminus. Later, in 1977 a Committee for Defence Planning (CDP) was set up under the Cabinet Secretary, comprising the three Chiefs of Staff and officials of the Planning Commission. In 1979, a Defence Planning and Implementation Committee was established under the Defence Secretary. It was at this stage that the Army, Navy and Air Force initiated the task of planning in right earnest.

None of the above structures, however, was able to manage planning exercises effectively and, therefore, in 1986, the Defence Planning Staff headed by an officer at the level lieutenant general level was created under the Chiefs of Staff Committee. Its basic task was to coordinate plans developed by Perspective Planning Directorates of the Army, Navy and Air Force.[13]

In the wake of the Kargil War, following a review of the national security set-up, a Group of Ministers (GoM) recommended restructuring of the Higher Defence Organisation. The GoM report observed serious weaknesses in the functioning of the COSC and in its ability to provide single-point military advice to the government, and relevant substantive inter-Service doctrinal, planning, policy and operational issues, adequately. Regarding the defence planning process, the report noted that it was greatly handicapped by the absence of a national security doctrine, commitment of funds beyond the financial year, lack of inter-Service prioritisation and the requisite flexibility. In the context of the COSC's deficiencies, the GoM's major recommendation was to create the institution of a Chief of Defence Staff (CDS) assisted by a Vice Chief of Defence Staff (VCDs), among other things, to "enhance efficiency and effectiveness of the planning process through intra and inter-Service prioritisation."

The recommendation for creating a CDS, however, has not so far been implemented for the reason that such an important decision needs to be taken only after broad based political consultations. However, the Integrated Defence Staff (IDS) has been created as an interim measure. There are, of course, misgivings among the armed forces also, as evident from a number of statements and writings. Understandably, the issue is highly contentious.

Whether or not, there is a Chief of Defence Staff, an effective system of inter-Service integration needs to be developed without any delay. Presently, the Chairman, COSC, is not in a position to oversee long-term defence planning or its sustained implementation. In his own Service, the Chairman, COSC continues to perform an important operational role. He, therefore, simply does not have the time to preside over the rigorous, long-term exercises needed for sound planning. The Chairman, COSC is intimately identified with his own Service and his choices regarding inter-Service prioritisation are unlikely to be seen as unbiased. The proponents of the institution of CDS argue that this problem can be overcome by him, as the position will be held by rotation among the different Services, at least to begin with. In any case, the CDS will not be returning to his own Service at the end of his term. It is also pointed out that the Chief of the IDS, is unable to function effectively as he has no teeth. It is further argued that the CDS is likely to help in planning of inter-Services' operations.

In future, technology will play a critical role. Technology-intensive systems needed for modern warfare are both expensive and have a long gestation. It is important, therefore, that planning for capability acquisition in all the three Services is an integrated process, keeping in mind the integrated strategic doctrine. An integrated approach for systems like reconnaissance satellites, cyber warfare strategies, space wars, advanced light helicopters, unmanned air vehicles, etc. would ensure the best value for money. It is in the above context that the Group of Ministers had stated:

> Under the existing system, each Service tends to advance its own capability without regard for inter-Service and even intra-Service prioritization. Accordingly, one of the most vital tasks that the CDS would be expected to perform is to facilitate efficiency and effectiveness in the planning/budgeting process to ensure the optimal and efficient use of available resources. This could be carried out through intra-Service and inter-Service prioritization of acquisitions and projects.[14]

The case of the 11th Defence Plan highlights the principal roadblock to timely finalisation of Defence Plans. The Parliament's Standing Committee on Defence has attributed the continuing delay to lack of agreement regarding the financial outlay for the plan: the Ministry of Finance has proposed a Defence Plan outlay, based on year-on-year increase in the range of 8-10 per cent whereas the MoD has asked for an annual average growth rate of 12.35 per cent. And the tussle for resources continues. This stalemate is not unique to the 11th Plan; earlier plans had met a similar fate. Ghosh observes that a discussion about Defence Plans basically boils down to "the extent of financial commitment the Finance Ministry was prepared to make and whether it was acceptable to the Ministry of Defence."[15]

The reluctance or inability of the Ministry of Defence/defence Services to finalise the plans keeping in view the realistic financial outlays is a recurrent problem. It is difficult for the Services to accept any ceiling as planning within such a ceiling, where some proposals considered vital would have to be rejected, becomes a complex process task for the Services or any other organisation for the matter. The Chairman, Chiefs of Staff Committee, does not have the requisite authority to decide on either intra or inter-Service prioritisation as it involves a reduction in the outlay of the affected Service. The difficult task of making choices and trade-offs is then passed on to the Ministry of Defence. Any major changes in the Services' plans by the Ministry naturally evoke strong resistance from the Services as civil servants of the Ministry do not normally have the relevant professional background. One of the problems mentioned by Admiral Arun Prakash (Retd) is "a generalist and transient bureaucracy (of the MoD) which lacks in-depth comprehension of defence security issues." Thus, decision- making with regard to Defence Plans can sometimes become a test of the civil-military relationship between the armed forces, and the civilian bureaucracy in the Ministries of Defence and Finance. It is argued that a Chief of Defence Staff would have both the professional expertise and the necessary authority to ensure inter-Service prioritisation within indicated financial ceilings.

A Defence Planning Board?

An alternative to the CDS which has been suggested and deserves some consideration is to set up a Defence Planning Board under the chairmanship of the Defence Minister, with a strong staff of professional experts.[16] The board could comprise retired Chiefs of Staff, scientists, industry representatives, a member from the Planning Commission, and other relevant experts. It can become effective only if it has a strong team of professional analysts able to analyse and evaluate the Services plans with the necessary expertise. Its task would be to look after planning on a whole-time basis in close collaboration with the Services, other relevant Ministries and agencies, the Planning Commission and the Ministry of Finance. However, such a board is likely to suffer from some handicaps. It may become detached from operational realities and lack credibility with Services Headquarters (HQ), field formations and the MoD. Perhaps, a carefully crafted mandate for the institution of CDS, which takes care of the specific circumstances of India and addresses aspects which are potentially problematic, can be a possible way forward.

Planning for Self-Reliance

India today has the dubious distinction of being among the top two importers of defence equipment. Self-reliance in defence is vital, not because of ideological considerations, but because it is a necessary condition for strategic autonomy. No nation, however, friendly, would be inclined to part with its cutting-edge strategic technologies and systems; these capabilities will have to be developed indigenously. Nor can one rely on uninterrupted supplies from foreign sources during the time of conflict. A nation's military capability would, in the ultimate analysis, depend on its mastery of strategic technologies. Therefore, investment in Research and Development (R&D) as well as production would need to be enhanced and the private sector more actively involved. Projects in defence R&D and production have a long gestation, and significant risks in terms of time/cost overruns and failures. Hence, they require much more careful planning as an integral part of Defence Plans. A committee set up under Dr. Vijay Kelkar, in its April 2005 report to the government has made a number of important recommendations which need to be integrated in the defence planning process. The logic for their speedy implementation is compelling. Implementation of the suggested measures is expected to lead to significant increase in defence industrial production and a higher growth rate in manufacturing. The increase in job creation was estimated between 1.2 to 2 lakh. Based on life-cycle cost analysis, the annual savings on maintenance and support were estimated by the committee in excess of Rs. 4,000 crore. Thus, time-bound implementation of the measures recommended in the report would lead to greater self-reliance (upto 70 per cent instead of 30 per cent at present), benefits in terms of R&D, technology spin-offs, higher industrial growth, higher exports, increased competition for better value for defence, besides providing strategic depth to defence production.

Planning for Human Resources

An important challenge for defence planning is to develop appropriate human resources for the future. The Indian Army, particularly, is regarded as manpower-intensive. The combatant of the future would have to be much better equipped and trained in a wide variety of tasks and prepared to meet wholly unanticipated contingencies. The world over, armed forces are improving their teeth-to-tail ratio and becoming leaner and more effective. Consequent to the 6th Pay Commission, the share of revenue expenditure in the defence budget has already increased to 61.31 per cent in 2009-10 from 54.5 per cent in the preceding year. The total wage bill has more than doubled from Rs. 21,892 crore in 2008-09 to Rs. 44,501 crore in 2009-10. This growing burden is likely to crowd out resources for modernisation. In this context, a former Chairman, Chiefs of Staff Committee, Admiral Arun Prakash (Retd) observes: "There is an urgent need to substitute manpower with technology (air-mobility, night-fighting capability, precision-guided weapons, surveillance and reconnaissance, network-centric warfare), and become lean." Given India's specific context and security environment, the scope for restructuring may be limited. Nevertheless, the issue of manpower costs needs to be revisited in the context of long-term defence planning.

There are two other issues relating to human resources which need to be addressed. First, for effective planning, the analytical capability of the defence Services and MoD needs to be significantly enhanced. Even among 'military professionals,' expertise in relevant areas like operations research and systems analysis is extremely limited. Significant improvements in planning are not possible unless conscious steps are taken to create or develop a staff having the requisite specialisation.

Secondly, there is need to promote greater specialisation among civilians in the MoD to enable them to make a substantive contribution to the defence planning process. Suggestions have been made in the past to carve out a special cadre of civil servants to staff ministries directly concerned with national security. A more practical alternative is intensive training and longer tenures in the Ministry of Defence. Selected officials from the armed forces can also be inducted in the ministry, provided they do not have to go back to their parent Service.

Conclusion

Defence requires investment of a significant share of the nation's resources. These resources need to be invested wisely as the attendant risks in terms of time and costs can be huge and in terms of the nation's security, unacceptable. Defence planning, therefore, deserves as much attention as national planning. While defence planning processes in India are now well-established and have seen many improvements in recent years, there are problems, besetting the system. Greater attention is needed in the areas of formulating a national security strategy in a systematic manner; effecting greater synergy between defence and national plans; capabilities-based planning, and greater integration among the Services. Reforms in planning processes, supported by sound analyses of specialists would facilitate the task of financial allocations for Defence Plans, which would be based on carefully assessed needs rather than any other ad hoc criteria. Sound Defence Plans, formally approved well in time, will lead to more efficient budget-making, timely utilisation of funds, value for money and speedy development of capabilities needed to safeguard India's security. The reforms envisaged cannot be implemented unless the issue of the required human resources for planning and decision-making in India's defence apparatus is simultaneously addressed. And, finally, for defence planning to be more effective in the future, there is need for greater transparency, consistent with national security. Sharing broad capability plans would facilitate greater synergy with the private sector, greater impetus to self-reliance through public-private partnership and strengthening of India's defence industrial base. Greater transparency, consistent with imperatives of security, will also ensure greater accountability in matters of national security.

Notes

1. Ministry of Defence, Standing Committee on Defence (2007-08), Twenty- Ninth Report, Lok Sabha Secretariat, New Delhi, April 2008, Para 2.49.
2. Report of the Comptroller and Auditor General of India for the year ended March 2006, N 4 of 2007.
3. Ministry of Defence, Standing Committee on Defence (1995-96), Tenth Lok Sabha, *Defence Policy, Planning and Management*, Lok Sabha Secretariat, New Delhi, April 1996.
4. Admiral Arun Prakash (Retd) in response to the author's questionnaire on the subject.

5. This questionnaire has been prepared on the basis of ideas presented by Alexander Nicoll in his paper "Challenges in Defence Planning", at the International Seminar on Defence Finance and Economics, New Delhi, November 13-15, 2006.
6. Richard K. Betts, "A Disciplined Defence: How to Regain Strategic Solvency", *Foreign Affairs*, 86(6), November-December 2007, p. 70
7. Ibid.
8. Ministry of Defence, Standing Committee on Defence, 'Defence Policy, Planning and Management', Sixth Report, Lok Sabha Secretariat, New Delhi, 1995-96, p. 37.
9. Betts, n.6, pp.79-80.
10. This problem has been underscored by Admiral Arun Prakash (Retd) in his replies to the author.
11. Paul K. Davis, ed., *New Challenges for Defense Planning: Rethinking How Much Is Enough* (Santa Monica: RAND, 1994); Idem, "Rethinking Defence Planning", John Brademas Centre for the Study of Congress, New York University, December 2007.
12. Admiral Arun Prakash's (Retd), reply to a questionnaire sent by the author
13. V.P. Malik and Vinod Anand, *Defence Planning: Problems and Prospects* (Manas Publications, 2006) pp. 24-26.
14. Report of the Group of Ministers on Reforming the National Security System, Government of India, New Delhi, 2001, p. 101.
15. Amiya Kumar Ghosh, *Defence Budgeting and Planning in India: The Way Forward* (New Delhi: Knowledge World, 2006) p. 224.
16. The suggestion has been made to the author in a discussion by Joint Secretary and Financial Adviser Shri Amit Cowshish dealing with budget and plans in the MoD.

Acknowledgements:

The author would like to express his gratitude to Admiral Arun Parkash (Retd) formerly Chairman, Chiefs of Staff Committee, Shri Vinod Misra, formerly Secretary Defence Finance and Shri Amit Cowshish, Addl. Financial Adviser and Joint Secretary Ministry of Defence, for their valuable insights; to Shri Laxman Kumar Behera, Associate Fellow, IDSA, and Shri Samuel Rajiv, Research Assistant, IDSA, for their help and inputs.

Defence Resource Management

□ Vinod K. Misra

Optimal resource management in defence offers myriad challenges and opportunities. The task of building up full spectrum capabilities to proactively and decisively respond to the needs of conventional, strategic and asymmetric warfare is truly daunting against the backdrop of rapid advances in defence technologies and the amalgam of platforms, sensors and armaments on offer or under development. Of course, India has its own unique perception of the blend of diverse yet mutually reinforcing capabilities that would have to be acquired in a time-frame of 15/20 years, based on medium and long-term threat assessments, force level comparisons with potential adversaries and the complementarity of the other elements of national security. Innovative solutions would have to be found if resource needs for the planned levels of capability acquisition in the desired time spans are deficient in relation to the likely resource base. Nonetheless, it is also apparent that concern for optimal allocations would continue to be paramount as inefficiencies, wastages, sub-optimal allocations and utilisation in high priority areas, thin spreading of resources, inadequate attention to efficient project management, lack of commitment control *et al* can seriously jeopardise the defence security build-up.

While comprehensive planning for a 15/20-year time horizon would lie at the root of a successful and focussed resource management process in defence, there are the continual imperatives of evolving a rational and balanced capability basket funded from the capital budget, on the one hand, and high readiness levels for fighting the next war in terms of operating, maintenance, ammunition and training costs for the standing armed forces funded from the revenue budget, on the other. Sub-optimal allocations for either the maintenance or the modernisation segment can have disastrous consequences for the preparedness levels of our armed forces. This brings in its wake abiding concerns for prioritisation, generation of choices for meeting the mission objectives and making an optimal choice, procurement and acquisition efficiencies, sharply focussed and cost efficient logistics, activity-based costing and cost consciousness, project management skills and a strong domestic capability for defence Research and Development (R&D), engineering and production. There are also constant improvements to be sought through purposeful monitoring of best practices and reforms practised worldwide in such areas and also facilitating, through a fair degree of appropriate transparency, and an informed debate in the public realm on vital defence concerns. The institutionalised association of think-tanks, defence analysts, scholars and thinkers and various supervisory agencies and organisations can have a truly transformational impact on optimal defence resource management. Some of these key issues are addressed in the succeeding paragraphs.

Resource Milieu

Allocations for defence currently account for around 2 per cent of the Gross Domestic Product (GDP) while in terms of share of the central government expenditure, the Budget Estimates for 2008-09 constitute approximately 14 per cent of the total central government spending. Given the compulsions for containing the fiscal deficit and eliminating the revenue deficit in terms of the commitments under the Fiscal Responsibility and Budget Management (FRBM) Act as also the escalating needs in terms of various subsidies, interest burden and social and infrastructural sectors spending, it is unlikely that defence allocations would experience a sharp upturn in the coming years unless we are faced with an emergent and crisis situation in relation to our security needs. Thus, we are likely to witness a steady and reasonable growth of the order of around 8 to 10 per cent per annum in nominal terms in the coming years. Even on this conservative basis of 10 per cent annual compounded growth, allocations over the next 10 years could aggregate to Rs.18.5 lakh crore. Also assuming that the current share of revenue and capital spending of around 55 : 45 between revenue and capital is about the optimal mix and would be sustained in future years, the availability for capital spending over the coming 10 years would aggregate to Rs.8.33 lakh crore or approximately $ 170 billion at current exchange rates. Nearly 85 per cent of this capital allocation goes towards modernisation of the defence Services while the balance 15 per cent is available for land, capital works, married accommodation projects and the like. It is, thus, obvious that sizeable funds would be available for financing the capability enhancement plans of the defence Services in the medium to long terms. Indeed, the pace of acquisition would need considerable stepping up to fulfill the envisioned targets for rapid modernisation. It is noteworthy that due to the inadequate pace of absorption of capital funds, allocations for defence have had to be curtailed by the Ministry of Finance at the Revised Estimates stage and there has been significant underutilisation of even RE funds in several years. It would be pertinent to note that significant additional funding, either distinctly or in an implied sense, would be available through the range of initiatives that could be coming to fruition during this period by way of offset investment inflows, public-private partnership/outsourcing, greater integration of the private sector in areas such as training, repairs, maintenance, overhaul, upgrades as well as supply chain management and targeting of lower costs through greater capacity utilisation by way of defence exports and concrete steps towards economy and efficiency in defence spending, potential for defence credits for large value acquisitions and the like.

Any bold reform impacting on defence manpower would also have significant potential for bringing down the substantial direct manpower and manpower related costs, including pensions, in respect of the defence Services.

Acquisitions and Procurement

Whether it is procurement of ordnance and general stores for meeting the maintenance needs of the standing forces or acquisition of complex, state-of-the-art weapon systems comprising platforms, sensors and armaments, there is large number of possibilities for bringing down costs and time-frames for induction and enhancing the efficacy of the entire effort and thereby serving the cause of capability enhancement in the shortest feasible time horizon. Some of these possibilities are briefly mentioned below:

- Acquisition costs are significantly dependent on quality specifications. On account of the high cost of incremental quality beyond the minimum need-based specifications, it is critical that qualitative requirements are set out carefully and are rooted in a mandatorily necessary quality threshold. Further, this alone is likely to provide for a fully competitive setting for acquisitions and consequently reasonable costs.
- Defence specific quality specifications across the board would continue to be an extremely expensive proposition. Apart from the high initial cost of acquisitions in situations where military sub-systems and systems have to conform to rigorous environmental standards, there is a significant cost dimension to through life maintenance of such systems. Consequently, military grade specifications have to increasingly become the exception rather than the norm. The availability of commercially-off-the-shelf technology items would impact extremely favourably on the life-cycle costs and maintainability of expensive weapon systems.
- The single-most delay-prone activity in the acquisition process is the technical and trial evaluation of the system planned to be acquired. This has also, more often than not, contributed to a single source situation as well. The current procurement procedure recognises that in the event technical evaluation of systems on offer leads to a single source situation, then the entire process has to be gone through afresh based on appropriately modified qualitative requirements which would yield a competitive setting. A similar formulation is

considered necessary in the event trial evaluation too brings us to a single source situation. It should generally be unacceptable that only one unique weapon system in the whole world meets the operational needs of the Indian armed forces. This apart, rigorous time-frames need to be prescribed for comprehensive trial evaluation in different regions, terrains and climatic conditions. Multi-disciplinary teams, which would have a certain degree of independence from those who formulated the qualitative requirements in the first place, would go a long way in providing the necessary transparency, accountability and time compression for this important facet of the acquisition process.

- Formal feedback from other customers of weapon systems on offer should be a mandatory element of the evaluation process. Weapon manufacturers/ suppliers should have no hesitation in making available to the Indian defence Services comprehensive inputs from other users on the reliability and maintainability aspects of the system on offer.
- Comprehensive Information Technology (IT) solutions have to be put in place for sharing data bases across the defence Services of common user revenue items as well as capital acquisitions.
- As a matter of conscious policy, the endeavour must be to develop long-term, reliable and stable relationships with a limited number of suppliers in the domestic market for the entire range of important defence stores based on a long range and transparent pricing basis. Only in the extreme event of such suppliers not abiding by the terms of the long-term price escalation basis, should the need arise to induct a matching number of additional suppliers for the item/s. This would nurture the necessary degree of long-term dependability of suppliers in an operational environment. Viewing each procurement decision afresh, may be several times in an year or even once every year, is neither cost-efficient nor a reliable arrangement.
- Elaborate IT-based systems would need to be put in place to be able to consciously evaluate, at regular intervals, whether the operational and maintenance costs, as assumed at the stage of the acquisition

decision based on life-cycle costs, are being borne true. While guarantees such as Mean Time Between Failures (MTBF), Total Technical Life (TTL), Time Between Overhaul (TBO), Mean Time To Repair (MTTR), Turnaround Time and the like are prescribed in the basic contract, it is necessary that appropriate monitoring of these parameters is carried out both from the point of view of assessing the true life-cycle costs as well as for benchmarking the performance of the concerned vendors.

Offsets

Defence offsets offer a unique opportunity for transforming defence capabilities and infrastructure in a reasonably rapid time-frame by leveraging our strengths as a significant defence importer.

The policy frame for offsets was first prescribed in the Defence Procurement Procedure (DPP) 2006 and has recently been updated in the latest 2008 version. With the projected defence imports of around $ 35 billion in the next five years' time horizon, offset inflows would aggregate to $ 10 billion even at the mandatory minimum level of a 30 per cent offset obligation (though a 50 per cent offset is potentially feasible in many of the more significant imports during this period and as has been sought in the Medium Multi-Role Combat Aircraft (MMRCA) context). While the existing policy does not seek to direct offset investments into any specific area, with the choice being left to the successful vendor to fulfill his offset obligation through any of the options comprising co-design and development and production, sourcing of specific items/assemblies/ systems from India, defence exports, infrastructure creation, repair/ maintenance/overhaul and the like, no prioritisation of the preferred destination for offset investments has yet been done. It is urgent that first of all the defence Services as well as the defence R&D and defence Public Sector Understandings (PSUs)/Ordnance Factories draw up full scale and focussed roadmaps listing their priority for attracting offset investments in key deficiency areas for development and production of weapon systems or augmenting the defence industrial base for meeting the life-cycle operations and maintenance needs of various weapon systems or for infrastructure covering training, operations, logistics, accommodation and repair, maintenance and overhaul of major systems in use with the defence Services. The suppliers of defence goods have to recognise that such investments in India would not only provide for reasonable returns on a sustained basis but would also lay the foundation for a long-term, mutually beneficial relationship with the defence Services. Offsets, thus, have the unique possibility of bringing about a transformation in defence related research, design, development, engineering, production and mid-life upgrade and overhaul of weapon systems as well as for filling the critical infrastructural gaps in the shortest feasible time horizon. With the coming into being of banking of offsets in DPP 2008, it should be possible to attract far more investments into the defence sector than would have been possible only on the basis of an offset contract concluded with the successful bidder. However, the time-frame of two years provided in the policy at present for banking purposes would have to be extended to a more reasonable four-five years, given the long lead time in setting up and operationising new ventures in India. There are also some issues regarding the maximum permissible foreign equity of 26 per cent in respect of joint ventures which would be getting formed in fulfillment of the offset provisions. Given the large value of offset obligations, foreign investors are likely to feel far more confident with a significantly higher equity participation. With the positive impact of induction of state-of- the-art technological know-how and know-why, the potential benefits by way of employment generation and the possibility of this contributing to a more meaningful export effort, the selective dilution of foreign equity participation norms would appear well merited. Similarly, except for some well-defined high technology areas, the government could also consider favourably the need to dispense with licensing norms in respect of Indian private sector entities keen to enter the defence sector.

Defence Industrial Base

The life-cycle cost encompassing expenditure on operations and maintenance over the life span of major weapon systems is estimated at several times the initial acquisition cost. Given the size of India's defence forces and the targeted level of strategic and operational capabilities, it would be clear that the predominant (around 70 per cent) import orientation in defence outlays on modernisation and maintenance is unsustainable in the long term. Increasingly, concepts of performance-based logistics and shedding responsibilities for all significant maintenance, repairs, overhaul and upgrade efforts to the original equipment manufacturers would take deeper roots. Also, judging by the erratic spares price pattern in respect of all spares consumption beyond the initial 2/3/5 years and the long lead time associated with procurement of spares or in getting repairs carried out abroad, all these point towards the need for empowering the domestic defence industrial base to make maintenance and modernisation a far more affordable proposition than it is likely to become before long. It is equally important that defence PSUs and Ordnance Factories as well as the defence R&D organisation do not continue to lead a sheltered existence and the undeniable advantages of a competitive setting are urgently brought into play in defence as well. Consequently, we have to seize the enormous opportunity offered to us through the potentially vast offset inflows as also the significantly altered policy framework permitting the larger integration of the vibrant private sector in India into matters of defence research, development and production. There are many outsourcing opportunities by way of shedding of core and non-core responsibilities by the defence Services to the Indian private sector. There are also enormous supply chain management efficiencies which could be achieved through larger farming out of logistics responsibilities to key entities in the private sector. Similarly, the rapid progress India has made in terms of capabilities of the Services sector, particularly in IT/ Telecom/ BPO/ KPO could be appropriately tapped for augmenting the capabilities of Indian defence. All this would also potentially lay strong foundations for Indian manufacturing entities to undertake defence exports in the medium to long term whereby it would not only contribute to the Indian Gross National Product (GNP) but also help us nurture long range defence oriented ties with several countries of interest to us in Latin America, Africa and Asia.

Manpower

Direct manpower and manpower related costs on account of elements such as rations, clothing, medical, housing, transportation and the like, account for an extremely sizeable percentage of the defence outlay in respect of the three defence Services. Consequently, any major reduction in resource needs of the defence Services can flow only through a significant down-sizing of the current manning levels. Apart from comprehensively reviewing the composition of the current force levels, force mix and force structuring from the point of view of modern warfare and the compulsions for more modular, adaptive and quicker responding operational units in the prevalent situation of both conventional and asymmetric warfare, there is also considerable scope for scaling down the manpower deployment on support services through outsourcing and greater integration of the capabilities of the Indian private sector for meeting defence needs. Any reduction in force levels would also have a corresponding impact on the sizeable pension bill (over Rs.15,000 crore for 2008-09) currently funded from the civil estimates of the Ministry of Defence (MoD). More significant reductions in manpower levels would, however, accrue only through a relook at the current terms of engagement and by bringing down the colour service to 7/10 years and having a large proportion of the officers' cadre made up by Short Service Commission. This would, of course, call for a well structured policy for absorption of armed forces personnel in the national mainstream, including jobs in the central police organisations, after they are discharged from the armed forces. It is pertinent though that notwithstanding very categorical recommendations made by the Sixth Pay Commission in this regard, it has not been possible to either accept or operationalise this far-reaching reform measure both for the officers as well as the Personnel Below Officer Rank (PBOR) categories. The objectives of a youthful and motivated profile for the cutting edges of the armed forces would be served only if the permanent cadre of officers and PBOR is drastically brought down. This would also offer richer promotion prospects for the officers and PBOR in the permanent employment of the armed forces.

All this, however, requires unprecedented will and courage at the highest levels in the armed forces to bite the bullet and set forth to comprehensively review manning needs in relation to core operational functions in the current scenario. This would call for threadbare analysis of the authorisations contained in WE (War Establishments)/ PE (Peace Establishments), which have, by and large, preserved a historical legacy. The underlying principles of a Zero-Based Budgeting exercise need to be applied to a one time, all encompassing and, subsequently, a periodical 5/10 years review of the continued validity and role of ongoing organisations/ structures, tasks and work centres. Similarly, a conscious exercise needs to be undertaken for outsourcing to the private sector, core and non-core tasks of the defence Services, which, in our perception, can be handled more cost-efficiently and reliably through private sector entities, leaving the armed forces free to focus on their truly core tasks.

Information Technology (IT)

Comprehensive use of IT capabilities holds the maximum potential for optimal resource management in defence. Unfortunately, we have a long way to go before these capabilities are fully realised. First of all, elaborate IT solutions provide for the kind of transparency, reliability, dependability, verifiability and accountability which the largely manual systems prevalent in the Services today can never match. Consequently, efforts towards ensuring a comprehensive data base and rational analysis for decision-making on a real-time basis, have, per force, been weak and sporadic. Any successful IT solution is expected to take full stock of all that can go wrong in the process of management of the vast resources of the defence forces. Likewise, given the spread of the defence Services across the entire country, communication networking on an all India and inter-Services and intra-Service basis is critical to efficient and time-bound decision-making in defence. This would call for an appropriate blend of in-house capabilities for defining the user requirement specifications, and outsourcing to leading IT entities in the country, the task of enunciating the system requirement specification and subsequent development of applications software. There is also the significant issue of network security and accessibility of these networks, data bases and applications. Even though some strides have been taken by the defence Services, it would be a long time before comprehensive and state-of-the-art capabilities are in place. Given the transformational strength of IT in diverse areas such as the pursuit of jointness, keeping an elaborate record of resource allocation and utilisation, targeting life-cycle costs of major weapon systems, achieving greater battlefield transparency, supply chain management, comparative cost control and providing a reasonable framework to prevent frauds, wastages, leakages and inefficiencies, it is about time that the government mandates a full scale realisation of all these capabilities in a time horizon of no more than 3 to 5 years through a rational mix of in-house capabilities and outsourcing.

Defence R & D

Around 6 per cent the defence budget is currently allocated in favour of the Defence Research and Development Organisation (DRDO). Nonetheless, it is also noteworthy that the import content, whether by way of capital acquisitions or for maintenance and upkeep of already inducted systems, is around 70 per cent. Considering that India has an ambitious and long-term mission for defence capability enhancement in the strategic and conventional warfare realms in a reasonable time-frame of around 15-20 years, the aforementioned situation is clearly unsustainable. Core capabilities for R&D, engineering for production and, finally, manufacturing have to be acquired and sustained in the country. Likewise, through life product support on operations and maintenance, which could be two to four times the initial acquisition cost, has to become a largely domestic capability- based endeavour.

A fillip to the creation of domestic defence R&D capabilities in the private sector is expected to be provided by the Defence Procurement Procedure promulgated by the Ministry of Defence in 2006/2008 and the significant offset investment inflows likely in the defence sector in the coming years. In so far as the DRDO is concerned, given the last mile issues, which have impacted adversely on the time and costs frames of DRDO's research projects, it is just as well that the organisation has already embarked upon a collaborative framework for the pursuit of key R&D and production capabilities in a more reasonable time and cost span. Of course, this effort has to be dovetailed with the long-term capability development plans of the defence Services, a thorough mapping of the key technological gaps in the domestic defence R&D capabilities and careful evaluation of the strengths of major defence R&D/production entities worldwide which would contribute to cost-efficient tie-ups. From the resource management perspective in defence, it is also important that we display adequate courage in short closing R&D projects which have gone on for an abnormally long period of time. Similarly, peer reviews of important projects have to be carried out not just initially at the stage of formulation of such projects but at certain defined milestones during the development phase in order that short closures are carried out as soon as the risks are found untenable in relation to time and costs parameters. Two more points need to be made in the DRDO context. One concerns the need for more concurrent transfer of technology to the identified production agency during the design and development phase, particularly from the point of view of achieving the necessary degree of efficiencies in engineering for production. The other concerns greater mapping of the vast potential in different manufacturing disciplines in the private sector and encouraging creation of capabilities with them as well rather than the reliance, thus, far on defence PSUs and Ordnance

Factories alone. It is also critical that in the process of collaborative tie-ups with major foreign R&D/ production entities, adequate work share for the Indian side in terms of "build to specifications" is secured rather than excessive reliance on "build to print" alone, where the contribution of the Indian organisations would be minimal in terms of acquisition of R&D and production strengths.

Jointness

Even though the Headquarters Integrated Defence Staff (HQ IDS) has been in existence for some time now (over five years), the underlying concepts of jointness in the context of the defence Services have yet to take significant roots. While it is true that each of the defence Services has yet a long way to go in terms of creation of an optimum blend of capabilities for the current warfare spectrum and to that extent is apprehensive that advocacy of the cause of jointness might imply larger allocations for other Service/s to its own detriment, the pursuit of jointness is a necessary goal if the benefits of scarce resources have to be optimised. This calls upon decision-makers to develop a viable range of competing options for meeting current and potential threats and developing capabilities and then making appropriate and prioritised choices on an intra-Service and inter-Services basis. All this would have to stem from concepts of joint doctrines, strategy, operations, tactics and training, at one end of the spectrum, to joint acquisitions/joint logistics/ joint repair and maintenance and overhaul organisations, at the other. Thus, even if resource levels for individual Services have to conform to the share pattern of recent years, it would still be possible to develop a more rigorous planning and resource induction/utilisation process from a joint Services perspective in the interest of optimal efficiency. An integrated, mutually reinforcing and holistic approach to realisation of current, state-of-the-art capabilities would contribute to a far more balanced basket of defence capabilities rather than the existing situation of stand-alone weapon systems availability with each Service. Similarly, a jointness orientation to issues such as supply chain management, outsourcing, training and the like offers extremely rewarding resource optimisation possibilities. This renewed emphasis should have nothing to do with the coming into being of the institution of the Chief of Defence Staff (CDS) as the three Services together with the MoD should be able to view the future capability spectrum from a common prism and endeavour to get the most out of a likely aggregate resource base for maintenance and modernisation of the defence Services.

Financial Management–Management Information System (MIS)

Notwithstanding the financial management structures and systems already in place, including an extremely comprehensive package of delegated powers for functionaries of the defence Services at the level of Service Headquarters/ Command Headquarters/lower echelons, significant reforms would be feasible in the matter of accounting, budgeting, internal audit and project management to bring about greater value for money in the utilisation of precious resources deployed in defence.

For instance, the underlying concepts of accrual accounting by way of comprehensive listing of capital and other assets and ensuring strict commitment control could be extended to defence in an urgent time-frame even though a formal switch over to the accrual system of accounting for the entire governmental accounts is likely to be a long drawn affair. Similarly, the basic principles of outcome budgeting by way of quantification of financial targets of defence spending and judging performance in terms of outcomes rather than outlays and expenditure could well be extended in several areas of defence where such quantification of physical progress is definitely feasible and, indeed, desirable. This would help underscore the key elements of transparency, accountability and result orientation which are essential for evaluating performance in the realm of defence.

Likewise, the essential tenets of Zero-Based Budgeting (ZBB) whereby the rationale for continuance of certain activities, processes and organisations could be evaluated afresh at regular intervals (say every five years rather than an annual review) could yield significant dividends by way of reprioritisation and funding core tasks fully.

Scheme/project/programme-based budgeting and expenditure review also hold significant potential in the defence context. However, this is dependent upon comprehensive computerisation whereby distinct elements of such project costs would get captured, aggregated and analysed in order to throw up variances with reference to the desired and pre-defined physical and financial milestones. Indeed, a major weakness in the defence context is the absence of cost data in vital areas such as inventories, activities/performance centres, repair/OH/upgrade costs where carried out through in-Service organisations, labour and material productivity, training and exercises, life-cycle costs of operations and maintenance and the like.

A renewed thrust is also needed to revamp internal audit in the defence arena. This would encompass an officer driven effort rooted in comprehensive computerisation of major activities and processes in defence, with an embedded element of integrated systems audit which would throw up, on a near real-time basis, variations and deviations from the desired norms. Internal audit clearly has the potential of being a concurrent and sustained exercise for providing timely inputs for mid-course corrections rather than a post facto non-conformance detection exercise which should largely remain the preserve of statutory audit by the Comptroller and Auditor General (C&AG). This, of course, also requires sustained efforts towards developing the requisite degree of professionalism among the key audit functionaries covering the elements of knowledge, skills and attitude and simultaneously an appropriate mindset at the highest executive levels to constantly value inputs provided by internal audit both for specific deficiencies and inadequacies in ongoing programmes as also for the purpose of appropriate changes and reforms in the policy framework.

Economy

Given the vast span of defence spending, ranging from pay and allowances, revenue works, transportation, rations, clothing, medical and ordnance stores to construction works and capital acquisitions and the complex regulatory framework in terms of delegated powers and accountability, there is no dearth of possibilities for economy in defence should there be a top driven, all pervasive sense of cost consciousness and austerity and a reasonable stress on accountability for outcomes. While the large number of audit paras in C&AG reports and the content of the Internal Audit Reports of the Defence Accounts Department give some clear pointers in this behalf, a brief enumeration of the areas which offer scope for economy in defence expenditure is contained below:

- Consumption of water and electricity in offices, cantonments, military stations, residential blocks.
- Economy in temporary duty and permanent moves.
- Economy in hiring of civil transport through greater reliance on speed post, courier services, container services and the like.
- Finding far more cost-efficient solutions to the present costly and difficult to monitor entitlements of railway warrants, concession vouches for travel by personnel and military credit notes for movement of stores.
- A serious relook at the continued relevance of organisations such as military farms in terms of their value addition and operational costs vs benefits
- Efficiency/economy in Petroleum, Oil, Lubricants (POL) usage; discontinuing the use of coal and firewood
- Avoiding cross-movements, repeated handling and storage
- Drastically scaling down scales of stocks for indigenous/PSU/ Ordanance Factory supplies items as well as imported items- based commercial off-the-shelf technology.

India's Nuclear Strategy: An Assessment

☐ **Manpreet Sethi**

In 1955, ten years into its existence as a Nuclear Weapon State (NWS), the US' nuclear relationship with its primary rival, the USSR, was securely premised on the concept of nuclear deterrence. The USA's demonstration in 1945 of the destructive power of the atom, and its use, or threat of use, to force another into submission had both attracted and scared the Soviets enough to reach the conclusion that they needed the same capability to stop the US from using it against them or their allies. The first Soviet nuclear test, therefore, broke the American nuclear monopoly and established *deterrence* as the name of the game in nuclear relations. Since then, the nuclear strategy of every nuclear weapon state aims at establishing and consistently enhancing the credibility of its deterrence by building the necessary military capability and associated infrastructure that can project a readiness to engage in nuclear war-fighting while actually seeking thereby to prevent deterrence breakdown.

In 2009, having completed a decade of its own existence as an NWS, India is in the process of establishing credible nuclear deterrence with its own nuclear armed adversaries. Existential deterrence is obtained once a country has demonstrated some sort of nuclear capability – developed indigenously (as by India) or acquired clandestinely (as by Pakistan). However, for nuclear deterrence to be more credible, and, hence, more sustainable, the national nuclear strategy must consciously define, acquire and refine three essential elements:

- A clear nuclear doctrine that provides the conceptual underpinning for the role of the weapon in the overall security strategy – whether it is considered a militarily usable tool of war, or a political instrument for enforcing deterrence. India's nuclear doctrine defines a political role of deterrence for its nuclear weapon.
- A credible capability of handling the adversary's use or threat of use of nuclear weapons. This requires development of necessary offensive and defensive capabilities based on whether the nation's doctrine subscribes to nuclear pre-emption or retaliation, a corresponding force structure, appropriate delivery systems, requisite numbers of weapons with reliable yield, and a robust and hardened command and control system.
- A communication of the doctrine and capability to the adversary in order to signal the seriousness of deterrence. While it is true that matters of detail of nuclear strategy must remain opaque, that there exists a well considered nuclear strategy that is capable of reliable functioning must be transparent. It is a task of the political and military leadership to enable this through nuclear signalling in a credible fashion with the optimal use of both threat and reassurance.

Evidently then, the game of nuclear deterrence requires nuclear states to abide by a set of rules. The nuclear strategy must be crafted to reflect subtlety and bluntness, soft confidence-building diplomacy and hard military capability, opacity and transparency, all at the same time.

Amongst the nuclear strategies of nuclear weapon states, India's nuclear strategy stands out for three reasons: firstly, for the nature of threats it deals with; secondly, for its doctrinal precepts that are different from the predominant paradigm; and thirdly, for the complex interlinkages of nuclear deterrence with conventional and sub-conventional warfare that it has to contend with. It is the purpose of this paper to explore and explain these three dimensions of India's nuclear strategy.

India's Nuclear Strategy: The Uniqueness of the Exercise

Having shed its nuclear ambiguity in 1998, India is engaged in the process of establishing a viable and stable deterrence relationship vis-à-vis its two nuclear armed neighbours, Pakistan and China. This is a unique exercise for more reasons than one.

First, quite unlike the case of the other NWS (except perhaps the USA and China), India seeks a workable deterrent posture against two real nuclear powers with two different kinds of nuclear doctrines and capabilities. The Pakistani nuclear doctrine advocates first use of nuclear weapons to deter a conventional Indian offensive as it indulges in sub-conventional conflict against India. Meanwhile, though India shares some doctrinal similarities with China, there exists an asymmetry in capabilities and legal nuclear status. Secondly, India seeks nuclear deterrence with countries with which it shares common borders that are yet to be delineated to the satisfaction of both sides. In fact, India has territorial disputes and a history of wars with both nations. Skirmishes at the border are routine and hence, the possibility of deterrence breakdown is high. Thirdly, both India's nuclear adversaries enjoy a robust mutual nuclear and missile proliferation relationship. In fact, in the history of development of nuclear weapons, China stands out as the only country to have provided its strategic ally, Pakistan, with weapons design, material, as well as having conducted a nuclear test on its own soil on its behalf. Fourthly, India's nuclear strategy grapples with the challenge of proxy wars fomented differently by both the adversaries under a similar doctrine of offensive defence. So, while Pakistan uses sub-conventional conflict with impunity from the shadows of its nuclear weapons, China uses Pakistan to indulge in proxy war against India. Stability at the nuclear level is, therefore, severely strained by instability at lower levels. Fifthly, the dangers of political instability, existing or potential, in both India's non-democratic adversaries do little to allay fears of nuclear war as a result of accident, miscalculation or unauthorised use. Lastly, both Pakistan and China are revisionist states. While Pakistan seeks to revise the borders with India, China too seeks a territorial revision besides one also in its status and prestige in the regional and international order, and both conceive a role for their nuclear weapons in these tasks. None of the other nuclear weapon states shares the same predicament as India's. In fact, unlike the other P-5 states that in this post Cold War era do not have a specific target for their nuclear weapons but hold on to their arsenals as general purpose insurance against future uncertainties, India has very real and specific threats that need to be addressed through nuclear deterrence.

India's Nuclear Doctrine: Some Major Attributes

In order to redress its distinctive dilemma, India has evolved a nuclear doctrine and strategy specific to its requirements. The distinctly Indian brand of nuclear deterrence begins with the basic premise that a nuclear war cannot serve a worthwhile cause and, hence, must not be fought. Extremely conscious of the horrendous damage in space and time that any use of nuclear weapons would cause, India clearly defines their use only for the limited task of safeguarding itself against nuclear coercion and blackmail. In fact, the Indian nuclear weapons are not meant to deter conventional war, unlike the case of Pakistan's or North Atlantic Treaty Organisation's (NATO's) nuclear strategy that openly envisages nuclear use against conventional weaponry. Instead, India's narrow articulation of nuclear weapons as a means of only nuclear deterrence actually reflects India's traditional abhorrence for nuclear weapons and the reluctant steps it took down this path due to the absence of any substantive progress on universal nuclear disarmament and the deteriorating regional security environment. As Prime Minister Vajpayee said soon after India's nuclear tests, New Delhi does not "intend to use these weapons for aggression or for mounting threats against any country; these are weapons of self-defence to ensure that India is not subjected to nuclear threats or coercion."[1]

A coherent document on India's nuclear doctrine was formulated just fifteen months after the conduct of nuclear tests in May 1998. In one of its earliest tasks, the first National Security Advisory Board (NSAB)[2] produced a draft that encapsulated recommendations on what the country's nuclear doctrine should be. On August 17, 1999, the body presented the draft nuclear doctrine to the National Security Council and through it to the government, which simultaneously made it available for public scrutiny and debate. [3] The national and the international community were likewise taken by surprise not only by the speed with which the task was undertaken, but also with the transparency that the then caretaker Indian government offered on a subject that is normally kept out of public purview.

India's nuclear doctrine[4] reflects three basic precepts: the acquisition of credible "minimum" deterrence; adoption of no first use (NFU); and the promise of *assured retaliation* to inflict unacceptable damage. With each one of these parameters, India has chosen to impose restraints and checks upon itself, and the nuances of each are explained in the paragraphs below.

Credible Minimum Deterrence

The rejection of the concept of nuclear war-fighting frees India from the need to match the nuclear arsenal of its adversary/(ies) weapon for weapon. It was stated by Kenneth Waltz several decades ago, "Forces designed for war-fighting have to be compared with each other. Forces designed for war-deterring need not be compared. The question is not whether one country has less than another, but whether it can do unacceptable damage to another."[5] With the principal role of India's nuclear force being to protect the nation from nuclear blackmail and coercion, instead of any desire to enforce compellence, or mount aggression, the country's policy-makers perceive the need for only "credible minimum deterrence" or a nuclear force sufficient to deter the adversary by the prospect of a counter-attack that would be punitive enough to be unacceptable. Official pronouncements, therefore, have refused to be drawn into quantifying the minimum deterrent. Rather, the actual size and composition of the nuclear arsenal is seen as a dynamic entity to be determined by the changes in threat perceptions and own and the adversary's technological capabilities. As Jaswant Singh, India's Foreign Minister in 1998, said,

> The minimum is not a fixed physical quantification. It is a policy approach dictated by, and determined in, the context of our security environment. There is no fixity. Therefore, as our security environment changes and alters and as new demands begin to be placed on it, our requirements too are bound to be evaluated.[6]

However, it is unlikely that the numbers will sway significantly. Most Indian strategic analysts agree that it would suffice to have between 100-250 weapons based on various calculations. The essential parameter for this determination would be the ability to inflict unacceptable damage on the enemy. Of course, this calls for a predetermination of what the enemy would consider unacceptable, which is possible only on the basis of an extensive and intensive study of the cultural, socio-political, and strategic factors affecting the likely response of the adversary to nuclear use. The ability of a country to stomach damage is a complex function of its strategic culture, political system, economic state of growth, and level of freedom enjoyed by the populace. For instance, during the 1950s, China's leader Mao Tse Tung described his country's damage acceptability threshold as very high. But a more developed and economically advanced China cannot be expected to ascribe to the same philosophy. As countries develop, they also become more vulnerable and less open to accepting damage. Economically backward or politically isolated nations, on the other hand, have little to lose and may be able to absorb more damage.

Therefore, on India's assessment of the level of punitive retaliation that an adversary will find unacceptable rests the ability to correctly calculate the number of nuclear weapons it must stockpile and those that must absolutely be made survivable for effective retaliation.

No First Use (NFU) Against Nuclear Weapon States

The central principle that logically flows out of the perception of nuclear weapons as political instruments of deterrence is the lack of necessity to use them first in a conflict. Doctrines that ascribe a war-fighting role to nuclear weapons need to adopt aggressive postures that envisage their first use. During the Cold War, the USA and USSR believed that a nuclear war could be fought and won, and, hence, went on adding numbers and newer delivery capabilities in order to maintain an edge over the other. For the Americans, crafted as their war strategy was on the Pearl Harbour experience, acting first and maintaining surprise was critical. Not surprisingly, therefore, Washington subscribed to Launch-on-Warning (LOW) and Launch-Under-Attack (LUA) postures because it believed that unless it was able to undertake a preemptive/surprise strike, it stood little chance of being able to destroy all Soviet targets as required by its elaborate war plans. And minimising, if not completely eliminating, the enemy's second strike capability was the primary task of the first strike, and, hence, the need for ever increasing numbers of warheads. NATO too adopted a first use doctrine to deter Soviet conventional might, a logic that Pakistan now uses in support of its nuclear doctrine of first use against India.

India's nuclear doctrine, in contrast, has freed itself of many of these requirements by basing its nuclear strategy on a *retaliation only* policy. Of course, there are several at the domestic level and outside who dismiss the NFU as nothing more than public posturing and not capable of offering any guarantee against first use if the need so arose. While this is true of any declaratory policy, the fact remains that an NFU gets translated into force postures differently than a first use doctrine and, therefore, provides a fair indication of a country's intentions. While a first use posture requires missiles to be on alert for LOW/LUA and the nuclear warheads to be mated or ready to be instantly mated with the delivery systems, NFU offers a more relaxed posture. It allows for a policy of recessed deterrence wherein nuclear weapons and delivery vehicles may be developed and built, but would be stored separately, ready to be assembled in the event of a crisis. In fact, several points of criticism of NFU do not stand up to scrutiny and it may even be argued, as is attempted in the following paragraphs, that NFU offers the best possible choice for India in the present circumstances.

First of all, it must be understood that the adoption of NFU does not in any way adversely impact India's ability to defend itself against nuclear weapons. Given that the country does not foresee any plausible, rational scenario for the actual use of nuclear weapons, and least of all where it might be compelled to use nuclear weapons first, not for coercion and nor for any territorial, expansionist ambitions, NFU appears most logical. By placing the onus of escalation on the adversary, while retaining the initiative of punitive nuclear retaliation, India steers away from nuclear brinkmanship. Meanwhile, by establishing the nuclear weapon as an instrument of punishment, India seeks to prevent deterrence from breaking down, and, thus, aims to minimise, if not prevent, the very use of the nuclear weapon. NFU actually encourages the possibility of "no use" instead of "sure use." Unless the adversary is completely irrational, has suicidal tendencies, or is utterly unmindful of national and international public opinion, the possibility of a nuclear war should not arise and the surety that India would not use nuclear weapons first, but would certainly cause unacceptable punishment through retaliation, further ensures that.

Secondly, it is also questioned whether India should retain NFU even if it gets to know that the adversary is preparing for a nuclear strike? Should not preemption then be the right step? The answer to this lies in understanding that even preparation is no guarantee of a nuclear strike. Rather, it may well be part of the adversary's strategy of "coercive diplomacy." It is not a coincidence that all the 51-odd incidents of threat of use of nuclear weapons actually were attempts at coercive diplomacy. Therefore, despite the apparent show of readiness, there will, more likely than not, still be a chance that nuclear weapons would not actually come into use. But, by striking first, India would end up inviting certain retaliation. In fact, even if the adversary's first use of nuclear weapon was planned to be a small demonstration strike, with India's preemptive nuclear strike, it would surely retaliate with all that is available. In such a scenario, India would have invited a much larger nuclear force upon itself. In fact, at such a time when a nuclear strike appears imminent, it would be better to clearly indicate own level of preparedness for nuclear retaliation in order to reinforce deterrence, and simultaneously seek international diplomacy to mount pressure on the adversary.

NFU is also the best answer to those strategic analysts that believe that nuclear weapons in India and Pakistan have led to a condition of instability since both sides have small nuclear forces and would be tempted to launch a disarming first strike in case of a crisis.[7] But India's NFU posture removes this temptation not just for itself, but also for the adversary. The NFU goes to alleviate Pakistani insecurity, which, in turn, is beneficial to India by relieving pressure on its leaders for launching a preemptive strike. If Pakistan was constantly under the fear that an Indian nuclear strike was imminent, its own temptation to use its nuclear force would be higher. Moreover, since the NFU necessitates greater emphasis on survivability in order to reduce vulnerability of own nuclear arsenal, the measures to enhance survivability mitigate the "use or lose" syndrome. This situation was best described by Robert McNamara in the context of the Soviet hardening their missile sites. He wrote,

> In a period of tension, I wanted the Soviet leaders to have confidence that those forces would survive an American attack and would be capable of retaliating effectively. Then they would not feel a pressure to use them preemptively… I had no desire to face, in a period of tension, an adversary who felt cornered, panicky and desperate and who might be tempted to move irrationally.[8]

Thirdly, by rejecting a first use stance, India has also removed the need for retaining nuclear forces on hair trigger alert, a situation not at all conducive to strategic stability, given the geographical realities of the neighbourhood. Having nuclear forces on alert not only raises the possibility of an accidental nuclear war based on a gross miscalculation, but also lowers the threshold of nuclear war in a crisis situation. In the case of India and Pakistan, this raises risks, given their proximity, low warning time and propensity to frequent border skirmishes. Therefore, India's NFU actually brings stability to the nuclear equation. It eases the risk of unintended or accidental attack since it precludes the need for delegation of authority for launch.

One situation, however, that could test India's NFU is a scenario where a Taliban type military man or some non-state actor were to take control of nuclear weapons in Pakistan after a period of political instability?[9] Two responses could arise in this case. Firstly, if the new player has assumed state power and become a political actor, then he would also have developed a stake in the political system and, hence, should be expected to display some sense of rationality in the use of nuclear weapons. Classical deterrence based on a retaliation policy can be expected to apply in such a scenario. But in case the non-state actor/organisation has got hold of nuclear weapons in a situation of political chaos, and threatens to use it against India in a suicide bomber mentality – in order to wreak nuclear havoc without worrying about the consequences of the same for its own state – then nuclear deterrence could become dysfunctional. In fact, this is the same problem being faced by nuclear doctrines worldwide because classical deterrence cannot apply against non-state actors since there are no tangible targets on which unacceptable damage can be inflicted as a means of deterrence.[10] To deal with cases such as this, India needs a multi-pronged strategy: one, it must declare that Weapons of Mass Destruction (WMD) terrorism would invoke retaliation against the state known to be sponsoring such activities, or which might have facilitated loss of weapon or material to cause an act of terror; secondly, it must participate in global/multinational endeavours aiming at controlling proliferation of dual use materials through strengthened export controls, and enhanced security and safety of nuclear arsenals; and, thirdly, it must focus on better intelligence and preparedness levels to mitigate a national disaster.

Evidently then, NFU can hold in a range of situations, and it offers several advantages for India, besides being morally the most correct stance to take. That is, if there can be anything morally correct about nuclear weapons at all. Nuclear weapons are not just any weapons that could or should be used indiscriminately. They are special in the sense of their immense destructive capability. And therein lies their value as deterrents. Also, the merit of India's NFU lies in challenging the long held nuclear theology of first use, as professed by the Western nuclear powers. Until now, this has been touted as the only possible approach to use nuclear weapons for national security. The Indian adoption of NFU offers a counter-view to the traditional aggressive and arms race generating doctrines. If this were to be accepted by all NWS, then the world might find itself on its way to a diminishing salience of nuclear weapons, perhaps their delegitimisation, and eventually their abolition.

Assured Retaliation

India's nuclear doctrine frees itself of the compulsion of immediate retaliation and bases deterrence instead on the *certainty* of retaliation that would be *punitive* in nature. As has been said, "The ability to retaliate with certainty is more important than the ability to retaliate with speed."[11] In fact, the time taken to retaliate would have to be dependent on technical realities such as time required to bring together the nuclear weapon and delivery vehicle, the nature of precise command and control and custody arrangements, the state of the country after having absorbed the first strike, as also other domestic and international political factors.

Accordingly, therefore, India's doctrine does not specify any time for retaliation. In fact, the NSAB draft did not even define the nature or magnitude of punishment, except for describing it as "punitive." Beyond this, it did not specify the character, extent or weight of retaliation. Instead, it followed the French logic that "the adversary must not be able to calculate what would be the reaction to this or that initiative he might take."[12] It only conveyed that retaliation would be certain and it would be devastating, irrespective of when or how it was inflicted.

However, while the draft nuclear doctrine mentioned "punitive retaliation", the 2003 official version changed it to "massive retaliation." But this has not necessarily enhanced the credibility of deterrence because it actually restricts the available response to an adversary's first strike to only an all out nuclear attack. This may appear too drastic for use except in extreme circumstances. While it may be argued that India's nuclear philosophy does not conceive nuclear use except in extreme circumstances, however, a case can be made for greater flexibility in response options. In fact, "punitive retaliation" is credible enough since it provides alternatives relative to the nature of strike and level of provocation. The US too in the 1960s had reached the conclusion that it should have a variety of response options other than massive retaliation against cities. This wisdom was obtained after several US officials, including Robert McNamara as Secretary of Defence, expressed their dissatisfaction with the inflexibility of a Standard Integrated Operation Plan (SIOP) that envisaged a preemptive first strike involving 3,423 weapons totalling 7,847 megatons against the Russians and Chinese in case of any conflict, irrespective of the provocation!![13] A more flexible response was, therefore, proposed that envisaged a substantial raising of the nuclear threshold for the critical initial responses to be made by conventional forces alone, keeping the use of nuclear weapons late and limited.

In the Indian context too, especially vis-à-vis Pakistan, as Ashley Tellis argues, by basing deterrence on massive retaliation that forecloses the option of a graduated response, India might end up encouraging its adversary to massively employ its own arsenal in the fear that India's "massive retaliation" to any nuclear use would anyway hold the possibility of disarming it. [14] Thus, India might invite a greater nuclear use upon itself than the enemy might actually have factored into its calculations. This would then rob India of the opportunity to exercise escalation dominance. Meanwhile, in relation with China, given its existing nuclear superiority and higher survivability quotient, Indian threats of massive retaliation after having suffered a first Chinese strike, hardly seem credible.

The credibility of nuclear deterrence premised on assured retaliation obviously depends on the survivability of a sufficient nuclear force that can mount a second/retaliatory strike with certainty. This necessitates the survival of not only the nuclear warhead, but also the delivery vehicle; the command and control set-up, including not just the primary decision-maker and his entire pre-determined chain of succession, but also the line of command up to the man in the field who is to execute the decision; secure communication systems; targeting coordinates; and, above all, the survival of the political will to stomach the horrendous damage that any nuclear strike would cause.[15] The last element can never be empirically assessed, though its survivability can be heightened through educating the political leadership about not only how deterrence works and the need to show adequate resolve, but also by ensuring the availability of other well informed advisers to help in times of crisis. Meanwhile, survivability of the other prerequisites can be enhanced through intelligent planning and adequate redundancy measures.

Traditionally, survivability has been assured through dispersion of nuclear forces, use of deception and by maintaining some sort of a relationship, not necessarily of parity, with the enemy's intelligence, surveillance, warheads and, more importantly, delivery capabilities. Dispersion is a function of mobility and the more mobile the components of the nuclear arsenal are, the greater is the chance of their survivability. But, managing mobility is not an easy task. The challenges include not just organising the frequent movement of actual nuclear warheads and delivery vehicles but also that of dummies so as to weave in adequate deception and camouflage into the survivability strategy, and ensuring that all elements can be brought together within a reasonable time-frame after a first strike has been suffered.

The determination of survivability also depends on the adversary's nuclear doctrine, force posture and strategy. A known counter-force or a counter-value targeting philosophy, or a mix of both, provides some indication of the likely targets to be prioritised and, hence, helps address some of the survivability issues. But these are complex challenges that require serious thinking and analysis in peace-time in order to enhance the credibility of deterrence.

Interlinkages Between Nuclear Deterrence and Lower Levels of Conflict

India's practice of nuclear deterrence deals with a unique predicament where its nuclear weapons, especially with Pakistan, are tasked to function in a fragile situation where Pakistan uses its nuclear weapons to defend its own conduct of sub-conventional conflict against a conventionally superior India. This distinct feature of the Indo-Pak nuclear relationship bears no comparison to any other. While it is true that the superpowers of the Cold War era deterred the conventional might of each other with their nuclear capability, they did not engage in activities that were meant to directly destabilise the other while taking refuge in their nuclear shield.

For India, therefore, the interlinkages among sub-conventional, conventional and nuclear war are far more complex. In fact, India's nuclear strategy has to grapple with the challenge of building credibility of its nuclear deterrence in such a way as to counter the adversary's attempt to blur the lines between conventional and nuclear war. While he threatens to lower the bar for nuclear deterrence breakdown in order to dissuade conventional attack from a superior force, India must not only raise the nuclear threshold, but also devise adequate conventional responses that can be safely executed in the situation of a nuclear overhang. Of course, this is easier said than done. As long as nuclear weapons exist, so does the potential danger of escalation to the nuclear level. Therefore, the true test of India's nuclear strategy lies as much in establishing mutual nuclear deterrence, as in tackling sub-conventional warfare with conventional tactics that operate in the presence of nuclear weapons, but without bringing them into play.

The shadow of nuclear weapons obviously imposes constraints on the range of military options and the nature of coercive force that India can indulge in. It demands greater caution so as to avoid the potential cost of miscalculation. On the one hand, the desire to win requires a demonstration of resolve and a willingness to fight. But, simultaneously, the fear of nuclear war demands caution and restraint in the use of force since the risks of an unplanned and uncontrollable escalation in the presence of nuclear weapons could be catastrophic for both.

Therefore, the conduct of war in the presence of nuclear weapons has to follow a different set of rules. At one level, in fact, nuclear weapons mean the end of classical conventional war of the kind envisaging acquisition of large swathes of territory, or a *blitzkrieg* to cause high attrition. Any such measure is certain to breach the threshold of the adversary's levels of tolerance, especially one with weaker conventional capabilities. This could compel him to resort to use of nuclear weapons, thereby leading to deterrence breakdown. So, if nuclear deterrence is to be maintained, then the war needs to be fought differently, in a manner where the risk of escalation to the nuclear level is minimised because the targets are so chosen as to not threaten the survival of the state or its critical elements. This obviously also has an impact on how victory and defeat can be defined in such a situation. In an all out conventional war, the difference between the victor and the vanquished is clearly evident based on the estimation of which side has suffered greater losses and damage. But, in limited wars, in which the level of destruction has to be carefully calibrated and imposed, this distinction is blurred.

An illustration of this is manifest in the experience of India and Pakistan during the Kargil crisis in 1999 and Operation Parakram in 2001-02. The manner in which the two episodes unfolded held several lessons for Pakistan, India and the larger international community on how use of force can be executed in the shadow of nuclear weapons. With Kargil, Pakistan realised that the acquisition of nuclear weapons had not provided it with a *carte blanche* on disruptive actions across the border. Rather, the presence of nuclear weapons placed clear limits on how far it could, or should go, so as not to breach the limits of Indian tolerance. Meanwhile, India also realised the constraints that the presence of nuclear weapons cast on its own exercise of military options. Despite the widely expressed opinion to strike against Pakistan once the identity of the Mujahideen as regular Pakistani soldiers was established beyond doubt, the political leaders imposed upon the military to undertake operations in such a way that the threat of escalation was minimised. Therefore, in an unprecedented gesture, the use of air power was limited to the Indian side of the Line of Control (LoC) in order to oust the illegal occupants of the heights. No strikes were authorised across the border, not even at the terrorist infrastructure known to exist in Pakistan Occupied Kashmir (POK).

India exhibited the ability to wage war with self-imposed limits. This proved to be as much a revelation to Pakistan as to the larger international community that had described this very region as the most dangerous flashpoint. The sense of responsibility and maturity in action displayed by India helped to shape a range of perceptions across several capitals. Indeed, India did feel the weight of nuclear weapons on its possible courses of action. But it also displayed that there was scope for retaliatory action that had to be intelligently discovered and exploited, and astutely meshed with politico-diplomatic measures best suited to the prevailing international political environment. In the future too, India's nuclear strategy will have to closely deal with lower levels of conflict.

Conclusion

India has premised its need for nuclear weapons on its desire to resist nuclear coercion or blackmail and, hence, espies their use only for self-defence. Accordingly, New Delhi has enunciated a nuclear doctrine that perceives a purely political role of deterrence for its nuclear weapons. Flowing therefrom, India's nuclear doctrine ascribes to a no first use posture since it conceives no role for its nuclear weapon in enforcing compellence or staging aggression. Rather, the weapon is only considered usable in retaliation to inflict unacceptable damage against the adversary's first use. In order to carry out this exercise, the doctrine aspires for a minimum nuclear deterrence whose credibility resides in its survivability.

Such a nuclear strategy provides the least risk option since it premises nuclear deterrence on a small arsenal that is not on hair trigger alert, and, hence, less prone to miscalculation or accidental use. Significantly, India's nuclear doctrine also accords due importance to the attainment of a nuclear weapon free world (NWFW) as the best insurance of Indian security. In an NWFW, India can be regionally more secure and globally better placed to pursue its objective of assuring strategic autonomy for its pursuit of economic and security objectives. But until such a world may be obtained, India's own brand of nuclear deterrence has to handle the challenges unique to the region.

In difficult regional circumstances such as the kind that India finds itself in, the possibility of conventional wars exists. Theoretically, therefore, there is also the possibility of escalation of a conventional conflict into an unwanted nuclear exchange. However, if India was to be self-deterred by this thought, it would mean complete erosion of both its conventional and nuclear deterrence capabilities. In order to deter war, India must maintain and project a high level of conventional capability that would be intelligently applied in a calibrated manner to keep the use of coercive force well below the assumed red lines of the adversary. In order to deter nuclear war, India must illustrate its ability to handle deterrence breakdown and retaliate against the adversary with enough capability and resolve to inflict damage that would impose a cost far beyond the value of the stake that made the first use of nuclear weapon against India thinkable.

Notes

1. *Suo Motu* Statement by Prime Minister Vajpayee in Parliament on May 27, 1998. As reproduced in *Strategic Digest*, July 1998.
2. The NSAB exists as an official body that is part of the National Security Council. It acts as a forum through which the government decision-making apparatus can draw on the advice and experience of appointed academics and retired civil servants and military officers. Members of the NSAB serve a term of one year.
3. National Security Advisory Board, *Draft Indian Nuclear Doctrine*, August 17, 1999. Available at http://www.meadev.nic.in

4. The NSAB doctrine as well as the Statement put out by the Cabinet Committee on Security, "Operationalisaion of Nuclear Doctrine," January 4, 2003, are considered the official doctrinal pronouncements.
5. Kenneth Waltz, as quoted by Gen Sundarji, *The Blind Men of Hindoostan* (UBS PD Ltd.), p. 68
6. Foreign Minister's Speech in Parliament on December 16, 1998. Downloaded from http://www.meadev.gov.in.
7. This has been forcefully argued by Sumit Ganguly and Kent L Bringer, "Nuclear Crisis Stability in South Asia," in Lowell Dittmer, ed., *South Asia's Nuclear Security Dilemma: India, Pakistan and China* (New York: M.E. Sharpe, 2005).
8. Robert McNamara, *Blundering into Disaster: Surviving the First Century of the Nuclear Age* (London: Bloomsbury, 1987)
9. The prospect of a nuclear capable state losing control of some of its weapons to terrorists was put forth as a real and immediate danger in the US Quadrennial Defence Review 2006 and has been maintained in official threat perceptions since then. American options in case of such an eventuality have been well brought out in Thomas Donnelly, "Choosing Among Bad Options: The Pakistani 'Loose Nukes' Conundrum," *National Security Outlook*, American Enterprise Institute for Public Policy Research, May 2006.
10. For a comprehensive examination of this, see Manpreet Sethi, "Current Trends in Nuclear Weapons Thinking and Strategies," *Asian Defence Review 2007* (New Delhi: Knowledge World, 2007).
11. Ashley Tellis, *India's Emerging Nuclear Posture: Between Recessed Deterrent and Ready Arsenal* (Santa Monica, CA: RAND, 2001), p. 326.
12. This is the logic of the French nuclear doctrine, as explained by Tellis, Ibid., p. 321.
13. For more on this, see DeGroot, *The Bomb: A History of Hell on Earth* (London: Pimlico, 2005), pp 268-69.
14. Tellis, n. 11, p. 340
15. Several Indian and foreign writers have doubted that the Indian leadership has a "killer instinct." For instance, Aditya Chibber writes, "The *vairagya* syndrome of renunciation has always robbed us of the killer instinct," in his *National Security Doctrine: An Indian Imperative* (New Delhi: Lancer, 1990), p. 85 ff. A fiction writer, Humphrey Hawksely, in his book *The Third Global War* too depicts a nuclear stricken Indian leadership unable to bring itself to retaliate with nuclear weapons.

Role of the Indian Army in National Security

□ V.P. Malik

It is difficult to deny that without adequate defence, a country cannot develop freely. Yet, it is equally impossible to deny that without development, there will not be a country worth defending.

Introduction

A growing recognition in the world that security issues cannot be looked at in isolation has led to widening of the security agenda by including a range of issues beyond the purely military dimension. A comprehensive security matrix comprises all major political, economic, environmental, societal and environmental issues together with the military dimension. Security problems, therefore, require more integrated national and international responses. Some inherent implications of the comprehensive security matrix are: (a) a more constructive form of security management; (b) improved rationalisation of the security agenda with a potential to improve national and international cooperation in human development; (c) it complements the new phenomena of globalisation and of economic reforms; and (d) commonly shared threats in different sectors of security will bring states together, eventually leading to greater cooperative security. This would particularly apply to issues like counter-terrorism, environmental, energy and economic activities.

The essential requirement, however, is that the comprehensive security concept must lead to bridging the gap between traditional development and security studies through cross-cutting policy agendas, and establish connections with related disciplines like international relations, regional studies, socio-politics and socio-political economy.

What are its implications for India's armed forces, the Army, in particular? It implies that as a major national institution, the armed forces not only have to defend the territorial integrity of the nation from external aggression and internal threats but also play a role in nation-building.

Army's Contribution to Nation-Building

I believe that the armed forces, the Army in particular, have been doing this ever since India became independent. The new comprehensive security concept is merely a recognition of this example for all developing nations. In support, let us take a few post-independence pages of India's nation-building history.

Institutional Strength. Army life, in its outlook and purpose, is heavily dependent on the traditions of service imbibed over years of blood-stained history. The Chetwode credo taught to all officers on commissioning is, "The safety, honour and welfare of your country come first, always and every time. The honour, welfare and comfort of the men you command come next. Your own ease, comfort and safety come last, always and every time." There is also the tradition of 'oath', considered as the bedrock of a profession which differentiates it from all other professions. Earlier, it was the concept of "Naam – Namak – Nishan: Be Honourable – True to Your Salt – uphold the Flag." Today, the oath is to the Constitution of India. The Army is a true reflection of the basic national concept of unity and diversity. Soldiers of all classes, creeds and religions serve loyally, effectively, and with total dedication: living together in barracks, eating from the same kitchens, speaking the same language, and observing each other's religious festivals. Secularism, discipline, integrity, loyalty, *espirit-de-corps* — these are essential values inculcated among Army personnel. Such a motivation makes them efficient and dedicated soldiers in service. Even after leaving the Army, they remain role models for others. Unaffected by divisive politics or casteist social activities, Army cantonments and stations have always been, and are, totally cosmopolitan: an oasis of national unity.

Apolitical Outlook. An important legacy of the Army and its sister Services has been their totally apolitical outlook. The oath, as stated earlier, is to the Constitution of India and to the constitutionally elected central government. The directions of the political parties or their hue and colour do not concern them. Unlike some armies in the neighbourhood, the Indian Army has stuck to the concept of its loyalty to the constitutionally elected government and, thus, enabled the nation to develop its unique democratic political ambience. It has supported democratic electoral process for free and fair elections in many insurgency affected states but stayed far away from the polling booths.

Supporting Consolidation of Nation-State. India's Army was put to test from the very first day of India's march to nation-building. The birth of the 'nation-state' that we see today was not such a smooth affair. The Army had to assist the civil authorities to control the communal frenzy that was unleashed by the artificial boundaries of partition. Hyderabad and Junagadh had to be militarily forced to integrate with India. In October 1947, when Maharaja Hari Singh of Kashmir signed the Instrument of Accession to India, the Army was inducted into Kashmir Valley and other parts of the state to thwart Pakistan's design of swallowing it. Consolidation of territorial frontiers was later carried forward by liberating Goa, Daman and Diu. Sikkim was integrated through a popular referendum in 1975. This was possible primarily due to the strong presence of the Indian Army in that state. The contribution to the consolidation of the nation goes on.

Integrating Border States. The Army played a significant role in integrating the border states and their people into the national mainstream. Deployment of the Army enabled governmental infrastructure to follow in remote areas of Mizoram, Manipur, Nagaland, Northeast Frontier Agency (NEFA)/Arunachal, Uttar Pradesh (UP) and Himachal Pradesh (HP) border with Tibet, and Jammu and Kashmir (J&K).

Ensuring Territorial Integrity. The main role of the Army is to defend the territorial integrity of the nation-state against external and internal threats. This is central to the concept of national security and paramount for all nation-building activities. History is witness to the fact that whenever a nation has lowered its security guard, external powers have been quick to exploit it to their advantage. In 1950s, we overlooked this important lesson of history and allowed the security apparatus to drift till the Chinese shook us in 1962. We had to relearn this lesson through an ignominious experience. Post-1962, we have had two 'brushes' against the Chinese: one at Nathu La in Sikkim in 1967 and the other at Wangdung in Arunachal Pradesh in 1986. The outcome made it clear that the Indian armed forces are alert, strong and determined to defend national territory. This feedback helped the Chinese and our leaders to open and pursue a dialogue, which is helping to improve our relations now. The Indo-Pakistan Wars in 1965, 1971 and in Kargil in 1999 are history today. Besides ensuring territorial integrity, they have had a positive impact on the security, *izzat*, and morale of the nation. Strong and well-trained armed forces are a deterrent to our potential adversaries. If this is achieved, developmental activities and nation-building can progress without external hindrance.

Internal Security. Nation-building is hardly possible where we fight and kill each other. I am referring to internal security and stability. Imagine a riot-torn Mumbai of 1992, when the economic capital of India came to a standstill. Communal and inter-caste riots, Naxalites, secessionist groups, militants and other anti-social elements, aided and abetted by foreign countries: internal security these days has become a very serious challenge to our national security. The demand, or the need, to use the Army for internal tasks which are primarily in the police and Home Department domain, has been increasing year after year. The positive impartiality of the Army when placed in such a situation is legendary. A soldier in his olive greens is looked upon as a source of confidence among the people. At a higher level, we are also fighting full-fledged insurgency. The Army has not allowed, nor shall allow, any attempt by any state or a section of the state to secede from the Republic of India. During such operations, the Army not only fights militants and anti-social elements but also reassures innocent people feeling insecure or neglected due to inadequate civil administration. Large scale civic action programmes are undertaken by the Army along with anti-terrorist operations. In Nagaland and Manipur, we have formed an Army Development Group for this purpose.

Disaster and Ecological Management. In addition to its employment in counter-insurgency and internal security duties, the Army is frequently called upon to render aid to civil authority for maintenance of law and order, flood and famine relief, and humanitarian aid during natural calamities such as the frequent cyclones on the Andhra and Orissa coast and the Latur earthquake, building of bridges and roads during emergencies. Floods in some parts of the country are a regular feature. And yet, when they occur, the civil administration in those areas is found ill-prepared or paralysed, and promptly sounds the bugle for Army assistance, as happened recently in the Bihar floods even after the formation of the National Disaster Relief Management Authority. In a train accident near Ferozabad some years ago, the affected people and locals refused to allow the police and other officials to remain at the site. The only people worthy of their confidence and trust were soldiers of 50 Para Brigade who had rushed to the site and did a commendable job. Lately, the Army has also been helping to sustain ecology through afforestation and other environmental protection missions with its Ecology Territorial Army (TA) battalions.

Cooperative Security and Image Abroad. Our troops, in support of the foreign policy of the government, have projected military power in Sri Lanka and Maldives. They have worked in UN Peace-Keeping missions in Korea, Congo, Gaza, Cambodia, Angola, Somalia, Rwanda, Namibia, Lebanon and so many other countries. Besides infantry battalions, artillery, engineers, signals and medical units and logistics personnel, the Army has contributed a large number of military observers and force commanders to missions in Asia, Africa and Latin America. The Army has also been part of international disaster relief operations in Bangladesh, Sri Lanka and Maldives. It has done the nation proud by its committed devotion, impartiality and efficiency. This aspect has received international recognition and would definitely help India one day in getting a seat in the UN Security Council.

New Threats and Challenges to National Security

Having observed the traditional role being played by the armed forces, the Army in particular, let us now look at the new threats, challenges and vulnerabilities that may affect India's national security in the 21st century, and would have direct implications on the role and preparedness of the Army. These threats and challenges include global shifts in economic and military power, rising uncertainties, emerging destabilising factors, and diverse, complex and invisible challenges whose precise nature is hard to predict. Hegemonic and power politics, impact of globalisation, resources and financial crises, failed and failing states, trans-national crime, global terrorism and spread of weapons of mass destruction are a few of the pressing challenges that face the country and may involve the defence Services at some stage.

Prolonged instabilities in the neighbourhood (Afghanistan, Pakistan, Nepal, Myanmar, Bangladesh, Sri Lanka), the emergence of China as a superpower, and territorial disputes with China and Pakistan coupled with the nexus between these two nations are also bound to have politico-security consequences for India. In addition to these external challenges, there are grave internal security challenges such as insurgencies in the Northeast, Jammu and Kashmir, and the Naxalite-Maoist movement. All of these pose a challenge to India's external and internal security. Non-military threats such as illegal migration, global warming, and security of oil and water resources, environmental security, and proliferation of small arms, large scale gun running and drug smuggling are also realities that need to be dealt with on an urgent basis. Even the Revolution in Military Affairs (RMA) poses a challenge. Unless we can catch up, India's national security would be affected adversely in the years to come.

Elongated Spectrum of Conflict

An assumption made by Admiral J.C. Wylie in *Military Strategy: A General Theory of Power* is, "Despite whatever effort there may be to prevent it, there may be a war." This assumption is neither provocative nor a justification for the existence of the armed forces in peace-time. Military history tells us that nations that neglect this historical determinism make themselves vulnerable to military surprise, defeat, and ignominy. The assumption, therefore, is a reminder to the strategists to visualise security threats, the possibility and nature of conflict (or war, when political negotiations no longer serve the purpose) and to always remain prepared for such an eventuality. Another basic assumption for war planning is that we cannot predict with certainty the pattern of conflict for which we prepare ourselves. It has seldom been possible to forecast the time, place, scope, intensity and general tenor of a conflict. India's conflicts with Pakistan and China, military involvement in Sri Lanka's ethnic conflict in 1987, and the recent wars in Iraq and Afghanistan are examples of the same. This is particularly true in the current and futuristic strategic scenario wherein the war potentials are more transparent and intentions more inscrutable. This assumption implies that our security plans should cater for the complete spectrum of conflict — a spectrum that will embrace any conflict situation that may conceivably arise. India's military strategies and doctrines must be flexible and non-committal, capable of application in any unforeseen circumstances. Planning for uncertainty is less dangerous than planning for certitude. With the paradigm shift in the nature of military and non-military security, the Army has a tougher job today to be prepared for this elongated spectrum of conflict, ranging from aid to civil authority, counter-terrorism, different levels of conventional war, to a war involving Weapons of Mass Destruction (WMD).

Future Roles and Missions

India's Ministry of Defence has spelt out the role for the armed forces that includes defending the country's borders as defined by law and enshrined in the Constitution and protecting the lives and property of its citizens against terrorism and insurgencies. It also aims at maintaining a secure, effective and credible minimum deterrent against the use or the threat of use of WMD, along with securing the country against restrictions on the transfer of material, equipment and technologies that have a bearing on India's security, particularly its defence indigenous research, development and production to meet the nation's requirements.

A futuristic mission would also include promotion of further cooperation and understanding with neighbouring countries and implementing mutually agreed confidence-building measures, working with the Non-Aligned Movement (NAM) countries to address key challenges before the international community and engaging in cooperative security initiatives such as the Association of Southeast Asian Nations (ASEAN) Regional Forum. There would be a need to pursue security and strategic dialogues with the major powers and follow a consistent and principled policy on disarmament and international security issues based on the principles of supreme national interest. However, it should always be kept in mind that changes in the security environment, grand strategy and military strategy dictate the military doctrines, shape and size of the armed forces.

Missions to ensure India's territorial integrity and national interests would include not only the ability to fight limited or full-scale conventional wars but also a capability for rapid deployment of forces to deal with border skirmishes and tri-Service task forces for out-of-area contingency missions and the security of sea lanes for India's trade. The tri-Service strategic forces are meant not just for deterrence but also for having the capability of a wide range of nuclear responses, options and ability to defend space assets. The armed forces also require capability to handle cyber, information and psychological warfare. Integration of air and sea power at strategic, operational and tactical levels would need to be appropriately addressed to deal successfully with the elongated spectrum of conflict.

In addition, the Army will require specially equipped and organised forces for counter-terrorism, counter-insurgency, proxy wars and other internal security missions.

Likely Nature of Conventional War

With the existence of nuclear weapons on the subcontinent, it is hard to see a large scale 1962, 1965 or 1971 type conventional war between India and Pakistan, or even with China. The probability of an all out, high intensity regular war between these nations will henceforth remain low, although it cannot be ruled out altogether. Even if a conventional war does break out, the war is likely to be limited in time, scope and space. It could even be limited in the quantum and pace of application of firepower. It will have to be conducted within the framework of carefully calibrated political goals and military moves that permit adequate control over escalation and disengagement. The likely duration of the war and time available to the armed forces to execute their missions and achieve politico-military goals would be crucial for their planning and conduct of operations. In a 'reactive' situation, like the Kargil War, this duration can be prolonged. However, the duration available will be much less if we decide to take the initiative.

There is a linkage between deterrence and escalation. Capability to wage a successful conventional and nuclear war is a necessary deterrent. A war may well remain limited because of a credible deterrence or escalation dominance. Escalation dominance means that one adversary has overwhelming military superiority at every level of violence. The other will then be deterred from using sub- conventional, conventional or nuclear war due to the ability of the first to wage a war with much greater chances of success. Limited conventional war, therefore, does not mean limited capabilities. It refers to the use of those capabilities.

In a limited conventional war, politico-diplomatic factors will play an important role. Careful and calibrated orchestration of military operations, diplomacy and domestic environment is essential for its successful outcome. Continuous control of the 'escalatory ladder' requires much closer politico-civil-military interaction. It is essential to keep the military leadership within the security and strategic decision-making loop and having a direct politico-military interface. During a conflict situation, all participants must remain in constant touch with the political leadership, as was done during the Kargil War.

Over the years, the range and lethality of weapons has increased and the time to target has reduced. The battlespace has expanded in all the three dimensions. In a digitised battlefield, timely access to intelligence can be matched with operational mobility with great effect. Enhanced mobility, long reach in targeting and effective command and control have obscured tactical and strategic boundaries. A small military action along the Line of Control or a terrorist act in the hinterland of the types that we have seen in the recent past, tends to become an issue for consideration and decision-making at the strategic level. It is a situation wherein a junior military officer is expected to understand political considerations, and the political leader to know the tactical and operational considerations.

The military implications on the ground are integrated capabilities, shallow objectives, rapid concentration and launch, multiple choices/thrust lines, short, sharp intense action, surprise, continuous surveillance, multiple choices/thrusts and shallow objectives, maximum use of forces (including Special Forces) and force multiplier, and ability to fight from a cold start.

A few important capabilities that require special attention are:

- **Use of Satellites for Surveillance and Communications**. Rapid and responsive military operations require timely and accurate reconnaissance reports, weather monitoring, precise navigation and long haul fail-safe communications. Indian aspirations for RMA would remain unfulfilled till we are ready for the Network-Centric

Warfare (NCW). The first requirement for which is a communication network that allows interoperability of the highest order among all the constituents of war fight machinery.

- **Use of Force Multipliers**. The concept of calculating conventional force levels to achieve an objective has been radically altered on account of force multipliers. A small well-equipped and better trained force can cause much more devastation and accomplish more than what was possible earlier.
- **Rapid Mobilisation**. The sooner an intervening force can arrive to influence the course of a military event, the smaller the chance that the conflict will devolve into a firepower intensive, wasteful slugging match. Rapid mobilisation outpaces the enemy and has the same advantage as surprise. The Army has already included the capability to fight from a cold start in its doctrine. We need not wait for full mobilisation of the entire theatre or border. It will also need a Rapid Reaction Force for border emergencies and out of area contingencies. Such a capability requires specially organised and trained formations, with the ability for cold start, adequate means for rapid transportation and capability to sustain till a build-up can take place, or its withdrawal.
- **Surveillance and Intelligence**. Without adequate intelligence and continuous surveillance, even the best of plans cannot succeed. We require a clear strategic, operational and tactical picture and assessment with the help of all possible technology and human intelligence. We have to obtain real- time intelligence and ensure that it reaches those who need it in time.
- **Helicopter Support**. Helicopters need to be integrated with fighting formations. Attack helicopters and long-range rocket artillery will be able to take on several difficult close air support fighter-bomber tasks.
- **Small-Scale Operations.** In a limited war, tempo and speed dictate that light, highly mobile infantry and commando forces make up the majority of the force. We need more Special Forces.

Role in Nuclear Deterrence

Due to the horrendous destructive power of nuclear weapons and their almost certain universal condemnation, the probability of their use would remain extremely low. But as long as there are nuclear weapons around, they *could* be used. Soldiers do not have the luxury to rule out such a possibility. Nuclear weapons are meant for deterrence and India's nuclear doctrine is based on credible minimum deterrence, no first use, civilian control and survivability of the warheads and delivery systems. The nuclear doctrine calls for integration of multi-disciplinary actions and coordination: the armed forces, foreign policy, atomic energy, Defence Research and Development Organisation (DRDO) and several other elements. We have an inter-Service Strategic Forces Command. The Services have yet to develop a joint operational doctrine on the employment of nuclear weapons. There is also a need to interface the nuclear capability with conventional capability and then reassess military strategy and force structuring. This is even more essential for the Army.

Changing the Attitude

Recent wars have involved a much greater level of integration of politics and military planning and execution. Even when diplomacy has run its course and a decision to employ the military is made, the political leadership seldom allows autonomous conduct of the war to the military. In practice, we are seeing a continuing erosion of the dividing lines between war and politics. In the new military conflicts environment, I believe that some of our large size combat organisations can be reduced in size and made more versatile and agile. It is time that we started thinking about greater combat effectiveness of our Special Forces, Combat Groups, Commands and Battle Groups and other equivalent formations. Having several large, unwieldy and expensive Strike Corps for conventional deterrence that tend to sit out of the war when it actually happens is not a cost-effective military strategy. The emphasis has to be more on quality and not on quantity.

Challenges of Internal Security

India has a unique centrality in the South Asian security zone. It has special ties with each of its neighbours, of ethnicity, language, culture, and common historical experience or shared access to vital natural resources. India's special ties with its neighbours tend to encourage the Indian secessionist groups in establishing safe sanctuaries across the borders in neighbouring states and in trans-border illegal migration, gun running and drug trafficking. Situated, as India is, between the Golden Crescent and the Golden Triangle, secessionist groups taking to violence find little difficulty in indulging in the drug trade and obtaining small arms. This problem may accentuate when we have greater free trade with our neighbours. There is a growing nexus among crime, insurgency and politics, with 45 per cent of India's geographical area covering 220 districts in the grip of insurgency today. In the last 20 years, over 65,000 people have been killed in terrorist violence and the insurgency problems of J&K and the Northeastern states are well known.

Strategically, India cannot afford to be perceived to be buckling down under internal security or externally induced terrorist pressures. That would be disastrous. Already, a large strength of the Army is deployed on these tasks. This needs to be reduced but that will be possible only if India can revamp its paramilitary, central and state police forces.

Military Diplomacy and Cooperative Security

In the comprehensive security matrix and the emerging security situations that we would be faced with, military diplomacy will be a potent instrument for advancing India's national interests and foreign policy objectives, particularly in the country's immediate and extended neighbourhood. We have good experience of peace- keeping, peace enforcement and disaster relief missions abroad. In recent years, the Indian military has shed its hesitant approach. But the endeavours are essentially in the fields of training and bilateral visits. Coopting the military leadership in strategic dialogues and consultative processes is essential. Military diplomacy as a strategic tool for engagement needs to be conducted at both bilateral and multilateral levels. There is a need for appropriate structures and organisations to use this important strategic tool which is lacking at present.

Media Management

Today is the age of information war where new forms of war reporting catch events at their source, when they are still history's raw material. It is not possible to resist the pressure to be transparent. The lesson learnt could be: don't try to seal all lips. Analysts, journalists, investors, employees, or members of the public consider knowledge of situational information to be their right, an entitlement rather than a luxury. During the Iraq War, the US government managed its communications effort from the very top. It appointed a Brigadier (Vincent Brooks), one of its brightest officers, to deal with daily briefings at Central Command Headquarters in Qatar. In Washington, the Secretary of Defence and Chairman of the Joint Chiefs of Staff held almost daily meetings with the Press. The information war structure in India suffers due to lack of trust among the political leadership, bureaucracy and military.

Shortage of Officers

The Army has always been proud of its unequalled tradition of selfless devotion to duty, sacrifice and valour. The RMA notwithstanding, the intangible but most awesome asset in conflicts and emergencies has always been the Indian soldier and his leader. The Indian soldier is a remarkable human being, one who is spiritually evolved, mentally stoic and sharp, physically hardy and skilled. Young officers have always displayed sterling qualities. They provide the immediate leadership, motivation, and the inner strength to the troops to overcome dangers and hardship in the execution of near-impossible tasks. During the Kargil War, we were short of tangible assets, but very strong on courage, determination, camaraderie, leadership and morale. The spirit was very strong. Currently, this asset is seriously threatened and seems to be on the verge of becoming extinct. There is an acute shortage of officers since young men and women with acceptable leadership potential are not joining the Army in adequate numbers, and those within, want to leave. Moreover, the rising deficiency of middle and junior officers has started impacting operational efficiency, administration and morale at the unit level. The manifestations of stress, strain and low morale are visible. This is an urgent challenge before the Army as well as the government

Conclusion

In the comprehensive security environment, the Army has an important role in ensuring security from external and internal threats as well as in nation-building. The ultimate test of nation- building is to win wars, if ever launched against the nation. As Clausewitz stated, "Wars are won by a trinity of the will of the government, the passion of the people and the competence of the military."

The emerging strategic environment is full of uncertainties. Military and non-military threats are diffused and less obvious. There is a need for the Army, Navy and Air Force to prepare contingency joint plans, which can be implemented at short notice/during the course of mobilisation. This requires a higher degree of jointness in defence and operational planning. We need to reorganise the networking system of the armed forces within, and with other government and non-government agencies that have an important role to play in a future war.

Politico-diplomatic factors will play an important role in future comprehensive security related issues. Careful and calibrated orchestration of military operations, diplomacy, and domestic political environment would be essential for the successful outcome of any mission. With conflicts becoming multi-dimensional, the armed forces require geo-strategically aware and specialised political guidance.

The Army needs to improve its management as well as attitude to achieve a faster decision-making process which should include faster response from the lower formations. Its human resource should not only possess essential traditional skills but also have the capability to assimilate technological advancements to analyse and act appropriately in the rapidly changing operational environment. It should divest elements which are clearly not needed in the future, while redesigning the remainder to play a more effective role in the comprehensive security environment. A status quo Army will not do. Dynamism tempered with sagacity, discretion and prudence is the need of the hour.

Salience of Aerospace Power

□ T.M. Asthana

Introduction

The art and science of warfare, through the ages, have made serious attempts to resort to technological excellence to create a favourable asymmetry. Be it the spear, the bow and arrow, the cannon or the rocket, the lethal distances increased and the precision of striking targets improved. All along, the intentions have been two-fold. Firstly, to increase the distance from where lethal blows may be delivered, and secondly, to achieve as clear a precision strike as possible. The challenge of the new strategic era is to selectively use the vast and unique capabilities of the armed forces to advance national interests in peace-time while maintaining readiness to fight and win when called upon. Exploiting the military capabilities of the third dimension had seemingly matured in World War II. The 20th century witnessed destructiveness and violence of war that resulted in the death of millions, and at times, widespread devastation across significant portions of the globe. The cost in human tragedy and economic waste has been enormous and few nations have escaped the impact of war. During this period, the role of technology and science in warfare has been particularly profound, which has pushed the theoretical limits of war to its absolute extreme – the potential to destroy life on the planet in a nuclear maelstrom. Man stood on the brink of absolute, total war for nearly forty years and only now is slowly stepping back from this edge.

Gen Eisenhower had emphatically stated, "The Normandy Invasion was based on a deep-seated faith in the power of the air force in overwhelming numbers to intervene in the land battle. Without the air force, without the aid of enemy air force out of the sky, without the power of enemy air to intervene in the land battle, that invasion would have been catastrophic and criminal." Ever since, airmen have tried to argue that air power should hold centre-stage. Over six decades have elapsed since Lord Trenchard wrote: "A strategic force can be defined as a military force capable of assuming the command of its own medium by its own combat resources. Until the advent of the airplane, the army and navy were valid expressions of the nation's ultimate military power on land and sea respectively. With the development of aircraft, however, that ceases to hold true. No longer the masters of their own mediums, those forces have lost their strategic significance. Conversely, the surface forces cannot on their own initiative interfere decisively with the functions of the air force. Consequently, the air force is the only strategic force, because it is the only force that can attain command of its own medium by its own combat resources. Thus, the air force has become the primary instrument of the nation's military strength."

Military operations have progressed from land, to land and sea, to land, sea and air, and now, we need to factor in space power, which has demonstrated its immense potential in an incremental manner. Indeed, we will continue to witness increments in land, sea and air power as well. Military operations will be compelled to synergistically employ them to advantage along with space power. It is important, therefore, for us to absorb the fact that space, or rather, aerospace power, to promote military operations, is here to stay. With the launch of the Sputnik on October 4, 1957, we entered the space age. Actually, the terminology "Aerospace Power" was coined in 1958 to encompass the continuous medium, including air and space. The true contributions of aerospace power were first demonstrated in Operation Desert Storm and the intrusive visual media made millions by spreading this awareness worldwide. Simultaneous engagements of strategic, operational and tactical target systems restricted the enemy's ability to recoup, realign or offer a counter. From this day onwards, it dawned that the integration of air and space powers together provides rich dividends. It is, hence, argued that aerospace power has to be unmistakably meshed into one.

If technological excellence provides the asymmetry to Tom, why not to Dick and Harry? "Elementary, my dear Watson"—it is the human interface at all levels that needs perfect synchronisation to deliver the goods. So, if your objective is aerospace power supremacy, do not ignore this major factor. Only then, will the salience of aerospace power befriend you.

Aerospace Power in Defence

In my personal opinion, the art, science and execution of defence are far more complex than offence. This is so because it is generally believed that defence is reactive to an aggressor's posture or action. (However, I don't wholeheartedly support this view since I believe that aerospace defence can also be offensive.) The world has today increasingly shunned offensive action primarily because witnessing our own brethren in body bags is a stigma attached to the nation, as also, conserving our hard earned economy is the top-most priority and, lastly, offensive action is costly in terms of both equipment and resources. In the interest of the nation, therefore, development of an impenetrable defence should remain the main agenda of the military. **Defence, hence, must form the essence of all military actions (including aerospace power) that will ensure a nation-state's least intervention in offensive action.**

India is the second most populous country of the world. To our credit, India is the most populous functional democracy that has traditionally demonstrated two brilliant attributes, i.e. resilience and hospitality. Also, history tells us that the inhabitants of this nation are valiant fighters for all causes that threaten our national security. We already have elements of aerospace defence in place. The subcontinent of India covers 3,287,590 sq km in area. It extends 3,000 km from north to south and 2,647 km from east to west. If we consider aerospace above this landmass, as well as the aerospace above our international waters and littoral interests, the areas and volumes reach mind-boggling figures. Just like we read in teachings of air power, so also, the basic aerospace defence criteria are to detect, identify, classify, track, intercept and destroy.

Perhaps the first responsibility of any air force to the nation is to prevent the enemy's air forces from attacking one's own surface forces, population, and support facilities in rear areas. The Indian Air Force (IAF) has historically been quite successful in this regard, such that, we now largely take the ability to operate in sanctuary from enemy attacks for granted. Since the recent upgradations, accretions (SU-30 and radars inclusive) and modifications to the IAF's inventory, the capabilities to defend against air attacks have nearly overmatched the air forces of our potential adversaries. Nevertheless, regional powers such as China are modernising their air attack capabilities, and new systems such as cruise missiles and beyond-visual-range air-to-air missiles are available, that, if employed competently, could challenge the capabilities of the present IAF to defend vital assets.

Aerospace defence will be a top priority mission as long as an enemy has the ability to threaten us with air, missile or space systems. Invariably, offensive operations will be conducted with aerospace defence operations to terminate any conflict as early as possible on terms favourable to us. Aerospace defence operations should be continuous and MUST not be conducted in isolation. Active aerospace defence measures are direct defensive actions. Passive aerospace defence measures include all measures, other than active aerospace defence, taken to minimise the effectiveness of hostile air, ballistic missiles, and space systems. Today, active aerospace defence operations are conducted using an assortment of weapon systems supported by secure and highly responsive Command, Control, Communications, Computers, Intelligence (C4I) systems. It has become popular to disparage air power/aerospace power and to argue, "It is not decisive in war." We do not claim either that aerospace power alone will be decisive in every instance, but it is the hardest hitting, longest reaching, capable of rapid response, and most flexible force that the nation possesses. It is difficult to imagine a future conflict of any major scope in which land or sea power could survive – much less be decisive – without aerospace power.

The contributions of space are good today, but they can reach unimaginable levels. Space offers immense potential in terms of information, surveillance and, consequently, intelligence. We have satellites that give us information, but we do not have dedicated defence satellites. It may be argued that dedication may be gross underutilisation of a satellite. I believe that among the three military Services, we are in a position to load the transponders of a satellite to capacity. If necessary, we may launch smaller satellites but they must be dedicated to the military. Today, we have satellites in orbit with the capacity to transmit intelligence data. In the near future, we will also have satellites capable of transmitting sub-metre resolution data. Space will ultimately be the largest repository of surveillance and intelligence information, which is the common denominator to ensure positive aerospace defence. It must be mentioned here that surveillance of the areas of interest is of paramount importance. Surveillance does not demand surgical and/or sub-metre resolution. Any doubts and suspicions that may arise after interpreting the surveillance inputs can always be verified by accurate reconnaissance, either by air or space power before initiating any defensive or offensive action. Geo-Stationary Orbit (GEO) satellites have been criticised for not being adept enough to provide meaningful inputs for surveillance, reconnaissance, navigation and targeting information, but there is no dispute that they are good communication facilitators. I am, however, convinced that GEO satellites will also be good enough for surveillance when it can be ensured that the quality of inputs provided by the interpreters is enhanced. Developing skills for identification and spotting of minor alterations, and mass relocations or movements can enhance this. They will always be good for early warning of all kinds and a special tool for aerospace defence. Above all, GEO satellites will provide continuous cover. We should, hence, also provide a fillip to our space-based assets by ensuring at least two GEO satellites are on station at pre-designated locations to provide surveillance cover for all our areas of interest in the west, north and east, 24 hours a day, throughout the year. These may be launched either together or in tandem, but, at the earliest. Preliminary research establishes that the cost of these two GEO satellites will be in the region of $1.4 billion.

It is natural for us to be concerned about the safety and survivability of these satellites, and, hence, space defence assumes significance. Whether we resort to active or passive or both forms of space defence only time will tell, but we must ensure that space defence is in place in an efficient manner. Above all, the plans must cater for adequate redundancies and alternates and BMD must also be factored into the overall plan of aerospace defence.

No single mission area is capable of providing complete protection from a determined aerospace attack against India. A combination of active and passive defence measures from all mission areas, integrated and coordinated by a robust and efficient C41 system is required to meet the stringent performance requirements demand for aerospace defense of India. **If you want to live in peace, create an impenetrable aerospace defence.**

Aerospace Power in Offence

Offence is the best form of defence, hence, in this section there will, at times, be reference to both these actions. Just as surveillance and reconnaissance, along with detection are considered prerequisites for defence, so also, they are the prerequisites for offensive action. *Jane's Defence Weekly* of August 29, 2007, carries a cover headline "Russian Long-Range Reconnaissance Assets Return to the Fray." Russia had stopped combat duty flights in 1992 following the end of the Cold War. It explains that the resumption was intended as a finale to the Shanghai Cooperation Organisation's counter-terrorism exercise "Peace Mission

2007." A small word of caution here – today, there are so many agencies, with their associated technologies, collecting information that it has created a situation of "information overload." In its raw form, information is no use for aerospace responsive actions. It has to be converted to useful and meaningful accurate intelligence before any offensive or defensive action can and should be taken. We are all aware of this fact, but there are occasions when actions are initiatated in haste, without proper intelligence evaluation. A lot has been accomplished by automation with digitisation, but the human element that does the final vetting, holds the key to efficient and appropriate proactive and responsive actions. The primary objective of offensive actions (including aerospace offensive) is to ensure that there is no collateral damage. If that cannot be ensured due to the prevailing circumstances, all attempts should be made by a responsible nation to ensure the minimum collateral damage.

Education remains a continuous stimulant for all activities, including military action. In the initial 40 years or so of air power application, it had become apparent that air power would effectively contribute not only to air operations but to land and sea operations as well. This basic tenet must be further analysed and plans must be generated to remain ahead of our potential adversaries in technological excellence. This would require dedicated and special munitions for specific tasks. Aerospace power practitioners need to be constantly involved in educating themselves and the concerned agencies in terms of the pros and cons, in order to select the most appropriate option from a variety of options that can be made available.

The destructive power of an explosive device is a linear function of its explosive and a squared function of its accuracy. Recent major changes in the accuracy of air-delivered weapons have enabled quite dramatic changes in their effectiveness. During World War II, the typical long-range bomber could deliver its load of bombs with a circular error probable of around 1,000 metres. This meant that it took upwards of 240 bombs to be confident of destroying a single 'hard target' such as a bridge or a command post. It also meant that aerial attacks on military targets would, perforce, result in heavy damage to people and facilities in the general vicinity of the targets. The situation changed little in Korea, or indeed, ten years after that in the early stages of the war in Vietnam. Only the invention of laser-guided bombs, first used in Vietnam in 1970, brought a marked improvement in the effectiveness and efficiency of aerial attacks on fixed targets. Today, laser guidance for air-delivered munitions has become both more accurate and more robust. Circular error probable is now down to about three metres for most modern platforms, and automation allows the pilots of single-seat aircraft to deliver laser-guided bombs. Perhaps more importantly, the advent of weapons guided by signals from the Global Positioning System (GPS) means that a modern aircraft can attack fixed targets with accuracies approaching those of laser- guided bombs in virtually all weather conditions. And the attack can be undertaken with little risk to people and facilities in the surrounding areas i.e. least minimum collateral damage. In 1971, a million pounds left the runway, and it cost about $10 million to destroy a bridge. In 1991, destroying a bridge took two Tomahawks with the same range as a F4 — 6,000 pounds left the runway, and it cost about $2.5 million. You saved money, no one was at risk, and you actually hit the target hard enough to destroy it. Today, the IAF has this capability, which will only get better with fresh accretions and better munitions in the pipeline.

The Iraq operations visually demonstrated (via the media) the incremental value of space power. Although space power remained a major facilitator to air power offensives as well defensive actions only, it is actually capable of much more. I have no intentions to introduce militarisation of space in this paper, but it is a fact that we need to respect. What is important to note is the fact that the continuum of air and space has provided dividends well beyond expectations.

Today, there are nearly 600 satellites in orbit and it has been predicted that in another 5 to 6 years, the figures will cross well above the 1,000 satellites in orbit. Internationally, it has dawned that exploitation of space assets is not only the monopoly of civilian facilities in terms of communications, information, etc, but has formed a major asset to militaries around the world in the form of aerospace power. The satellites in orbit used in recent wars have exceeded 100 in number in each case. Hence, in any future conflict, reliance will be heavy on use of satellites and air power by itself will always graduate to aerospace power. The Outer Space Treaty (OST) is still respected wherein only soft kills are visualised at the interference levels only in respect of the adversary's satellites. The day is not far when hard kills of satellites will become the norm, and we have seen that Anti-Satellite (ASAT) weapons have been developed. There are frantic efforts to develop nano and micro satellites and provide almost similar facilities by resorting to miniaturisation. Whenever this converts to reality, the hard kill option is also likely to be developed. In addition, by launching small satellites in greater numbers, there will be dispersion of space assets, as well as the clear possibility of catering to redundancies. Above all, the cost of launching smaller satellites will be far more affordable. We, therefore, see that just as for air power, terminologies like "air defence" and "counter-air operations" have matured, so also terminologies of "aerospace defence" and "counter-aerospace operations" will materialise. In a similar fashion, "aerospace superiority" will be bandied around. It is an established fact that offensive action has remained the first option of all military operations, and will also remain the first option for aerospace operations.

On June 21, 2004, humanity achieved another first when a 62-year-old American called Michael Melville reached an altitude higher than 62 miles and became the first citizen to fly a craft into what is, technically, space. The vehicle he piloted, "Space Ship One", was designed to win the US $10 million Ansari X Prize. The concept of hypersonic aircraft seems like an unlikely dream today but research is in progress. Theories suggest that rockets will propel aircraft into, and out of, space. The "Scramjet" or "Supersonic Combustion Ramjet" could provide the propulsion in space — these are basically "Pulse Detonation Wave" engines utilising liquid methane directed onto hot fuselage. This would permit hypersonic aircraft to attain speeds up to Mach 10 at the lower levels of Low Earth Orbit (LEO). These aircraft could be used as weapon platforms including ASAT weapons, as also, to deploy satellites. In addition, Unmanned Aerial Vehicles (UAVs) too are making a niche for themselves through miniaturisation and advanced propulsion techniques, enabling them to operate at near LEO levels with enhanced endurance and loiter. A major factor that had troubled the air power platforms was survivability. Today, assisted by stealth technology, Electronic Counter-Measures (ECM) on board and long-range cruise missiles munitions permit a far greater survivability to modern combat aircraft. Thus, we see that aerospace power will not only comprise air breathing and space platforms, but also near space platforms. Indeed, aerospace is set to grow into a more and more potent asset for military power. Consequently, it follows that the offensive component of aerospace power will emerge as the major contributor to the overall military power of a nation in offensive actions.

Aerospace Power as a Deterrent

The security of India will depend (to a large extent) on the effective execution of aerospace power in both defensive and offensive operations against an assortment of conventional weapons and weapons of mass destruction. The principal objective of aerospace power is the protection of our country, its people and other valuable assets.

> *For it is not profusion of riches or excess of luxury that can influence our enemies to court or respect us. This can only be affected by fear of our arms.*
>
> – Vegetius
>
> *De Re Militari*, circa 4th century

Aerospace forces accomplish several purposes in both conflict and peace. The introduction of aircraft and space systems has provided revolutionary changes to a nation's ability to fight. One needs to understand specific missions that aircraft and spacecraft can accomplish or support before employing them. Aerospace forces can conduct deterrence, denial, coercion, decapitation and humanitarian missions. Perhaps the most important mission of these missions is the mission of deterrence. These forces make a potential enemy think twice before launching a preemptive, be it a nuclear or conventional, strike. The speed, range and flexibility of aerospace forces give a nation a decided advantage in achieving conventional deterrent value. Aircraft that are ready to bomb targets at a moment's notice also help to stop another nation from taking certain action because of a swift, decisive reaction. Aircraft can demonstrate deterrent value by providing a visible display of combat power if they fly near an enemy's border or conduct training exercises in plain sight of an adversary.

Once combat action starts, aerospace forces can coerce an offending nation to adopt a certain course of action. Coercion involves the use of force to punish the transgression of a foe in the hope of altering the nation's will. Although coercion may use attacks on physical targets, its mission goal is to change the behaviour of the nation, organisation, or group of people through psychological means. However, a significant issue regarding coercion is the problem of escalation. Does the coercive power increase the level of attack? Is there a ceiling to the escalation of force? All these factors need to be considered. In 1999, the North Atlantic Treaty Organisation (NATO) conducted Operation Allied Force, an air campaign against Serbian forces in Kosovo and Yugoslavia. NATO's aerospace forces attacked a number of targets that affected the living conditions of the Serbian populace to end the conflict. The air action added internal pressure on the Serbians to accede to NATO's demands by creating public dissensions among the Serbian populace and reducing the country's war-making capacity, as well as will. Aerospace forces have the ability to contain the escalation through precision, avoiding unnecessary civilian casualties and collateral damage. It will be argued that we cannot create a similar asymmetry, but the lesson to be learnt is, **"If deterrence fails, apply coercion of the aerospace forces."** It is for certain that deterrence will have an impact in the future also.

Aerospace forces can, and have also, contributed immensely to humanitarian needs like evacuation, supply of food, medicines, etc. Such operations conducted in a swift and efficient manner make a potential adversary pause and think seriously. A nation that can provide humanitarian aid in quick time, at relatively large distances, in quality and quantity, is also capable of inflicting punitive action at a moment's notice. Such a capacity for humanitarian aid is a clean, decisive and telling demonstration of aerospace forces, to any potential adversary, of the nation's deterrent value, which, if not respected, could well be converted to coercive value.

We may safely arrive at the conclusion that the IAF will certainly be superior to the Pakistan Air Force (PAF) in aerospace power capability for quite a few years, given the present trends. The deterrent value of India's aerospace power would hold since Pakistan has virtually no indigenous space capability. In respect of our other potential adversary, suffice to say that the October 2008 Vol 6, No.2 *Force* magazine carries a specific article on the IAF for this subject. The cover has a headline of "Indian Air Force on the China Front." This article has a sub-heading of "Only the IAF can Provide Dissuasive Deterrence."

Network-Centric Warfare (NCW)

No missive on any military subject can be complete today without mention of NCW. Recent successful wars have given birth to some terminologies. It is not as if, we, in this profession did not practise these aspects, but now we have terminologies to go by. Operation Desert Storm (Iraq) produced the Revolution in Military Affairs (RMA); Kosovo produced Effects-Based Operations (EBO) and Information Warfare (IW); Iraq (2003) produced Network- Centric Warfare (NCW). The objective of NCW is to ensure that the Combined Operational Picture (COP) is available at all levels, which is relevant and real-time. The basic principle of NCW is to ensure continuous availability of relevant, timely, and accurate information, which is converted to applicable intelligence. This real-time intelligence is today the backbone of all military operations, including aerospace operations. EBO in the 21st century enabled by NCW is a methodology for planning, executing and assessing military operations designed to attain specific effects, which translate information advantage into combat power to achieve the desired national security outcomes. Intelligence superiority is the central theme. This is of utmost importance to aerospace operations, and hence, the world over, there is a race to better the reaction time-frames. NCW has a war winning and war deterrent capability reducing the Orient, Observe, Decide, Act (OODA) loop. There is a demand for aerospace dominance to achieve reliable, efficient and survivable NCW capability.

The Human Capital

Much of this paper has focussed on hardware. While the aircraft, spacecraft, missiles, satellites, munitions, and other equipment that the IAF operates are the most visible manifestations of capability, they are not the most important. Rather, the most significant factors determining the capabilities of the force are the qualities of the people who operate and maintain the equipment and who command the units the force comprises. Without doubt, the people serving in the IAF today are highly capable. Nevertheless, sustained and deliberate efforts will always be needed to maintain this.

A nation may have all the physical elements necessary to organise and function as an aerospace force, but it is useless unless it is given the right direction. Piloting a combat aircraft is a complex task. Maintenance of an aerospace force requires the existence of a technological base. The people involved must have sufficient motivation, education and training to keep aircraft and space systems at their optimum performance levels. The nation must also maintain a sound educational system to instill the proper levels of knowledge to produce engineering technicians and individuals capable of operating and maintaining advanced technology for exploiting it. Aircraft and satellites cannot function by themselves. Although crew can fly the aircraft, it relies on a number of diverse complex support systems while in the air or on ground. Aerospace power is not just about platforms.

A nation could use its aerospace power for independent military actions. Better still, the country could use this aerospace power as a catalyst in support of its surface forces as an equal partner to all of its sister forces' objectives. Before the nation decides to use its aerospace power, it needs to provide direction and guidance. It is generally agreed that absence of this clear perspective may make a country acquire attractive but unnecessary aircraft and space systems that might not give it the appropriate forces to win a war in the national interest.

Conclusion

To support and achieve India's goals, the nation must have a force able to deter and/or decisively defeat a range of potential adversaries. Forces, inclusive of aerospace forces, must be lethal and flexible and must have strategic reach, yet must also protect the nation against potential attacks, large and small. Personally, I am convinced that the IAF has some of these assets, and for those that are not on its inventory, plans are in the advanced stages. With such an envious contribution on offer by aerospace power, it would indeed be foolish to ignore this asset. We need to plan not only for "Aerospace Superiority", but for "Aerospace Dominance." Only a precise, well-conceived plan will deliver aerospace dominance, for which we need to create the hardware and software of aerospace assets. Viewed practically or even pessimistically, if aerospace dominance cannot be achieved, at least, aerospace superiority will be ensured. Building up the aerospace power capability that we visualise is a multi-pronged task, which needs a multi-pronged approach. It is imperative that a single agency handles this aspect with total dedication. We need to ensure that we gain the prescribed aerospace capability (with gradual efforts to continuously improve on it) in a systematic and practical plan. I propose the obvious i.e. **"The Indian Air Force must assume the total responsibility as the executing agency for the nation to establish the requirements (assets) and stipulate priorities for a well honed aerospace power for the nation."** Acquiring such an offensive, defensive and deterrent capability of aerospace power must be transparent internationally and to our potential adversaries in particular.

India's Maritime Strategy for The Decades Ahead

□ **Arun Prakash**

The Historical Underpinning

Disembarking at Jakarta airport in 2005, I marvelled at the sight of two airliners parked alongside each other on the apron, displaying the logos of *"Garuda Airways"* and *"Jatayu Airlines"*; both names derived from mythical birds in the Hindu tradition. While *Jatayu's* sacrifice forms an evocative episode in the *Ramayana*, *Garuda* is described as Lord Vishnu's mount, and today forms the state emblem of both Indonesia and Thailand. Over the next few days, the Indonesian Navy repeatedly surprised me. The motto of the Service, *"Jalaseva Jayamahe",* is Sanskrit for "On the Sea we are Glorious"; the sailors of the Eastern Fleet wore, on their uniforms, a unit insignia depicting the flaming wheel of the *Brahma-Astra*. Their submarines had names like: *Pashupati, Chakra* and *Trishul*.

All this may come as a surprise to Indians, because very few of us are aware of the deep and ancient cultural linkages that bind us to Southeast (SE) Asia, and even the knowledgeable may not be aware that these linkages could have only been established and sustained through intense maritime intercourse between India and Siam, Kampuchea, Java, Sumatra and Bali, once collectively known as *Suvaranabhumi.*

Through mythology, ancient literature, folklore and artifacts, historians claim to have traced wide-ranging Indian maritime activity as far back as circa 2500 BCE, to the Indus Valley civilisation. But about 300-400 years before the birth of Christ, and perhaps a century after the Peloponnesian Wars between Athens and Sparta, we find a brilliant Indian treatise on administration, economics and diplomacy, the *Arthashastra,* which provides more concrete evidence.

The *Arthashastra,* written by Kautilya, Prime Minister in the court of Emperor Chandragupta, reflects the acute maritime consciousness of the highly progressive Mauryan state. It describes the elaborate organisation in place for dealing with matters relating to shipping, ports, navigation and maritime safety under a Minister who headed an Admiralty Board[1]. From the growth of huge Hindu Kingdoms right across present day SE Asia, it is evident that the Andhra, Pallava, Pandya and Chola dynasties, which succeeded the Mauryas, sustained the intensity of mercantile and cultural traffic with this region by sea. This hoary Indian maritime tradition survived up to about the 13th century and thereafter waned rapidly with the advent of Central Asian invaders into northern India.

The date May 20, 1498, represents a defining moment in our history, for it was on this date that Vasco da Gama landed on Indian soil[2]. At this juncture, the Lodhi dynasty ruled North India, while the South was divided between the Bahamini and Vijaynagram Kingdoms. Neither of them was blessed with a maritime vision, much less a navy.

Just over 100 years later, English Captain Hawkins obtained from Emperor Jehangir, a permit to undertake peaceful trade in Surat. This was the beginning of a European "scramble for loot, trading links and political advantage"[3] which was to enslave India for two centuries. Myopic Indian rulers had abandoned the seas and left them for the colonial powers to dominate, not realising that they were paving the way for their own subjugation.

This somewhat lengthy prologue was necessary to establish the historical moorings for modern India's maritime ambitions by putting across three essential points to the reader:

- That both by geographic disposition and historical tradition, India has been a maritime nation, but collective amnesia persuaded its inhabitants for centuries that the Himalayas were the country's sole frontier.
- It was this blinkered vision that rendered us defenceless at sea and virtually invited invasion and domination by the European maritime powers.
- These facts should remain etched in the national psyche, along with the resolve not to let our "maritime guard" down, ever again.

Changing Maritime Perceptions

India's continental mindset and landward orientation persisted for many years after independence. For this reason, the fortunes of the Indian Navy (IN) tended to fluctuate with the vagaries of the annual defence budget[4]; its precarious existence earning it the sobriquet of "Cinderella Service." With the dawn of economic liberalisation, on the one hand, and acceptance of *realpolitik,* on the other, the last two decades have, however, brought about a radical change in perceptions.

Having shed the bonds of its traditional "Hindu rate of growth", India has been pitched into the mainstream of the international trading system and locked tightly into the world economic network. Foreign trade makes a substantial contribution to India's Gross Domestic Product (GDP), and over 97 percent of it is transported by sea. In addition to foreign trade, it is also undersea gas, oil and minerals as well as vast fishery resources that have sharply focussed attention on the seas around us. A major share of our maritime trade is devoted to the import of energy and raw materials which are vital to the economy. Since the Sea Lanes of Communication (SLOCs) have now become the arteries which carry the lifeblood of India's globally-linked economy, our merchant fleet, ports, shipbuilding industry and other constituents of maritime power have acquired unprecedented criticality.

The status of the IN as the centre-piece of the nation's maritime power has at last been clearly understood by the intelligentsia as well as the security establishment. However, maritime security remains a sub-set of national security, and that is a vexed subject which I will dwell upon briefly.

India's National Security Paradigm

We are all aware of the Utopian attitude towards national security that engulfed the nation in the years after independence. Since India was committed to non-violence, the armed forces were considered "superfluous." We were perhaps on the verge of beating our "swords into ploughshares"[5], when the Chinese brought us to our senses in 1962 with a good hard knock in the Northeast Frontier Agency (NEFA) and Ladakh.

Regrettably, we did not learn from this traumatic experience, because just three years later, in 1965, we were once again taken by surprise in Jammu and Kashmir (J&K). Again, in the mid and late 1980s, we were caught napping as Pakistan fanned the embers of discontent into full blown insurgencies, first in Punjab, and a few years later, J&K. Then, in 1999, we faced the ignominy of having to admit a massive intelligence failure because the Pakistanis had sneaked in to occupy the Kargil heights[6].

Our ambivalent attitude to national security has repeatedly lulled us into the false belief that there is peace all around and defence expenditure can be cut back. This, in spite of repeated demonstrations during the past six decades that developmental plans for the people cannot be implemented unless the nation can create a secure environment for itself. Such conditions can be achieved only when it is clear to everyone in the neighbourhood that India is a strong, confident nation with a surplus of security assets, and the ability to cope with all possible threats.

The story, however, does not end there, because it is not just a question of capabilities – you must have the resolve and the ability to use them when the need arises. Which brings me to another factor: a strategic vision. The fact that we as a nation have lacked a tradition of long-term strategic thinking has been repeatedly demonstrated by our ad hoc and maladroit handling of episodes like the Rubayia Mufti kidnapping, the IC-814 hijacking, and the armed forces mobilisation of 2001.

Our effete responses to cross-border terrorism, Naxalite activity and the repeated bombing of our cities have given us the deserved reputation of a "soft state." This by itself constitutes a provocation and an invitation for further interference and aggression by malignant forces.

The Intellectual Moorings of Maritime Security

Much of the responsibility for this situation, in which neither national aims and vital interests are clearly articulated, nor security policies defined, has to be borne squarely by India's political establishment. The minutiae of sheer survival occupy so much of the time and mental space of our politicians, that they have none to spare for national security, especially since it is not seen as an electoral issue that will garner votes. Major decisions on security issues often remain in limbo, or are taken on the inputs of the bureaucracy and selected technocrats, often to the nation's detriment.

This strategic vacuum has tended to deny the IN a contextual frame of reference within the overall national security paradigm (such as it is). In the late 1990s, the American analyst George Tanham wrote in a monograph on Indian strategic thought: *"Status and symbolism matters greatly in Indian society…and Indian Admirals may need no justification or rationale for a powerful navy other than that India's greatness mandates it.*"[7] Others like Ashley Tellis have often attributed grandiose maritime designs and alleged hegemonistic intentions on the part of India.

Despite all the rhetoric that one hears of the three armed forces, it is only the Navy that is inherently "trans-national" in nature because of the medium that it operates in, and the spirit of international laws that govern the high seas. A warship or a naval task force can remain off a foreign shore for an indefinite period without infringing any law or territorial rights, provided it stays outside the 12-mile limit. As the 2004 tsunami clearly demonstrated, IN units were already in foreign waters on their mercy mission even before the formal government approval had been issued.

Under such circumstances, it was inappropriate that the IN should continue to extemporise in the absence of higher directives. The intellectual challenge of drawing up a clear-cut roadmap to synergise our national maritime endeavours, was taken up, and in October 2005, the Service issued a document titled *"Freedom of the Seas...India's Maritime Strategy."* This became the companion volume to the *Maritime Doctrine* which had been issued a year earlier.

The IOR Security Environment

The Indian Ocean Region (IOR), at whose focal point India lies, is replete with unique features, many of which contain the seeds of conflict. For example, it is home to one-third of the world's population, and the world's fastest growing economies coexist with some of the poorest. It is the largest repository of not just hydrocarbons, but also many strategic materials.

Countries of this region are afflicted with problems of poverty, fundamentalism and insurgency, and most of them are under military dictatorship or authoritarian rule. Consequently, the region has seen much instability, and even violence, in the post-Cold War era. The Horn of Africa, Bay of Bengal and Malacca Strait are witness to frequent incidents of lawlessness, including piracy, hijacking and human trafficking.

Sitting in a strategic geographical location astride major shipping lanes from the Persian Gulf to the Malacca Strait, India's location in the IOR is particularly fortuitous. A long peninsular coastline studded with deep-water ports, a well-endowed Exclusive Economic Zone (EEZ), a rich hinterland and island territories on both seaboards complete this happy picture. Our dependence on the sea for food, energy and minerals will grow in the coming years, not just because of necessity, but also because evolving technology will bring its exploitation within the realms of possibility. This will place added demands on security resources for protection of our maritime assets.

With a large merchant fleet, which has just crossed 9 million Gross Registered Tonnes (GRT), India ranks 15th amongst seafaring nations. This fleet, operating out of 12 major and 184 minor Indian ports, can carry a little less than a sixth of our seaborne trade, and has much scope for expansion. India's share of total world trade has been hovering around just one percent, and the government is aiming to double it by 2009.[8]

Annually, over 100,000 merchantmen transit the waters of the Indian Ocean, carrying cargo worth about a trillion dollars. Both east-bound and west-bound shipping has to pass through a number of choke points where it is vulnerable to interdiction or interference by state and non-state entities. Any disruption in the supply of energy, or commodities, would send prices skyrocketing, and destabilise industries as well as economies worldwide.

The vast IOR is also a stage where a number of other issues are played out. Some of them have serious security implications, and these need discussion here.

Energy Security

Apart from the major SLOCs that criss-cross its waters, the proximity of the world's primary oil producing region – the Persian Gulf – and the oil traffic emanating from it, create an energy security issue of international significance in the IOR.

In addition to greedy speculators and "inside traders", the galloping energy demands of both China and India seem to have contributed to a phase of permanently high oil prices. The world held its breath as prices edged towards $100 a barrel – but there was deafening silence all round when it actually happened, and consumers meekly absorbed the price rise. July 2008 saw a historic high of US $147 per barrel, but as the world steeled itself for the $200 mark, prices mercifully started declining.

The world's first "oil shock" had come in 1973, and in the 35 years since then, prices have steadily risen by almost 50 times. The oil producing nations will continue to "make hay while the sun shines" and it is futile to expect a respite. Till such time that viable alternate energy resources are found (oil companies permitting) oil will continue to rule the economies – and destinies – of non-Organisation of Petroleum Exporting Countries (OPEC) nations.

Any threat or impediment to the free flow of oil could have catastrophic repercussions worldwide. The combination of rising oil prices and foodgrain shortages, created by the quest for bio-energy, has created intensified inflation and badly hurt the Third World. Ensuring the unhindered flow of oil from this region is, therefore, a major economic issue with maritime connotations – one that is driving countries to befriend India because of our critical location astride the sea lanes from the Persian Gulf to the Malacca Strait.

As far as India herself is concerned, an assured and steady supply of energy is vital to sustain its current economic trajectory, and sustain hopes of wiping out poverty by the third or fourth decade of this millennium. Our current dependence on imported energy of over 75 percent is expected to reach 85 percent by 2025, and the country will become the world's largest oil/gas importer by 2050.

An inevitable development in this context has been the acquisition of oil and gas fields across the globe by ONGC Videsh. While investments worth billions of dollars have been made in these overseas assets extending from Sakhalin in the Russian Far East across Central Asia and Africa to South America, little thought has been given to their protection, which will certainly have maritime security connotations[9].

Terrorism and Low Intensity Threats

One of the unfortunate fallouts of political turmoil in our region is economic backwardness, which provides fertile ground for religious fundamentalism and breeds militancy. Organisations like the Liberation of Tigers of Tamil Eelam (LTTE), Al Qaeda and the Jemmah Islamiyah, manifestations of this phenomenon, find both recruits and financial sustenance in our neighbourhood, and use the sea routes for their nefarious activities.

Given all the catalysts, it should surprise no one that, the fountainhead of terrorism and the trans-national nexus of blatant nuclear and missile technology proliferation lies in our immediate neighbourhood in Pakistan. One determined individual, Dr. A.Q. Khan, with the sly or unwitting support of half a dozen countries, has done irreparable harm to the international non-proliferation regime and added to regional instability. These are threats that will continue to pose a major hazard not just to the region, but the world at large.

Low intensity maritime conflict waged by non-state entities is also a grim reality in our region. Such conflicts cut across state boundaries and require a multinational response. The IN has been engaged in internal security duties in the Palk Bay, along our maritime boundary with Sri Lanka, since the late Eighties, and on the west coast since 1993, to prevent the clandestine entry of arms and explosives, and to check the movement of terrorists.

In the coming decades, the challenge of threats such as piracy, gun-running and drug smuggling will only grow, requiring the cooperative use of maritime forces to counter them. Joint patrols are already being conducted with Indonesia, Thailand and Sri Lanka. We are now engaged in building cooperation with key maritime nations to build up our maritime domain awareness.

The Strategic Players

The strategy that we are discussing, essentially spans a time-frame of 10-15 years. Of the many players in the IOR, we need to focus on just three starting with the USA, and see what kind of influence they will cast in this time-frame.

The United States

The US considers itself an Asia-Pacific power for all practical purposes, and must, therefore, figure as a major factor in India's calculus. The three abiding US interests or preoccupations in this region are: safeguarding the hydrocarbon resources of the Middle East and Central Asia, the containment of China to protect the autonomy of Taiwan, and combating the mounting challenge of militant Islam.

Currently, America's resources and attention are concentrated on the ongoing operations in Iraq, and the requirement to keep the terrorist hubs in Pakistan and Afghanistan under check. Monitoring, and finding ways to circumscribe, the nuclear ambitions of North Korea and Iran are the other two issues that have critical significance for the USA today.

It is now becoming obvious that while she can try and set agendas to suit her interests, the USA cannot single-handedly, implement them worldwide. Signs of overstretch are showing, and a lack of visible success in Iraq has hardened domestic opinion. A radically new approach may become necessary, and concepts like the "1,000-ship Navy", floated by a former US CNO[10] seem to sound the right note.

In this context, India's vital national interests would benefit if we are in a position to exert some influence on the course of events in our neighbourhood. But for that to happen, we must bring ourselves to accept a degree of participation, direct or indirect, in the affairs of our strategic "near abroad." For far too long has our attitude towards events in neighbouring Afghanistan or participation in the Proliferation Security Initiative (PSI) in the Indian Ocean been marked by ambivalence and diffidence, which do no good for our security.

It is possible that the withdrawal of the Communists from India's ruling coalition and the (impending) approval of the 123 Agreement by the US Congress may accelerate the progress of Indo-US cooperation.

China

Regarding China, which though not on the littoral, looms menacingly over the IOR as a rapidly emerging entity with her sights set firmly on superpower status, in the context under discussion, there are just four major points to be noted.

Firstly, there is competition between China and India in the economic and military spheres. No matter how far we lag behind, since both are Asian powers, it is a historical inevitability that they will have to compete for the same strategic space, and may even come into conflict with each other.

Secondly, with Sino-Indian bilateral trade having crossed the US$ 30 billion mark, China is well on the way to becoming our largest trading partner. But trade could become a Trojan horse if it lulls us into complacency. While Indians may suffer memory lapses, the Chinese remain firmly focussed on their claim to 100,000 sq km of Indian territory, in addition to illegal occupation of Aksai Chin. The fact that China has armed Pakistan to the teeth with conventional and nuclear weaponry is a clear declaration of deeply malign intent vis-à-vis India.

Thirdly, China has deliberately ensured the strategic encirclement of India, by providing military and economic aid to countries all around us. Having anchored her Indian Ocean strategy in Pakistan, China has threaded Iran, Saudi Arabia, the Horn of Africa, Seychelles, Maldives, Sri Lanka, Myanmar and Bangladesh into a "string of pearls" in order to secure potential footholds all over the region.

Finally, the past few years have seen China allocating increasing resources to finance a maritime build-up which includes a large diesel submarine force, amphibious shipping, surface escorts and naval aviation. This force accretion can possibly be ascribed to her contingency plans for Taiwan and her "access denial" strategy vis-à-vis the US Navy. Of more serious concern to India is the growing Chinese force of armed ballistic missiles as well as attack nuclear submarines (SSBNs and SSNs respectively) as well as the aircraft carrier reportedly under construction. Entry of the People's Liberation Army (PLA) Navy into the Indian Ocean would seriously upset the balance of power in the region[11].

Pakistan

Much as we try to ensure that Pakistan does not dominate our radar scope, it is certainly in our interest that our troublesome neighbour should remain a stable and integral nation, and outgrow the sense of insecurity which has haunted her since partition. Although democracy has made yet another comeback in Pakistan, it remains hostage to many antagonistic forces and internal contradictions, as this nation reaps the bitter harvest of the religious fundamentalism it has nurtured for years.

Notwithstanding domestic turmoil, however, the one constant in Pakistan's policies will be its continuing endeavour to fish in India's troubled waters through a sustained campaign targeting our integrity and secular structure, with a covetous eye on Kashmir.

Quite apart from the mischief Pakistan may attempt on her own, we need to remember that she is a critical tool in China's anti-India strategy. In this context, the Gwadar deep sea port, built with Chinese assistance on the coast of Baluchistan, is the thin end of the wedge. Situated at the mouth of the Persian Gulf, it could be an ideal base for future operations of PLA Navy ships and nuclear submarines.

China's assistance to Pakistan in the field of nuclear weapons has already forced India on the strategic back–foot. Conventional Chinese weaponry, including warships, is also freely available to Pakistan at affordable prices. As a subaltern maritime power, Pakistan has adopted the classical sea denial strategy vis-à-vis India. This will be executed by posing a "threat in being" to Indian naval units, merchant shipping and energy traffic in the North Arabian Sea by means of missile armed submarines and aircraft, mine warfare and clandestine attacks. This threat could assume truly menacing proportions if the Pakistan Navy and PLA Navy decide to collude.

Peace-time Maritime Strategy

Although our statesmen seem oblivious of it, the IN has kept its sights firmly on Lord Palmerton's dictum about the preeminence of national interests over transient friends or adversaries. As part of its doctrinal initiatives Naval Headquarters (NHQ) has produced a *Maritime Capabilities Perspective Plan,* which outlines its force requirements in the strategic scenario likely to prevail till the end of the next decade. The plans seeks to "right-size" and transform the Navy into a technology-intensive, manpower-lean and net-worked force, ready to operate in a joint environment. The focus has been deliberately shifted from "number of units" to the "capabilities" that need to be delivered in war and peace.

Unlike air forces which continue to delude themselves that wars can be won from the skies, the IN has pragmatically adopted Julian Corbett's historical tenet that major issues between nations have rarely been decided by naval action alone. Wars have largely been won by what the army can do against the enemy's territory and national life, or by what the fleet can do to help the army attain its aims[12]. The IN, has therefore, decided that its strategy and equipment policies will be drawn up in a manner so as to have a clear linkage with the conduct of operations on land.

The Navy's peace-time strategy must encompass the vital components highlighted in the succeeding paragraphs, in order to ensure that the not just the Service, but also the battlespace is prepared for any eventuality.

Shipbuilding Programmes

The IN aims for a stabilised order of battle of about 150-170 ships and submarines, and possibly 250-300 aircraft and helicopters. Timely replacement of ageing platforms and obsolescent equipment will be the Navy's biggest challenges.

Amongst the 40 odd vessels currently on order, or in different stages of construction in various public sector shipyards are: the 37,500 ton indigenous aircraft carrier (or IAC); Scorpene class submarines, follow-on destroyers of the Delhi class; stealth frigates; landing ships (tank); anti-submarine warfare corvettes; fast attack craft; and offshore patrol vessels.

In addition, the former Russian Navy carrier *Gorshkov* (renamed INS *Vikramaditya)* is under repair and modernisation in Severdovinsk, and due to be commissioned by 2012. Three frigates are also under construction in Russia and an oil tanker in Italy.

The picture is, however, not as rosy as it may appear and naval force levels have already dipped below the minimum approved figure of 140 ships. The Russians have brazenly reneged on costs as well as delivery schedules of ships, in violation of solemn agreements. Regrettably, our own shipyards, never known for their productivity, are years behind in construction and crores ahead of cost estimates.

One of the more serious challenges before the Navy's leadership will be to stay a step ahead of the wily Russians, and somehow motivate Indian shipyards to rise to the occasion.

Foreign Cooperation

Navies are fortunate in having a serious peace-time operational roles to discharge because of the medium in which they operate. It is said that the oceans truly make neighbours of distant nations; therefore, while the Foreign Service forms the frontline of our international interface, the IN can form a very useful second-line, provided the Ministry of External Affairs (MEA) shows vision and imagination.

In the maritime context, "foreign cooperation" has wide connotations and covers a whole gamut of activities. While naval exercises establish navy-to-navy rapport and interoperability, joint patrolling builds confidence, and port calls and flag-showing deployments enhance bilateral goodwill and understanding.

Countries in our immediate neighbourhood, many of them island nations, seek maritime security, sometimes, through direct naval presence, but more often through urgent requests for material aid, training assistance and advice. The slow pace of decision-making in the Ministry of Defence (MoD), and often the insecurity of the MEA have served to stall many a Naval Headquarters (NHQ) proposal and driven a potential ally into the arms of other willing donors of assistance. These are self-inflicted injuries which the IN must try and fend off.

The best way of making friends is to help someone in times of need. After its sterling performance during the 2004 tsunami[13], it will be the unstated expectation of our neighbours that the IN will promptly come to their assistance in times of natural calamity, and the Service must prepare itself accordingly.

Making long-term plans for foreign cooperation and building lasting bridges of friendship with our immediate maritime neighbours, particularly Myanmar, Thailand, Sri Lanka, Maldives, Mauritius,

Seychelles, Indonesia, Malaysia, Vietnam and Oman will pay great dividends in the long term.

Networked Operations

Today, the IN is in an unusual situation where it has weapons of formidable capability which outrange its sensor performance. It also has platforms which operate on the sea, under water and in the air, over a vast geographic area extending from the Persian Gulf to the Malacca Strait. During peace-time, the commander at sea is keen to have a composite three-dimensional picture of maritime activity, not only in his own area of responsibility, but also in other distant locations of concern.

In operational situations, his sole objective will be to locate, identify and destroy the enemy. In such a scenario, the ability of units to network closely and exchange intelligence as well as sensor and weapon data with each other, assumes crucial significance. Only then can the commander hope to work "inside" the adversary's decision-making loop.

The secure, high speed link for such a network will come from a dedicated geo-stationary communication satellite whose footprint covers the IN's complete area of interest, from Africa to the Malacca Strait. Since a foreign satellite would neither be affordable, nor meet our security requirements, the Indian maritime communication satellite will have to be an Indian Space Research Organisation (ISRO) product.

While powerful platforms with potent weapons and sensors may provide the required muscle to the IN, it is networking which will be the catalyst to lift the IN into the big league of navies. Networking will no doubt be a formidable technological challenge, but the IN is ready to pick up the gauntlet.

Transformation

After debate and discussion lasting almost a decade, the armed forces of advanced nations came to terms with the putative Revolution in Military Affairs or RMA, and have now internalised it. The profound changes that have accompanied this process, go beyond just leveraging of technological advances, and encompass a deep cultural change in many spheres of thinking, training, working and fighting.

In our context, the IN has only now reached the threshold of accelerated growth, in a fast changing geo-strategic and technological environment. Rapid induction of high-tech platforms and advanced capabilities is scheduled over the next decade or so, and will be accompanied by the augmentation of shore support facilities and modernisation of infrastructure. The rapidity and scale of change taking place will be such that it may just bypass some (especially the senior leadership) in the Service, or cause disorientation amongst others, thereby affecting operational effectiveness. The change over to networked operations will comprise the first big test in this context.

In order to ensure that the change is orderly, free of trauma, and applied universally, a comprehensive process would need to be initiated by the IN. Essentially, it would mean adoption of a broad philosophy encompassing three main features:

- that the Navy acknowledges the need for change.
- that it is prepared to bring it about in a phased and organised manner, and
- that it would prefer to adopt a transformational rather than an incremental approach to achieve it[14].

Maritime Strategy in War

Conventional Deterrence

While we have dwelt at some length on the Navy's peace-time strategy because peace fortunately prevails most of the time, it must, however, be borne in mind that the prevalence of peace is an indicator that deterrence is working. Should deterrence fail, war will surely follow, and war is what navies train and prepare for. An essential element of this preparation for war is the availability of a maritime strategy to guide planning and deployment of forces.

With conventional maritime challenges on the decline, the IN too has considered it necessary to make a small shift in emphasis from open-ocean warfare, to crisis-response and combat in the littoral. As stated earlier, with our long land borders, we must acknowledge the importance of the land-battle, and try to influence it; no matter how indirectly, or from how far. The US and Coalition Navies have played a major role in the Balkans, Iraq and Afghanistan through the instrumentality of air power, land attack missiles (ship and submarine launched), and Special Forces. The IN too contemplates expeditionary warfare in support of the Indian Army's operations, wherever required.

The IN strategy encompasses the resolute and judicious deployment of maritime forces in both direct and indirect operations. Indirectly, by interdicting the enemy's foreign trade and energy lifeline in an attempt to starve his industry, and economy, and bringing his military machine to a halt. And directly, by intervention through missile attacks and air strikes from the sea, undertaking amphibious operations or inserting Special Forces to create an impact on the land battle. Maritime forces in all three dimensions are weapon intensive, and must seek all-arms synergy, as well as repeated engagements with the enemy to inflict debilitating attrition on him.

The littoral being a hazardous environment, it would be essential to initiate a set of sequenced operations in order to sanitise the battlespace and create favourable conditions, prior to launching an offensive. Attaining information dominance, sea control, and a favourable air situation would be some of the essential prerequisites for success[15].

Close networking of our balanced trio of surface, sub-surface, and air forces over a wide geographical spread, through a satellite backbone, is one of the key requirements of our new maritime strategy.

Nuclear Deterrence

India's nuclear doctrine clearly envisages, and is based on, a deterrent in the form of a triad with land-based, aircraft-borne, and submarine launched weapons. Of this triad, we only have the first two in our inventory at present. Our nuclear doctrine abjures first use, and is quite clear that nuclear weapons are not meant for war-fighting. It seeks to achieve deterrence by convincing the enemy of the futility of a nuclear first strike, because the response would be so devastating that it would render the strike meaningless.

The only platform which can claim to be virtually invulnerable to attack, and ready for instant response is the nuclear propelled submarine, armed with nuclear tipped missiles of intercontinental range (ICBMs). It is, therefore, imperative for India to create a force of 3-4 SSBNs, with a similar number of SSNs for their protection.

If media reports are to be given credence, a Russian Akula class SSN is being sought on lease by India[16]. However, an SSN will not contribute anything to nuclear deterrence, and will at best be a training platform for our nuclear submariners.

The future of the IN indubitably lies in the strategic underwater deterrent rather than in its surface component. The two technology hurdles that our nuclear and rocket scientists need to master in this context are:

- The indigenous design and manufacture of a workable submarine nuclear propulsion plant, for the SSBN as well as SSN; and
- The indigenous design and of an underwater launched ICBM of 3,000-5,000 km range.

These are both technological challenges of monumental complexity. While some headway has presumably been made both in the realm of reactor design as well as missile development, there is a long hard road ahead. Both projects need to be funded and pursued with unremitting resolve and a sense of purpose; qualities that politicians are not known for, but will have to be infused by the naval leadership.

Conclusion

While nature has ensured that India's geographical configuration makes her as reliant on the seas as any island nation, geo-political imperatives, as well as the nation's current economic trajectory serve to emphasise her maritime dependence. Our colonial past has conditioned us in a manner that not only do we wear the mantle of authority with considerable unease, we also exercise power with great timidity. Some would argue that it has been our diffidence, and lack of resolve that has often tempted our adversaries into adventurism.

Technological change and the shifting balance of power may have led to Mahanian templates fading into irrelevance, but navies continue to have their hands full with roles and activities extending over the full spectrum of peace as well as conflict. There is little point in spending vast amounts of money in building a powerful instrument of state policy and then not learning how to use it to the country's advantage.

In 1941, when Admiral Cunningham, the Commander-in-Chief (C-in-C) Mediterranean, was planning the evacuation of Allied forces by Royal Navy warships after the fall of Crete, he was warned about the grave peril the ships would face from intense Luftwaffe bombing. His response is now history: "It takes three years to build a ship; but it takes 300 years to rebuild a tradition. The Navy will go." [17]

These are words of significance for the IN. This compact, integrated and technologically advanced 21st century combat force will be the product of the perseverance and labours of our visionary predecessors. But as Cunningham observed, the physical creation or acquisition of platforms as well as infrastructure is the easy part; it is the creation of a tradition and a lofty ethos that requires our undivided attention.

As she reaches out for great power status, it is imperative for India to first become a robust and confident maritime power. Such an endeavour must encompass not just the 50,000 men and women who serve under the Indian White Ensign, but the whole nation; India needs to reawaken an awareness of her ancient maritime heritage in the minds of the people, and instill maritime consciousness in the political establishment, officialdom and the intelligentsia. Above all, we must inspire a spirit of adventure and affinity for the sea amongst the nation's youth.

Notes

1. K.M. Pannikkar *India and the Indian Ocean* (London: George Allen & Unwin Ltd., 1945), pp.17-23.
2. RAdm Satyindra Singh, *Blueprint to Bluewater* (New Delhi: Lancer International, 1992), pp.5-13.
3. Paul M. Kennedy, *The Rise and Fall of British Naval Mastery* (London: The Ashfield Press, 1983), p.17.
4. Jaswant Singh, *Defending India* (Bangalore: MacMillan India Ltd., 1999), pp.113-128.
5. "...and they shall beat their swords into ploughshares and their spears into pruning hooks; nations shall not lift up sword against nation, nor shall they learn war anymore." Isaiah, 2:2-4. The Holy Bible; King James Version.
6. K. Subrahmanyam, et al. *The Kargil Review Committee Report* (New Delhi: Sage Publications, 2000) p.102.
7. George K. Tanham, *Indian Strategic Thought; An Interpretive Essay* (RAND, 1999), p.60.
8. *Indian Maritime Doctrine*. IHQ MoD (Navy); 2004, pp.63-71.
9. Alexander Nicoll and Jessica Delaney, eds., *India's Energy Insecurity*, IISS Strategic Comments, November 9, 2007, www.iiss.org.stratcom.
10. John Morgan Jr, and Charles Martoglio, *Global Maritime Network*, US Naval Institute Proceedings, November 2005.
11. Dipanjan Roy Chaudhury, *China: Boosting Maritime Capabilities in the Indian Ocean*, August 23, 2007, Worldpress.org. http://worldpress.org/Asia/2098.cfm.
12. Julian S. Corbett, *Some Principles of Maritime Strategy* (London: Longmans Green & Co., 1911), pp.13-14.
13. Vijay Sakhuja, "Indian Navy Diplomacy: Post Tsunami," February 8, 2005. http://www.ipcs.org/Military_articles2.jsm?action.
14. *The Future Indian Navy: Strategic Guidance for Transformation*, IHQ MoD: 2006.
15. *India's Maritime Military Strategy*, IHQ MoD (Navy), 2007, pp.60-61.
16. Sandeep Unnithan, "Indigenous *N-Sub in two Years: Navy Chief*," September 7, 2007. http://indiatoday.digitaltoday.in/index.php
17. General Ismay, *Memoirs* (London: Heinmann, 1960), p.205.

Need for Smart Counter-Terrorism

□ B. Raman

The attack in Mumbai from November 26 to 29, 2008, by a group of 10 Pakistani terrorists belonging to the Lashkar-e-Tayyeba (LeT) has been the subject of study by the intelligence and security agencies of many countries in order to examine whether the *modus operandi* (MO) used by the terrorists in Mumbai called for any changes in the counter-terrorism strategies adopted by them. The US Senate Committee on Homeland Security has also held a detailed hearing in order to understand how and why the terrorists succeeded in Mumbai and how to prevent such incidents in the US. It was terrorism of a conventional nature rendered smarter by modern communications equipment and a good understanding of the way modern media operates. Counter-terrorism failed in Mumbai because it was not as smart as the terrorists. Smart counter-terrorism is the need of the hour. That is the main lesson from Mumbai.

The Mumbai attack has caused concern right across the international counter-terrorism community not because the terrorists used a new MO, which they had not used in the past, but because they used an old MO, with destruction multiplier effect provided by modern communications equipment and lessons drawn from the commando courses of regular armed forces.

There were 163 fatalities in the sea-borne commando-style attack in Mumbai. Only five of them were caused by explosives. The remaining 158 were caused by hand-held weapons (assault rifles and hand-grenades). There had been commando-style attacks with hand-held weapons by terrorists in the Indian territory even in the past, but most of those attacks were against static security guards outside important buildings such as the Parliament House in New Delhi, the US Consulate in Kolkata, a temple in Ahmedabad, etc.

The Mumbai attack of November 2008 was the first act of mass casualty terrorism by the *jehadi* terrorists against innocent civilians using hand-held weapons. The previous two acts of mass casualty terrorism with fatalities of more than 150 were carried out with timed Improvised Explosive Devices (IEDs) – in March 1993 and in July 2006, both in Mumbai.

The increasing use of IEDs by the terrorists since 9/11 had led to strict anti-explosive checks even by private establishments. The killing with IEDs tends to be indiscriminate with no way of pre-determining who should be killed. Moreover, the publicity earned from IED attacks tends to be of short duration. As was seen during the attack on the Parliament House in December, 2001, the visual impact of TV-transmitted images of attacks with hand-held weapons as they are taking place tends to be more dramatic. In an attack with hand-held weapons, the terrorists can pre-determine whom they want to kill.

In Mumbai, 72 people were killed in the terrorist attacks in two hotels and in the Nariman House where a Jewish religious-cum-cultural centre is located, and 86 persons in public places such as the main railway terminus, a hospital, a café, etc. The attacks in the public places by two terrorists on the move lasted less than an hour, but caused more fatalities. The static armed confrontations in the hotels and the Nariman House lasted about 60 hours, but caused fewer fatalities. The static armed confrontations got the terrorists more publicity than the attacks by the two terrorists on the move in public places. By the time TV, radio and other media crew came to know what was happening in the public places and rushed there, the attacks were already over. In the hotels and the Nariman House, the media crew were able to provide a live coverage of almost the entire confrontation. Margaret Thatcher, the former British Prime Minister, had once described undue publicity as the oxygen of terrorists. The terrorists in Mumbai had 60 hours of uninterrupted oxygen supply.

Within a few hours of the start of the confrontation, the security staff of the hotels reportedly switched off the cable transmissions to the rooms. The terrorists were, therefore, not in a position to watch on the TV what was happening outside, but their mobile communications enabled them to get updates on the deployments of the security forces outside from their controllers in Pakistan who, like the rest of the world, were able to watch on their TVs what was happening outside. This could have been prevented only by jamming all mobile telephones. Such jamming could have proved to be counter-productive. Of course, it would have prevented the terrorists from getting guidance and updates from their controllers in Pakistan. At the same time, it might have prevented the security agencies from assessing the mood and intentions of the terrorists and could have come in the way of any communications with the terrorists if the security agencies wanted to keep them engaged in a conversation till they were ready to raid.

The Mumbai attack poses the following questions for examination by all the security agencies of the world:

- Presently, the security set-ups of private establishments have security gadgets such as door-frame metal detectors, anti-explosive devices, closed-circuit TV, etc, but they do not have armed guards. It would not be possible for the police to provide armed guards to all private establishments. How to strengthen the physical security of vulnerable private establishments and protect them from forced intrusions by terrorists wielding hand-held weapons?
- What kind of media control will be necessary and feasible in situations of the type witnessed in Mumbai? This question had also figured after the Black September terrorist attack on Israeli athletes during the Munich Olympics of 1972. Since then, the position has become more difficult due to the mushrooming of private TV channels and private FM radio stations.

- How to ensure that mobile telephones do not unwittingly become a facilitator of ongoing terrorist strikes without creating operational handicaps for the security agencies

The LeT terrorists, who attacked Mumbai, had a three-point agenda:

- An anti-Indian agenda to create fears in the minds of foreign businessmen about the security of life and property in India and in the minds of the Indian public about the competence of the Indian security agencies to protect them.
- An anti-Israeli and an anti-Jewish agenda whose objectives coincided with those of Al Qaeda.
- An anti-US and an anti-North Atlantic Treaty Organisation (NATO) agenda, whose objectives coincided with those of Al Qaeda and the Afghan Taliban. Of the 25 foreigners killed, nine were either Israelis or Jewish persons, 12 were from countries which have contributed troops to the NATO force in Afghanistan and four were from other countries.

All these agendas coincide with the agenda of the global *jehad* as waged by the International Islamic Front (IIF) for Jehad Against the Crusaders and the Jewish People, formed by Al Qaeda in 1998. From 1998 till April 2006, Osama bin Laden projected the global *jehad* as directed against the Crusaders (Christians) and the Jewish people. In an audio message disseminated by him in April, 2006, after the visit of the then US President George Bush to India, he expanded the objectives of the global *jehad* and projected it as directed against the Crusaders, the Jewish people and the Hindus. The Mumbai attack targeted these three proclaimed adversaries of the IIF, of which the LeT is a member.

Since 2003, there have been indications that following a weakening of the command and control of Al Qaeda because of the US military operations in Afghanistan, the LeT had started playing the role of coordinator of the IIF on behalf of Al Qaeda. The Mumbai attack brought out the increased capabilities of the LeT for the planning and execution of simultaneous commando-style attacks against multiple targets. The LeT now poses a serious threat not only to India as it was doing in the past, but to other countries as well. It is a new and major threat to international peace and security which has to be fought by the united efforts of the international community.

Since the terrorist attack lasted 60 hours and the lives of the nationals of many countries were in danger, the intelligence agencies of India, Israel, the US and the UK – and possibly of other countries too – were monitoring through technical means the conversations of the terrorists holed up in the two hotels and in the Jewish centre with each other and with their controllers in Pakistan. Thus, a substantial volume of independent technical intelligence exists – collected by the intelligence agencies of these countries independently of each other.

All this independent evidence clearly showed that the terrorist attack was mounted by the LeT from the Pakistani territory with the help of 10 Pakistanis specially recruited and trained for this operation in training camps in Pakistan Occupied Kashmir and then in Karachi. As other Pakistani governments had done in the past, the present government, headed by President Asif Ali Zardari, too has avoided extending mutual legal assistance to India as required by the conventions followed by the Interpol and by the UN Security Council Resolution No.1373 adopted unanimously by the UN General Assembly after the 9/11 terrorist strikes in the US. It first even denied that the terrorist captured alive by the Mumbai police is a Pakistani national. Under pressure from the US, it reluctantly admitted that he is a Pakistani national, but continued to question the credibility of the evidence collected by India. It made clear that there is no question of handing over any Pakistani national to India for trial.

The Mumbai attack and the role of Pakistan in it demonstrated the inability of the Indian state to retaliate against an act of mass casualty terrorism mounted from Pakistan through means other than a direct military strike, which may not be advisable against a neighbour except in extreme circumstances. States such as the US and Israel maintain a covert action capability for use against terrorists operating from the territory of another country when they consider direct military action as inadvisable or as not feasible. India has had no covert action capability since 1997 when the limited capability that it had was wound up as a unilateral gesture to Pakistan. This has proved to be an extremely unwise step.

Smart counter-terrorism has four components – prevention through timely and precise intelligence, prevention through effective physical security, crisis or consequence management to limit the damage if prevention fails, and a capability for deniable retaliation if the terrorists operate from the territory of another state. In Mumbai, intelligence was available, but considered inadequate by the police and the Navy/Coast Guard; physical security by the police and the security establishments of the targeted places was deficient, coastal surveillance by the police and the Coast Guard was weak, the consequence management by the National Security Guards (NSG) and others was criticised as tardy and lacking in coordination; and deniable retaliatory capability was not available. In their testimonies before the US Senate Committee on Homeland Security, non-governmental US security experts said that the success of the terrorists in Mumbai demonstrated the weak state of India's preventive capability and the non-existence of a retaliatory capability. Unless immediate action is taken to remove the deficiencies, more terrorist attacks of a serious nature cannot be avoided.

P. Chidambaram, who took over as the new Home Minister after the Mumbai attack, has already initiated certain measures such as enhancing the powers of the police, setting up a National Investigation Agency (NIA) to investigate certain types of terrorism cases and creation of regional hubs of the NSG in order to reduce delays in response as had allegedly occurred in Mumbai. These are the starting blocks of a revamped counter-terrorism strategy and apparatus, but much more needs to be done to keep in step with the evolution of counter-terrorism strategies and systems in other countries such as the US, the UK, Spain, etc, which had suffered serious terrorist attacks since 2001.

Just as terrorists are constantly evolving in their thinking and ideology, in their educational background and skills, and in their *modus operandi*, so too the counter-terrorism strategy of the state actors has also been evolving to meet the threat posed by them.

Before 1967, counter-terrorism was seen largely as the responsibility of the police and the civilian intelligence agencies. After the terrorist organisations took to aviation terrorism involving aircraft hijackings and blowing up aircraft in mid-air as part of their *modus operandi*, the need for special intervention forces trained by the Army was felt. After a surge in acts of terrorism against Israeli nationals and interests in Israel and outside after the Arab-Israeli War of 1967, counter-terrorism in Israel acquired an increasingly military dimension with the role of the police subordinated to that of the armed forces.

This trend towards the increasing militarisation of counter-terrorism acquired a further momentum after vehicle-borne suicide bombers, suspected to be from the Hezbollah, blew themselves up outside the barracks of the US Marines and French paratroopers then deployed as part of an international peace-keeping force in Beirut, killing 241 US servicemen and 58 French paratroopers on October 23,1983. It was after this incident that the US started talking of a strategy to combat terrorism instead of a strategy to wage a campaign against terrorism. Al Qaeda's attack against the US naval ship, the USS *Cole* in Aden in October 2000, and the subsequent discovery of the plans of Al Qaeda to indulge in acts of maritime terrorism in ports and in choke points such as the Strait of Gibraltar and the Malacca Strait to disrupt international trade and the flow of energy supplies and to damage the global economy, gave a naval dimension to counter-terrorism.

Even long before 9/11, counter-terrorism had acquired a scientific and technological dimension due to the increasing use by terrorists of IEDs, this dimension was restricted to detecting the presence of IEDs and neutralising them. This Science and Technology (S&T) dimension has since grown in importance due to the attempts of Al Qaeda to acquire Weapons of Mass Destruction (WMD) material and its proclaimed readiness to use them, if necessary, to protect Islam. This dimension has further expanded due to apprehended threats to critical information infrastructure that could arise from terrorists or hackers helping terrorists, who are adept in the use of information technology for destructive purposes.

Before 1967, terrorism was largely a uni-idimensional threat to individual lives and property. It has since evolved into a multi-dimensional threat to the lives of large numbers of people, to the economy and to the critical information infrastructure. It is no longer viewed as a purely police responsibility. It is the responsibility of the police, the armed forces, the scientific and technological community and the experts in consequence management such as psychologists, fire brigades and medical personnel and experts in disaster relief and rehabilitation. How to ensure coordinated and well-synchronised action by the different elements of the counter-terrorism community and what kind of counter-terrorism architecture is required are the questions constantly engaging the attention of national security managers of countries affected by terrorism.

Combating terrorism military-style evolved into a war against terrorism after the 9/11 terrorist strikes in the US homeland. This had three implications. Firstly, a no-forces barred approach in combating terrorism – whether it be the army, the air force, the navy, the police or the special forces; secondly, an enhanced leadership role for the armed forces in the war against terrorism; and thirdly, a new criminal justice system to deal with terrorists that not only provided for special laws and special courts, but also enabled the armed forces to deal with foreign terrorists operating against US nationals and interests as war criminals liable to be detained in special military camps such as the one in the Guantanamo Bay and to be tried by military tribunals and not by civil courts. President Barack Obama has been trying to reverse some of these practices and has initiated action to close the Guantanamo Bay detention centre within a year and to transfer the responsibility for trial to normal courts from military tribunals.

Keeping pace with this evolution of a new strategy to combat terrorism, there has been a simultaneous evolution of the counter-terrorism architecture with the addition of many new elements to this architecture. The two most important elements in the US are the Department of Homeland Security (DHS) and the National Counter-Terrorism Centre. The DHS acts as the nodal point for coordinating all physical security measures against terrorism and all crisis management measures to deal with situations arising from successful acts of terrorism in US territory or on its borders as well as with natural disasters. While the Department of Defence created in 1947 is responsible for all policy-making and coordination relating to US military operations abroad, whether against a state or a non-state adversary, the DHS is responsible for all policy-making and inter-departmental coordination relating to internal security and natural disasters. A Homeland Security Council in the White House performs an advisory and policy-making role in respect of internal security and natural disasters.

The Homeland Security Council is structurally similar to the National Security Council, with a Secretariat of its own, which is headed by an official, who is designated as the Adviser to the President for Homeland Security and Counter-Terrorism. Its meetings are chaired by the President and attended by various Cabinet members having responsibilities relating to internal security.

In August 2004, Bush established the National Counter-Terrorism Centre (NCTC) to serve as the primary organisation for integrating and analysing all intelligence pertaining to terrorism and counter-terrorism (CT) and to conduct strategic operational planning by integrating all instruments of national power. In December 2004, the Congress incorporated the NCTC in the Intelligence Reform and Terrorism Prevention Act (IRTPA) and placed the NCTC under the supervision of the Director of National Intelligence, a newly-created post to coordinate and supervise the functioning of all intelligence agencies of the US.

In the UK, as in the past, the police and the MI-5, the security service, continue to have a preeminent role in counter-terrorism of a classical nature such as acts involving the use of hand-held weapons and IEDs. The armed forces and the S&T community play an enhanced role only in respect of likely terrorist strikes involving WMD material, aviation and maritime terrorism and terrorism through the internet.

A long-term counter-terrorism strategy called CONTEST formulated in 2003, has four components: prevention, pursuit, protection and preparation. Prevention refers to the role of the political leadership in preventing British citizens and residents in the UK from joining terrorist organisations through appropriate measures for redressing grievances and for countering the ideology of the terrorists. Pursuit refers to the responsibility of the intelligence and security services and the police to collect preventive intelligence regarding planned terrorist operations and to disrupt the functioning of terrorist organisations through physical security measures and successful investigation and prosecution of terrorist incidents. Protection refers to the physical security measures required to prevent acts of terrorism based on threat or vulnerability perceptions. Preparation refers to the various agencies being in a state of readiness to meet the consequences of an act of terrorism. This is what we in India call crisis management.

Between 9/11 and July 2005, in the UK too, as in the US, the military dimension of counter-terrorism tended to acquire a greater importance than before due to the perception that the main threat to the UK would be from foreign-based Al Qaeda elements. This perception changed after the July 2005 terrorist strikes in London by four suicide bombers, who had grown up in the UK. The Intelligence and Security Committee, a Parliamentary oversight committee that reports to the Prime Minister on the performance of the intelligence agencies, which enquired into the failure to prevent the July 2005, attacks, concluded that the police and the security agencies had failed to adjust sufficiently quickly to the growth of domestic terrorism. It said: "We remain concerned that across the whole of the counter-terrorism community the development of the home-grown threat and the radicalisation of British citizens were not fully understood or applied to strategic thinking."

The counter-terrorism strategy and architecture evolved in the UK emphasise the role of the police working under the overall supervision of the Home Secretary. A lesson drawn by the British from the July 2005 terrorist strikes in London is that no counter-terrorism strategy will be effective unless it is supported by the community from which the terrorists have arisen. The importance of police-Muslim community relations for preventing the radicalisation of the youth and for de-radicalising those already radicalised and of police-business community relations in order to motivate and help the business community to protect itself from terrorist strikes on soft targets are now two of the important components of the British counter-terrorism strategy.

Among the new elements in the British counter-terrorism architecture, one could mention the National Counter-Terrorism Security Office (NaCTSO). The NaCTSO, which is funded and operated by the Association of Chief Police Officers, works on the "protect and prepare" strand of the government's counter-terrorism strategy. Its aims have been defined as follows:

- raise awareness of the terrorist threat, and spread the word about measures that can be taken to reduce risks and mitigate the effects of an attack;
- coordinate security advice through the Counter-Terrorism Security Adviser (CTSA) network and monitor its effectiveness;
- build relationships between communities, police and government agencies; and
- contribute to the national and international counter-terrorism policy.

It trains, tasks and coordinates a nationwide network of centrally funded, specialist police advisers known as CTSAs. The primary role of these advisers is to provide help, advice and guidance on all aspects of counter-terrorism security to the public. It has developed and published guides on physical security against terrorism in sporting stadia and arenas, shopping centres and bars, pubs and clubs. It has undertaken the preparation of similar guides for other soft targets.

The Israeli Counter-Terrorism Strategy has three components: defensive, operative and punitive. Defensive and operative refer to prevention through timely and precise intelligence and operations to disrupt planned terrorist strikes and punitive refers to retaliation by the state against terrorist organisations and their foreign state or non-state sponsors. No intimidation by terrorists, no succumbing to pressure by terrorists, making the terrorists and their sponsors pay heavily for their acts of terrorism, protection of the lives and property of Israeli citizens at any price and a refusal to be paralysed into inaction against terrorists due to fears of adverse reactions from the international community are the basic principles underlining the Israeli counter-terrorism strategy.

A Global Counter-Terrorism Strategy adopted by the UN General Assembly on September 8, 2006, laid down that any plan of action against terrorism should have the following four components:

- Measures to address conditions which could be conducive to the spread of terrorism.
- Measures to prevent and combat terrorism.
- Measures to build counter-terrorism capacities and to promote international cooperation.
- Measures to protect human rights and to enforce the rule of law.

Whereas other democracies such as those of the US, the UK and Israel have been facing only terrorism of one or two kinds, India has been facing terrorism of multiple origin with varied objectives and different areas of operation. Our intelligence agencies and security forces have been facing cross-border terrorism and hinterland terrorism; urban *jehadi* terrorism and rural Maoist terrorism; ideological terrorism, religious terrorism and ethnic or separatist terrorism; indigenous *jehadi* and pan-Islamic *jehadi* terrorism; and indigenous and Pakistan and Bangladesh sponsored terrorism. The likelihood of maritime terrorism and WMD threats from Al Qaeda based in Pakistan's tribal belt and cyber terrorism from Information Technology (IT)-literate terrorists have added to the complexity of the scenario.

Against this background, India's counter-terrorism strategy has to have a common core, with principles applicable to all terrorism and separate modules tailor-made and suited to the different kinds of terrorism that we have been facing. The principles of this common core, some of which are in force even now, are:

- The police would be the weapon of first resort in dealing with hinterland terrorism of all kinds and the Army would be the weapon of only last resort.
- In dealing with cross-border terrorism in Jammu and Kashmir (J&K) and with the United Liberation Front of Assam (ULFA) and the tribal insurgents in the Northeast, the Army would have the leadership role – with the police operating in the interior areas and the Army operating nearer the borders. The paramilitary forces would be available for assistance to the police as well as the Army.
- Intelligence collection against hinterland terrorism would be the joint responsibility of the Intelligence Bureau (IB) and the state police and in the border states of the IB, the police and the military intelligence. Intelligence collection regarding the external ramifications of all terrorist organisations would be the responsibility of the Research and Analysis Wing (R&AW).
- Physical security against hinterland terrorism would be the joint responsibility of the state police and the central security forces such as the Central Industrial Security Force (CISF). In the border areas, it will be the joint responsibility of the Army, the paramilitary forces and the police.
- The new mutations of terrorism, which could strike India one day, such as WMD, maritime and cyber terrorism have to be dealt with jointly by the armed forces, the scientific community and the police – with the Army having the leadership role in respect of WMD terrorism, the Navy/Coast Guard in respect of maritime terrorism and an appropriate S&T organisation in respect of cyber terrorism.

- While dealing with *jehadi* terrorism calls for the strengthening of urban policing, dealing with Maoist terrorism cannot be effective without strengthening the rural policing.
- While we should follow a no-holds barred approach to crush terrorists from Pakistan and Bangladesh operating in our territory, our strategy in respect of our own nationals who have taken to terrorism should be nuanced with a mix of the political and security strands.
- While we should avoid the pitfalls of over-militarisation or Americanisation of our counter-terrorism strategy, which would be counter-productive in our country with the second largest Muslim population in the world, and with our location in the midst of the Islamic world, we should not hesitate to adopt with suitable modifications the best counter-terrorism practices from the US, the UK and Israel. Among practices worthy of emulation, one could mention empowering the police with special laws, the creation of a central agency for coordinated investigation and prosecution of terrorism cases, strict immigration control, strong action to stop illegal immigration and to expel illegal immigrants, action to stop the flow of funds to the terrorists from any source – internal and external – and the adoption of the concept of an integrated counter-terrorism staff for an integrated analysis of all terrorism-related intelligence and joint action on them. All agencies having counter-terrorism responsibilities should be represented in the staff.

The evolution of our counter-terrorism strategy has been in fits and starts as and when we faced a new kind of terrorism or a crisis situation. Similarly, our counter-terrorism community too has grown up in a haphazard manner. Our approach to terrorism has been more tactical than strategic, more influenced by short-term thinking than long-term projections. The time has come to set up a dedicated task force to recommend a comprehensive counter-terrorism strategy. The strategy has to be community-based to draw the support of all communities, political consensus-based to draw the support of all political parties and should provide for a close interaction with the private sector to benefit from its expertise and capabilities and to motivate it to protect itself in soft areas.

In 2004, the government of Dr. Manmohan Singh created two posts of National Security Advisers (NSA) – one for external security, which was held by the late J.N. Dixit, and the other for internal security, which was held by Shri M.K. Narayanan. After the death of Dixit in January 2005, the government reverted to the previous practice of having a single NSA to deal with internal and external security, a post that was held by Shri Narayanan who has now been replaced by Shri Shiv Shankar Menon. A reversion to the 2004 practice of having an NSA exclusively for internal security is necessary for improving our counter-terrorism management.

Another important step should be the reorganisation of the Ministry of Home Affairs (MHA) of the Government of India. Counter-terrorism is one of its many responsibilities. While the trend in other countries has been towards having a single ministry or department to deal exclusively with counter-terrorism, our MHA has resisted this trend.

In any unified command and control for counter-terrorism, the ministry responsible for counter-terrorism has to play a pivotal role. The importance of having a single leader for dealing exclusively with internal security, without being burdened with other responsibilities was realised by Rajiv Gandhi and Narasimha Rao. Instead of bifurcating the MHA, Rajiv Gandhi created a post of Minister of State for Internal Security in the MHA to handle all operational matters, including waging a joint campaign against terrorism by the Centre and the states. This continued under Narasimha Rao. The time has come to create an independent Ministry of Internal Security.

Inadequacies in our intelligence agencies have remained unidentified and unaddressed. Every successful terrorist strike speaks of an intelligence failure. There is a lack of coordination not only among the agencies at the Centre, but also between the central agencies and those of the state police. How to improve the quantity and the quality of the intelligence flow? How to ensure better coordination at the Centre and with the states? Important questions such as these were addressed by the Special Task Force for Revamping the Intelligence Apparatus headed by Shri G.C. Saxena, former head of the R&AW, appointed by the Atal Behari Vajpayee government in 2000.

The implementation of its recommendations has not had the desired impact on the ground situation. Why? What further measures are needed? These issues have to be urgently addressed by a dedicated task force on terrorism-related intelligence capabilities.

Preventive physical security is the responsibility of central police forces and the police of different states. While the capability at the Centre has improved, it has improved in certain states and declined in certain others. A strong physical security capability can thwart a terrorist strike even in the absence of intelligence. A weak capability may not be able to prevent it even if intelligence is available. Identification of weaknesses in our physical security set-up and action to remove them must receive priority.

Successful investigation and prosecution deter future terrorist strikes. Poor investigation and prosecution encourage terrorism. India has a poor record in successful prosecutions. Effective coordination of the police in all the states, the creation of a national data base to which the police of different states can have direct access and the quick sharing of the results of the enquiries and investigations through this data base could improve our record in investigation and prosecution.

How to prevent attacks on soft targets? This has been a dilemma for all states. Israel, which sees many attacks on soft targets by Palestinian suicide bombers, follows a policy of reprisal attacks by the state on the leaders of the suspected organisations after every attack on a soft target, in order to demonstrate to the terrorists that their attacks on soft targets will not be cost free. It is able to do it because the targets chosen by the state agencies for reprisal attacks are located in the areas under the control of the Palestinian Authority. It does not indulge in reprisal attacks in its own territory. Despite such reprisal attacks by the state agencies, Israel has not been able to stop attacks on soft targets. There is no short-term solution to attacks on soft targets except improvement in the capability of our intelligence agencies to collect timely preventive intelligence. Gradual attrition of organisations indulging in such attacks through arrests and neutralisation of their leaders could be a medium and long-term solution. That too would require precise intelligence, which is not always available.

Suicide terrorism is a lethal strategic weapon, to which no state has been able to find an effective response. While suicide terrorism against hard, heavily protected targets can be prevented through strict access control, suicide terrorism against soft targets is difficult to prevent unless the suicide terrorist is accidentally detected or the explosive device fails to function.

About 80 percent of the acts of suicide terrorism are carried out with explosives. Strict explosives control in order to prevent them from falling into the hands of terrorists can make the problem of suicide terrorism more manageable, but the increasing use of commonly available materials such as nitrogenous fertilisers, cosmetics used by women, etc by the terrorists for fabricating explosives has added to the difficulties of the counter-terrorism agencies in preventing explosive substances from falling into the hands of terrorists. Yet, how to tighten up controls over the purchase, sale and acquisition of explosives and substances capable of being converted into explosives is a question which needs serious attention. While considerable attention has been paid to devising measures to prevent the proliferation of small arms and ammunition, similar attention has not been paid to explosive substances.

Strategic threat analysis has undergone a significant change since 9/11. Before 9/11, analysis and assessment of threat perceptions were based on actual intelligence or information available with the intelligence and security agencies. 9/11 brought home to policy-makers the difficulties faced by intelligence agencies, however well-endowed they might be, in penetrating terrorist organisations to find out details of their thinking and planning. This realisation has underlined the importance of analysts serving policy-makers constantly identifying national security vulnerabilities, which might attract the attention of terrorists, and suggesting options and actions to deny opportunities for attacks to the terrorists. Vulnerability analysis has become as important as threat analysis.

Strategic analysts can no longer confine themselves to an analysis and assessment of strategic developments of a conventional nature arising from state actors, but should pay equal attention to the strategic impact of non-state actors, such as international or trans-national terrorists, crime mafia groups and nuclear proliferators on global security, in general, and our own national security, in particular..

India's record in dealing with terrorism and insurgency is not as negative as it is often projected to be. We have had a successful record in Punjab, Nagaland (partial), Mizoram, Tripura and in Tamil Nadu in dealing with the terrorism of Al Umma. Even in J&K, the ground situation is showing signs of definite improvement.

However, there are two kinds of terrorism/insurgency where our record has been poor till now – the *jehadi* kind, which is essentially an urban phenomenon outside J&K, and the Maoist (Naxalite) kind, which is essentially a rural phenomenon. We have succeeded where the terrorism or insurgency was a regional phenomenon and was confined to a narrow area. We have not succeeded where the threat was pan-Indian in nature, with the network extending its presence to many states in the north and the south.

A pan-Indian threat requires a coordinated pan-Indian response at the political and professional levels. Unfortunately, the multiplicity of political parties, the era of coalitions and the tendency in our country to over-politicise terrorism come in the way of a pan-Indian political response. The tendency of the intelligence agencies and the police of different states to keep each other in the dark about what they know and not to admit to each other as to what they do not know comes in the way of a pan-Indian professional response. There has been a plethora of reports and recommendations on the need for better sharing and coordination, but without any effect on our agencies and the police.

The agencies and the police are largely responsible for the absence of a coordinated professional response, but the political leadership at the Centre and in different states cannot escape their share of responsibility. A determined political leader, who has the national interests in mind, can use a whip and make the agencies and the police cooperate. A political leader whose policies and actions are motivated by partisan and not national interests will come in the way of professional cooperation.

Any cure to the problem of *jehadi* and Maoist terrorism has to start at the political level. A political leader has to play a dual role. He has to help the professionals in taking firm action against the terrorists – whatever be their community and ideology. He has to give them whatever tools they need. At the same time, he has to identify the circumstances and perceptions which drive young Muslims to take to *jehadi* terrorism and young tribals to take to Maoist terrorism. Anger is one of the common root causes of all terrorism. Unless this anger is addressed, professional handling of the threat alone, however effective, cannot bring about an enduring end to this threat.

An effective political handling has to start with a detailed analysis of the causes of anger and action to deal with them. Our young Muslims, who are taking to *jehadi* terrorism, are not bothered by issues such as lack of education and unemployment, reservation for Muslims, etc. They are angry at what they consider to be the unfairness to the Muslims, which, according to them, is widely prevalent in India. Unsatisfactory political handling of the Muslim youth by all political parties is an aggravating cause of the threat from *jehadi* terrorism.

Similarly, it is the absence of meaningful land reforms and perceptions of suppression of the tribals by the non-tribals and the administration, which is an important cause of the tribal anger in central India. It is the responsibility of the political class and the society as a whole to address this. They do not do so and keep nursing an illusion that more and more money, men and equipment for the agencies and the police will end this problem. It won't.

The way we kick around the problem of terrorism like a football blaming everybody else except ourselves can be seen in the TV debates and media columns. The same arguments are repeated without worrying over their validity.

Flow of human intelligence about *jehadi* terrorism is weak because of the post-9/11 phenomenon of global Islamic solidarity and the adversarial relationship between the agencies and the police, on the one side, and the Muslim community, on the other. Feelings of Islamic solidarity prevent even law-abiding Muslims from volunteering to the agencies and the police information about their co-religionists who have taken to terrorism, and from assisting the police in their investigation. The adversarial relationship has resulted in mutual demonisation. How to come out of this syndrome is a matter for serious consideration not only by the police and the agencies, but also by the political class and the civil society, including the media.

Once we allow terrorism and insurgencies of different kinds to make their appearance in our society, it takes a long time to deal with them. We took 19 years to deal with the Naga insurgency, another 19 years to deal with the Mizo insurgency, 14 years to deal with Khalistani terrorism and about 10 years to deal with Al Umma. The French took 19 years to deal with the terrorism of Carlos and his group. Even after 41 years of vigorous implementation of a no-holds-barred counter-terrorism strategy, Israel is still grappling with the terrorism of the Palestinians and Hezbollah. The British took over 20 years to bring the Irish Republican Army under control.

The attitude of our political class to terrorism is ambivalent. On the one hand, it is worried – rightly – over this growing threat. On the other, it continues to view this as a vote-catcher. Every political party has been firm in demanding action against terrorism when it is out of power. It becomes soft when it comes to power. That is the bane of our counter-terrorism. Only voter pressure can force the political class to stop exploiting terrorism as an electoral weapon and start dealing with it as a major threat to national security, which should unite the political class and the civil society.

Finally, the *jehadi* terrorism in our territory has been able to thrive because of the support from the intelligence agencies of Pakistan and Bangladesh. Our anxiety for improved relations with them has been coming in the way of any deterrence to their continued use of terrorism against India. The deterrence has to be in the form of an effective covert action capability, which we should be prepared to use against the terrorist infrastructure in Pakistani and Bangladeshi territory, if left with no other option. The covert action capability, which was reportedly wound up in 1997 out of a misplaced sense of generosity to Pakistan, has to be revived.

Governance and Left Wing Extremism

□ Prakash Singh

The universally accepted features of good governance, according to the Planning Commission, are "the exercise of legitimate political power and formulation and implementation of policies and programmes that are equitable, transparent, non-discriminatory, socially sensitive, participatory, and above all accountable to the people at large." There could, however, be aspects of governance that are contextually driven and geared to address the local concerns.

Good governance helps secure human well-being and sustained development. It is, as Kofi Annan, former Secretary General of the United Nations, said, "The single most important factor in eradicating poverty and promoting development." On the other hand, poor governance erodes individual capabilities as well as institutional and community capacities to meet the needs of sustenance.

Addressing the Chief Secretaries and the State Police Chiefs a few years ago, the then Prime Minister said that one of the fundamental reasons for the ills of insurgency, extremism and crimes affecting internal security was the lack of good governance, especially at the cutting edge level. According to Madhav Godbole, former Union Home Secretary, "Our whole administrative apparatus is in shambles and the organised and highly qualified civil services based on open competitive examinations are on the brink of extinction."

There are remote areas in the country where there is hardly any governance. Abujhmarh in Narainpur district of Chhattisgarh is one such area. It has a tribal population of 27,000, inhabiting some 260 far-flung villages over a sprawling area of 4,000 sq km. The tribals here are primarily the Maria; they are the most backward tribals between the rivers Ganga and Godavari. Abujhmarh remains cut off from the rest of the civilised world for about six months a year. Surprisingly, the area has not been surveyed to date and has hardly any revenue or police presence on a regular basis. No wonder, the Naxals have made it one of their strongholds. Even in areas which are not so much in the interior, the absence of adequate public intervention, especially in education, health and employment has allowed the non-state actors to push their agenda among the people.

Apart from poor governance, several other factors have contributed to the growth of left wing extremism in the country. Poverty continues to be a huge problem. At the beginning of the new millennium, according to the Planning Commission document on the Tenth Five-Year Plan (2002-2007), 260 million people in the country did not have incomes to access a consumption basket which defines the poverty line. It is a huge figure – and, unfortunately, it takes a quantum jump if we accept the latest yardstick by two leading development banks to define poverty. The Manila-based Asian Development Bank (ADB) has proposed scrapping the \$ 1 per day poverty measure popularised by the World Bank in 1990 as an estimate of the per-person cost of procuring the 2,100 calories a day deemed necessary for human health. The ADB's new Asian Poverty Line raises the bar to \$ 1.35 per day. The World Bank, on the other hand, prefers raising the global poverty line to \$ 1.25. Even if we take the World Bank yardstick, the total number of poor jumps to 455 million; it would shoot up further to 622 million if we accept the ADB's definition. It is a depressing scenario. About half the population of the country would appear to be living below the poverty line.

Land reforms have unfortunately not received adequate attention in the states except in West Bengal. The Tenth Five-Year Plan admitted that "the record of most states in implementing the existing laws is dismal." Unemployment figures also showed worsening. According to the Economic

Survey 2007-08, the incidence of unemployment on current daily status basis increased from 7.31 per cent in 1999-00 to 8.28 per cent in 2004-05.

The present situation of tribals in the country was beautifully analysed by an Expert Group on *Prevention of Alienation of Tribal Land and its Restoration* constituted by the Ministry of Rural Development, Government of India. The Group found that "the socio-economic condition of the tribals is beset with severe complexities and problems" and that "the tribal societies, with few exceptions, are deficient in terms of social development and the tribal people suffer from capability poverty." The Expert Group also noted that the quality of life of the tribals had failed to improve despite the constitutional amendments. It noted with anguish that land belonging to the tribal people was being alienated in all the states despite the existence of an umbrella of protective legislation.

It is this combination of factors – poor governance reflected in acute poverty, neglect of land reforms, mounting unemployment and the alienation of tribals – which has contributed to a sense of grievance and frustration among large sections of the people. They felt that the peaceful political process would not bring about the necessary changes because vested interests controlled the levers of power, regulated the wheels of industry and had a feudal stranglehold over the predominantly agrarian economy. An armed struggle was the only way out. Thus, was born the left wing extremism in India.

Early Years: The Charu Phase

The Naxalite movement derived its name from a small village Naxalbari on the tri-junction of India, Nepal and what was then East Pakistan, where tribals took up arms against the oppression of the landlords in 1967. Charu Mazumdar's ideology captured the imagination of the people. His assessment was that "every corner of India is like a volcano" about to erupt, that "there is the possibility of a tremendous upsurge in India", and he, therefore, called upon the cadres to start as many points of armed struggle as possible. The movement spread like wildfire to different parts of the country. Some of the finest brains and the cream of India's youth in certain areas left their homes and colleges to chase the dream of a new world, a new social order.

Drawing inspiration from the Maoist ideology, the movement had a meteoric phase for about two years from the formation of the party till the end of June 1971. The ripples starting from Naxalbari spread in ever-widening circles to practically all parts of the country. The dominant strand of the movement was the annihilation of class enemies. It was viewed as a "higher form of class struggle and the beginning of guerrilla war." "Expand anywhere and everywhere" was Charu's message. Such expansions were particularly noticed in Srikakulam in Andhra Pradesh, Debra-Gopiballavpur in West Bengal, Mushahari in Bihar and Palia in Lakhimpur district of Uttar Pradesh (UP).

The government saw in the Naxalite movement the emergence of a threat not only to law and order but to the very existence of the democratic structure of the country. It, therefore, organised joint operations by the Army and the police in the bordering districts of West Bengal, Bihar and Orissa which were particularly affected by Naxalite depredations. The operations were undertaken from July 1 to August 15, 1971, and were code-named Operation Steeplechase. The broad strategy of the security forces was to surround as large an area as possible and seal the routes of entry and exit. The Army formed the outer cordon and the Central Reserve Police Force (CRPF) the inner ring. The local police, which was generally accompanied by a magistrate, carried out thorough search of the area. Suspected Naxalites were arrested, while their illicit weapons, ammunition and explosives were seized. These operations covered Midnapur, Purulia, Burdwan and Birbhum districts of West Bengal; Singhbhum, Dhanbad and Santhal Parganas of Bihar, and Mayurbhanj of Orissa.

The operation achieved the desired results, though not to the extent anticipated by the administration. The organisational apparatus of the Naxalites in the aforesaid districts was thrown out of gear and the party activists fled from their known hideouts to other places in search of safety. Violence registered a drop. Incidents of arms-snatching fell down. Above all, it restored the confidence of the people in the strength of the administration. Charu Mazumdar was also arrested by the Calcutta police detectives on July 16, 1972. A few days later, he died. Charu's death marked the end of a phase in the Naxalite movement. The period following his death witnessed divisions and fragmentations in the movement.

Revival in Andhra Pradesh: The Kondapalli Phase

The formation of the People's War Group (PWG) in Andhra Pradesh in 1980 under the leadership of Kondapalli Seetharamaiah gave a new lease of life to the movement. The PWG's programme included:

- redistribution of land;
- enforcing payment of minimum wages to the farm labour;
- imposing taxes and penalties;
- holding people's courts;
- destroying government property;
- attacking policemen; and
- enforcing a social code.

The PWG is believed to have redistributed nearly half a million acres of land across Andhra Pradesh. Its activists also insisted on a hike in the daily minimum wages and the annual fee for *jeetagadu* (year-long labour). The poorer sections found that what the politicians had been talking about and the government promising year after year could be translated into reality only with the intervention of the Naxalites. *Gorakala doras* (Lord of the Bushes) is how the Naxalites came to be known in the interior areas.

The revolutionary writers of the *Jana Natya Mandali*, the cultural front of the PWG, greatly helped in preparing the environment in which the Naxalite ideology found ready acceptance. Its moving spirit was – and continues to be — Gummadi Vittal Rao, better known as Gaddar. He is a balladeer who fights the establishment with the power of his songs. The People's War Group gradually spread its organisational network to the coastal and Rayalaseema districts in the state. It extended its tentacles to the adjoining areas of Maharashtra, Madhya Pradesh and Orissa and made a dent even in the bordering districts of Karnataka and certain pockets of Tamil Nadu.

The Andhra Pradesh government banned the PWG and its six front organisations in 1992. At the same time, the state police, assisted by the central paramilitary forces, undertook well organised counter-insurgency operations. As a result, 248 Naxalites were liquidated and 3,434 activists were apprehended in 1992. The arrest of Kondapally Seetharamaiah and other important leaders meant a further setback to the PWG. There was demoralisation among the ranks and about 8,500 Naxals surrendered before the authorities.

In Bihar, the Maoist Communist Centre of India (MCCI), another major Naxalite formation, perpetrated acts of violence. Its organisational network extended to most of the central Bihar districts. However, what began as a fight for social and economic justice, gradually degenerated into a caste conflict with a veneer of class struggle. The MCC ran virtually a parallel judicial system in certain pockets. These were described as *Jan Adalat* or People's Court where they would even shorten an accused by six inches – behead him, in other words!

Present State of the Movement

The present phase – we could also call it the third phase — of the movement commenced with the holding of the Ninth Congress of the People's War Group in 2007. The Congress "reaffirmed the general line of New Democratic Revolution with agrarian revolution as its axis and protracted people's war as the path of the Indian revolution", and resolved to "advance the people's war throughout the country, further strengthen the people's army, deepen the mass base of the party and wage a broad-based militant mass movement against the neo-liberal policies of globalisation, liberalisation, privatisation."

Violence has ever since taken a high trajectory, as the following figures show:

Total Incidents		Deaths
2001	1,208	564
2002	1,465	482
2003	1,597	515
2004	1,533	566
2005	1,608	677
2006	1,509	678
2007	1.565	696

The Ministry of Home Affairs, Government of India, admits the spread of the Naxalite movement to thirteen states of the Union, namely, Andhra Pradesh, Bihar, Chhattisgarh, Jharkhand, Madhya Pradesh, Maharashtra, Orissa, Uttar Pradesh, West Bengal, Kerala, Karnataka, Haryana and Tamil Nadu.

Main Features

- The disturbing features of the movement are:
- Spread over a large geographical area.
- Increase in potential for violence.
- Unification of PWG and MCCI.
- Plan to have a Red Corridor.
- Nexus with other extremist groups.

The Government of India has expressed concern over the spread of the Naxalite movement over a huge geographical area. The Prime Minister described the Naxalite movement as the single biggest threat to the internal security of the country. According to the Institute for Conflict Management, the movement has actually spread to 194 districts in 18 states.

The Naxals' potential for violence has increased substantially with their acquisition of sophisticated weapons and expertise in the use of Improvised Explosive Devices (IEDs). They are said to be in possession of at least 6,500 regular weapons including AK-47 rifles and Self-Loading Rifles (SLRs). They have built this arsenal essentially by looting weapons from police/landlords, purchasing them from smugglers, acquiring from insurgent groups like the National Socialist Council of Nagaland – Isak-Muivah [NSCN (I-M)] and the United Liberation Front of Assam (ULFA) and also obtaining some weapons from Nepal.

The movement got a tremendous boost when its two major components, the PWG and the MCCI, decided to merge on March 21, 2004, though a formal announcement was made only on October 14, 2004. The unified party was called the Communist Party of India (Maoist). The merger, apart from augmenting the support base of the movement, gave it the character of a pan-India revolutionary group. The Naxals' plan to have a Compact Revolutionary Zone stretching from the Indo-Nepal border to the Dandakaranya region is likely to get a fillip with the unification of their ranks.

The Naxalite nexus with the other extremist organisations has added to the complexity of the problem. There are indications that the PWG cadres received training in the handling of weapons and IEDs from some ex-Liberation Tigers of Tamil Eelam (LTTE) cadres. They have some understanding with the NSCN (I-M) also to support each other's cause. Some batches of CPML-Party Unity appear to have received arms training under the guidance of ULFA. The Communist Party of India (Maoist) has fraternal relations with the Communist Party of Nepal (Maoist) also.

The nature of Naxal violence has undergone a qualitative change in the recent years. Small scale isolated attacks have been replaced by large scale, well organised attacks on the government apparatus. The following incidents, which amounted to challenging the authority of the state, demonstrate the growing confidence of the Naxals:

- The attempt to assassinate the Chief Minister of Andhra Pradesh, N. Chandrababu Naidu (October 1, 2003), on a road between Tirupati and Tirumala in Chittoor district while he was proceeding to attend the Brahmotsavam celebrations.

- The raid in Koraput district of Orissa (February 7, 2004), when the extremists overran several government establishments and decamped with about 200 weapons, including SLRs and carbines and about 60,000 rounds of ammunition.
- The attack on the Jehanabad District Jail in Bihar (November 13, 2005), when about 200 cadres of the united Communist Party of India (Maoist) assisted by about 800 sympathisers freed 389 prisoners who included quite a few Naxals, while 20 activists of the Ranvir Sena were taken away, of whom nine were subsequently murdered.
- The interception and capture of a train in Latehar district of Jharkhand (March 13, 2006), when 628 Dn. Passenger was stopped between Mughalsarai and Barkana stations.
- The attack on a police base camp at Rani Bodli village of Bijapur district in Chhattisgarh (March 15, 2007) in which 55 persons including 16 personnel of the Chhattisgarh Armed Force and 39 Special Police Officers (SPOs) were killed.
- The killing of the son of former Chief Minister, Babu Lal Marandi, along with 18 other persons while they were watching a cultural programme in Giridih district of Jharkhand (October 27, 2007).
- The attack on Nayagarh town in Orissa (February 15, 2008), when the Naxals overran three police stations and killed 13 policemen and two civilians and decamped with 1,100 weapons, though 400 rifles were subsequently recovered by the police.
- The attack on the combined Andhra Pradesh and Orissa police parties in Chitrakonda reservoir of Malkangiri District in Orissa (June 29, 2008), resulting in the killing/drowning of 35 security forces personnel belonging to the elite "Greyhound" force of Andhra police.

State Response

The government has a 14-point plan to deal with the problem. The salient features of the plan are:

- Deal sternly with the Naxals indulging in violence.

- Address the problem simultaneously on the political, security and development fronts in a holistic manner.
- Ensure inter-state coordination in dealing with the problem.
- Accord priority to faster socio-economic development in the Naxal affected or prone areas.
- Supplement the efforts and resources of the affected states on both security and development fronts.
- Promote local resistance groups against the Naxals.
- Use mass media to highlight the futility of Naxal violence and the loss of life and property caused by it.
- Have a proper surrender and rehabilitation policy for the Naxals.
- Affected states not to have any peace dialogue with the Naxal groups unless the latter agree to give up violence and arms.

The following administrative measures and arrangements have also been initiated at the highest level:

- Security Related Expenditure Scheme (SRE): The SRE scheme envisages reimbursing the expenditure incurred by the state on ammunition, training, upgradation of police posts, etc. At present, 76 districts in 9 states badly affected by Naxal violence are covered by this scheme.
- Strengthening of law enforcement: This includes raising India Reserve Battalions to strengthen the security apparatus at the state level and also releasing funds under the police modernisation scheme to the states to modernise their police forces in terms of weaponry, communication equipment and other infrastructure.
- Backward Districts Initiative (BDI) and Backward Regions Grant Fund (BRGF): The government has included 55 Naxal affected districts in 9 states under the BDI component of the Rashtriya Sam Vikas Yojana (RSVY). The BRGF scheme covers a total of 250 districts and is administered by the Ministry of Panchayati Raj. It should accelerate socio-economic development in these 250 districts.

- Task Force: A Task Force has been constituted in the Home Ministry to deliberate upon the steps needed to deal with Naxalism more effectively and in a coordinated manner. The members of the Task Force comprise nodal officers of the Naxal affected states and representatives of the IB, CRPF and the SSB.
- Coordination Centre: A Coordination Centre was set up in 1998 headed by the Union Home Secretary with Chief Secretaries and Directors General of Police (DGPs) of Naxal affected states as its members. It reviews and coordinates the steps taken by the states to control Naxal activities.
- Empowered Group of Ministers: At a meeting of the Chief Ministers held on September 5, 2006, it was decided to set up an Empowered Group of Ministers (EGoM) headed by the Home Minister and comprising select Union Ministers and Chief Ministers to closely monitor the spread of Naxalism and evolve effective strategies to deal with the problem.

Recent Initiatives

Rehabilitation and Resettlement Policy, 2007: The Government of India announced a new rehabilitation policy on October 11, 2007, to make the displacement of people for industrial growth a less painful experience. Land in return for land for displaced families, preference in project jobs to at least a member of each family, vocational training, scholarships for children and housing benefits, including houses to the affected families in rural and urban areas, are some of the benefits under the new policy.

Forest Rights Act, 2006: The Scheduled Tribes and Other Traditional Forest Dwellers (Recognition of Forest Rights) Act, 2006, (popularly known as the Forest Rights Act) is a significant step in recognising and vesting the forest rights of Scheduled Tribes and other traditional forest dwellers who have been residing in such forests for generations but whose rights could not be recorded. It provides a framework for recording the forest rights so vested.

National Rural Employment Guarantee Act (NREGA), 2006: The NREGA is the largest ever employment programme visualised in human history. It holds out the "prospect of transforming the livelihoods of the poorest and heralding a revolution in rural governance in India." However, as brought out by the report of Comptroller and Auditor General (CAG), there are "significant deficiencies" in implementation of the scheme. There is lack of adequate administrative and technical manpower at the block and Gram Panchayat levels. This affects the preparation of plans, scrutiny, approval, monitoring and measurement of works, and maintenance of the stipulated records at the block and Gram Panchayat levels.

People's Support – Salwa Judum

Mobilising the support of the people is absolutely essential to weaken the support base of the Naxals. The political parties, with the possible exception of the CPM, are not playing their role in this regard. The representatives of major parties have virtually abdicated their responsibility of engaging in any kind of anti-Naxal propaganda in their constituencies. In fact, most of them have sought shelter in the safety of the urban centres or the state capital

The circumstances in which Salwa Judum evolved in the state of Chhattisgarh must be clarified in this context. The Naxals were, to start with, welcomed by the Bastar tribals because they were harassed by corrupt revenue, police and forest officials and were exploited by the traders from plains areas who never gave them a fair price for their products. The Naxals appeared as the benefactors, protecting and upholding their interests. However, in due course, as the Naxals entrenched themselves in the region, they started showing insensitivity to the feelings of tribals. They interfered with the social customs and cultural practices of the local tribals. *Ghotuls* were closed. Weekly bazaars were looted. Traditional celebrations at the time of marriage were discouraged. Images of Budhadev (Lord Shiva) were damaged and the tribals were asked to worship Mao only. Village priests were driven away. All this deeply hurt the tribals. There was a strong feeling of resentment. The Naxals did not allow the tribals to pluck *tendu* leaves also. This was a regular source of income for them and every family earned Rs.10,000 to 15,000 from the trade. This was denied to them. The resultant economic hardship proved to be the proverbial last straw. Enough was enough, the tribals felt.

It was against this background that the tribals rebelled against the Naxals. Large groups of people held rallies where they expressed their vehement opposition to the aggressions of the Naxals. This was the beginning of Salwa Judum, reflecting the resentment of the tribals against the activities of Naxals interfering with their social customs, cultural practices and hurting their economic interests. Mahendra Karma, a Congress (I) leader, gave them the leadership. It was a spontaneous movement, though it is a fact that at present, the camps are being maintained and financed by the state government. There is a concerted campaign by the human rights groups that Salwa Judum should be wound up.

Peace Talks

Peace talks were held between the People's War Group and the state government of Andhra Pradesh during June-July 2002 at the initiative of the Committee of Concerned Citizens. Three rounds of talks were held but there was no agreement on the substantive issues.

Peace talks were again held from October 15 to 18, 2004, at Hyderabad. The Naxals presented an 11-point charter of demands. The most important point related to land reforms. Again, there could be no agreement.

The government's Status Paper on the Naxal problem appropriately mentions a holistic approach and lays emphasis on accelerated socio-economic development of the backward areas. However, clause 4 (v) of the Status Paper states that "there will be no peace dialogue by the affected states with the Naxal groups unless the latter agree to give up violence and arms." This is incomprehensible and is inconsistent with the government's stand vis-à-vis other militant groups in the country.

The government has been having peace talks with the Naga rebels of the NSCN (IM) faction for more than ten years even though the rebels have not only not surrendered their weapons but continue to build up their arsenal. In relation to ULFA also, the government is prepared to have a dialogue without insisting on the insurgents surrendering their weapons. In Jammu and Kashmir (J&K), the government has more than once conveyed its willingness to hold talks with any group which is prepared to come to the negotiating table.

Why a different approach to the Naxals? The relevant clause in the Status Paper deserves a second look. The government must keep the option of talks open without giving any impression of weakness or conceding any unreasonable demands.

Prospects

The government plans to combat the Naxal problem appear generally sound on paper. However, there is a huge gap between the formulation of policies and their implementation at the ground level. Corruption, rampant at all levels, is upsetting all calculations. As conceded by the Expert Group formed by the Planning Commission to study "Development Challenges in Extremist Affected Areas," there is a "failure of governance which has multiple dimensions". The poverty level is depressing. Land reforms, unfortunately, remain "a romantic theme for the intellectual, a populist slogan for the politician, and a persistent source of hope for the landless." Unemployment figures continue to rise. Tribals are an alienated lot despite all the laws to protect their interests. There are huge disparities in different segments. As stated by Shankar Acharya, a former Chief Economic Adviser to the Government of India:

> Perhaps most worrying for the long run are the increasing dualisms between rich versus poor, organised versus informal sectors, urban versus rural, employed versus jobless, fast growing versus backward regions and skilled versus unskilled labour. The party may still be swinging ….

but it is not for everybody. How long before the widening economic and social disparities fuel serious social strife, which shreds the remnants of reasonable governance and impedes the overall momentum of development?

The factors which gave rise to Naxalism – the extent of poverty, uneven development, poor governance, neglect of land reforms, rising unemployment and tribals getting a raw deal — are, unfortunately, very much present today also. Unless these basic issues are sincerely addressed, a security-centric approach by itself would not be enough to deal with the problem.

Salience of Information in Comprehensive Security

□ **Wajahat Habibullah**

Mahatma Gandhi, the father of our nation, spoke of the Swaraj for which he fought in the following terms:

> The real Swaraj will come not by the acquisition of authority by a few but by the acquisition of capacity by all to resist authority when abused.

Evolution of Right to Information and Democracy

The Right to Information Act, 2005, in its statement of objectives recognised that "democracy requires an informed citizenry and transparency of information, which are vital to its functioning and also to contain corruption and to hold Governments and their instrumentalities accountable to the governed."

Freedom of access to information is increasingly regarded as the signature theme of democracy. Its evolution as such may be dated from 1766. In that year, Sweden, of which today's Finland was then part, included the freedom of information in Sweden's Constitution. But it was only nearly two centuries later that the concept began to take hold in what were then the Western democracies. In 1951, Finland enacted a law on the Public Character of Official Documents. The USA enacted its Freedom of Information Act in 1966, which, by an amendment of 1974, placed the onus of justifying restriction of access clearly upon government. This law places time limits for responding to requests and provides for access to all non-secret information disclosable through a principle of 'severability', also adopted in our law, by which even otherwise exempt information can be 'severed' to require disclosure of that part that is not so exempt. Disciplinary action – though not financial penalty – is mandated against officials for wrongful non-disclosure.

Denmark and Norway followed suit in the 1970s. The UK enacted a Freedom of Information Act in 2000, which became enforceable in 2005, allowing its government five years to suitably equip itself, and Mexico followed with a Public Government Information Law in 2002.

In South Asia, Pakistan, then under military rule but with democratic pretensions, was the first to enact a Freedom of Information Ordinance in 2002. Nepal followed our initiative in calling theirs a Right to Information (RTI) Act adopted by a democratic government in 2007. Indonesia followed with a Freedom of Information Act in 2008, with Bangladesh being the most recent entrant with an RTI Ordinance of October 2008, drawing extensively upon our Act, as that country's government paved the way towards restoration of democratic rule.

Kofi Annan former UN General Secretary has succinctly described the power of information:

> The great democratizing power of information has given us all the chance to effect change and alleviate poverty in ways we cannot even imagine today. Our task, your task … is to make that change real for those in need, wherever they may be. With information on our side, with knowledge of a potential for all, the path to poverty can be reversed.

India's Information Revolution

In our own country, we have over the years moved towards an information revolution. India's Constitution, in its Fundamental Rights, carries Article 19(1) (a), the Freedom of Expression, which the Courts have held to include the right to information, thus, accounting for the naming of India's legislation as a 'right' and not merely 'freedom' which had been the term used in relation to this legislation in nations across the world hitherto.

In the 1970s, under the government's declared policy of *garibi hatao*, ambitious poverty alleviation programmes were launched across the country. But by the early 1980s, it had started to become clear that the returns were not keeping pace and were by no means commensurate with the investment made. Almost in tandem, unnoticed by many in its early years, a revolution in information technology had begun to gather pace by the late 1980s. This was accompanied by a withdrawal of government monopoly over information and broadcasting in the 1990s. These factors opened the ground to the initiatives of civil society, most notably by the Mazdoor Kissan Shakti Sangathan (MKSS) in Rajasthan led by the Garboesque Aruna Roy, a Tamil and former civil servant, who threw up the relative comforts of service in government to give herself wholly to serving the peasantry. With the opening of the free media came the Freedom of Information Act piloted through Parliament by Arun Jaitley of the National Democratic Alliance (NDA) government in 2002, but never enforced. The United People's Alliance (UPA) government, in its very infancy, then revised the law and today we have the Right to Information Act, 2005.

When presenting the Bill for the Right to Information in Parliament on May 11, 2005, the Prime Minister of India, Dr. Manmohan Singh said, "I believe that the passage of this Bill will see the dawn of a new era in our processes of government, an era of performance and efficiency, an era which will ensure that benefits of growth flow to all sections of our people, an era which will eliminate the scourge of corruption, an era which will bring the common man's concern to the heart of all processes of governance, an era which will truly fulfill the hopes of the founding fathers of our Republic."

But if, as I have said, India's Constitution provides for a right to freedom, how can I describe it as a part of Article 19(1) (a)? The Supreme Court has, in repeated judgments, so held, the most significant in terms of its consequences being in the State of UP vs. Raj Singh – 1975, where Mathew J. on behalf of the Bench held as follows:

> In a government of responsibility like ours, where all agents of the public must be responsible for their conduct, there can be but few secrets. The people of this country have a right to know every public act, every thing that is done in a public way, by their public functionaries. To cover with the veil of secrecy the common routine business, is not in the interest of the public.

What then is information and how is it defined in the Act?

> Information means any material in any form, including records, documents, memos, e-mails, opinions advices, press releases, circulars, orders, logbooks, contracts, reports, papers, samples, models, data material held in any electronic form and information relating to any private body which can be accessed by a public authority under any other law for the time being in force.[1]

The key concepts are, therefore, transparency and accountability in the working of every public authority, the right of any citizen of India to request access to information and the corresponding duty of the government to meet the request, except the information exempted under Section 8 and Departments excluded from coverage under Section 24, listed in the Second Schedule. It is the duty of the government to proactively make available key information to all[2]. But this Act is not the responsibility of government alone. It brings a heavy responsibility to bear upon all sections of civil society, notably the citizenry, Non-Governmental Organisations (NGOs), and the media.

Liberty vs Security?

So we can conclude that the freedom of which our forbears dreamt has now come closer to fruition. But those concerned with national security might well ask: what is the bearing this has on national security? We need, therefore, to recognise clearly what we mean by 'comprehensive' security. Do we mean the security of our military installations, of our economic infrastructure or of our physical structures? All these are without doubt essential to what we refer and different constituents thereof. But in the ultimate analysis, it can hardly be denied that national security is synonymous with the security of the people of our country, which all these institutions serve. And if that is conceded, it needs no argument to state that if the people are the objective in comprehensively securing the nation, it is the people who must share responsibility for so ensuring it.

In the course of presentations at the Chief Ministers' Conference on April 15, 2005, it became increasingly clear that local self-government required to be developed as the primary instrument to counter a disturbingly growing welter of terrorism nurtured by discontent; 157 districts were identified by the National Security Adviser as Naxalite entrenched. According to the Intelligence Bureau (IB), Jharkhand, which had been with impunity circumventing the laws on *Panchayati Raj*, accounted for a whopping 25 percent of the incidents of terrorist violence in the country. *Jan Adalats* in Jharkhand and Chattisgarh and *Praja Adalats* in Andhra Pradesh were actually dispensing law and awarding death sentences. All this is but evidence of a growing gulf between government and people.

The Right to Information Act fundamentally restructures the debate on governance from what should be revealed to what must be kept secret and undoubtedly reflects the potency of India's vibrant democracy, a concept never bettered as a means to involve people in their own governance. It thus seeks to supplant the colonial Official Secrets Act of 1923 that had been hitherto the touchstone of confidentiality in government. The Central Information Commission can, under Section 19(8) (a) (iii), require every public authority to "publish certain information or categories of information" under the Act. Should the public authority not comply, Section 19(8) (c) gives the Commission the power to impose any of the penalties provided under this Act.

Primary among the obligations the Act brings to every public authority is laid down in Section 4, sub-section (1)(a) which states, "Every public authority shall – (a) maintain all its records duly catalogued and indexed in a manner and the form which facilitates the right to information under this Act and ensure that all records that are appropriate to be computerised are, within a reasonable time and subject to availability of resources, computerised and connected through a network all over the country on different systems so that access to such records is facilitated." What then is a public authority?

Public Authorities

A public authority means any authority or institution of self-government established or constituted:

- By or under the Constitution.
- By any other law made by Parliament.
- By any other law made by a State Legislature.
- By notification issued or order made by the appropriate government and includes any:
  - body owned, controlled or substantially financed;
  - non-government organisation substantially financed directly or indirectly by funds provided by the appropriate government.

From such authorities, what can be accessed is what is defined as the "right to information," which means that every public authority shall provide access to any citizen of India, which includes the right to:

- Inspect works, documents, records.
- Take notes, extracts or certified copies of documents or records.
- Take certified samples of material
- Obtain information in the form of printouts, diskettes, floppies, tapes, video cassettes or in any other electronic mode or through printouts.[3]

But to the sceptic, the full freedom provided by this Act might be seen as giving licence to malcontents to use it to promote anarchy to supplant orderly government. It is my contention in this essay that this Act will in the long run become a major tool in strengthening security and, therefore, that the very reverse will be the inevitable result of the implementation of this Act.

It is, of course, necessary that certain information held in trust by a public authority not be placed in the public domain to ensure that the government continues to provide security and balanced development, which, after all is what any public will seek from those it has empowered by the process of election in a democracy. Therefore, Section 8 of the Act holds that the following are exempt from disclosure:

- Information, disclosure of which would prejudicially affect the sovereignty and integrity of India, the security, strategic, scientific or economic interest of the state, relations with a foreign state or lead to incitement of an offence.
- Information that has been expressly forbidden to be published by any court of law or tribunal or the disclosure of which may constitute contempt of court.
- Information, the disclosure of which would cause a breach of privilege of the Parliament or State Legislature.
- Information including commercial confidence, trade secrets or intellectual property, the disclosure of which would harm the competitive position of a third party, unless the competent authority is satisfied that larger public interest warrants the disclosure of such information.
- Information available to a person in his fiduciary relationship unless the competent authority is satisfied that the larger public interest warrants the disclosure of such information
- Information received in confidence from a foreign government.
- Information, which would impede the process of investigation or apprehension or prosecution of offenders.
- Cabinet papers, including records of deliberations of the Council of Ministers, Secretaries and other officers.

- Information, which relates to personal information the disclosure of which has no relationship to any public activity or interest, or which, would cause unwarranted invasion of the privacy of the individual.
- Notwithstanding any of these exemptions or the Officials Secrets Act, 1923, a public authority may allow access to information, if public interest in disclosure outweighs the harm to the protected interests – Section 8(2).
- Infringe copyright, except of the state (Section 9).
- Where practicable, part of the record can be released under Section 10.
- Intelligence and security agencies exempt under Section 24, except in cases where corruption or human rights violation is alleged.
- Third party information to be released after giving notice to the third party.

Most exempt information is, however, expected to be released after 20 years, with some exceptions, particularly those weighing on national security. Besides, under the RTI Act, 2005, notwithstanding anything in the Official Secrets Act, nor any of the exemptions described above, a public authority may allow access to information, if public interest in disclosure outweighs the harm to the protected interests, thus, putting the public's interest foremost. The RTI Act then takes precedence over the Official Secrets Act, an offspring of colonial India, which treated the public as an adversary of the government.

But other than this, withholding any information knowingly or providing access while violating the time limit mandated under the Act, will bring with it penalties imposable by the Information Commission on Public Information Officers (PIO) or officers asked to assist the PIO, who have failed so to do. For unreasonable delay, this amounts to Rs. 250 per day up to Rs. 25,000; for illegitimate refusal to accept an application, malafide denial, knowingly providing false information, destruction of information, etc Rs. 25,000 and recommendations for departmental action for persistent or serious violations. There is also compensation for damages[4] – Section 19(9) (b). There is, however, no criminal liability, and immunity from legal action for action taken in good faith.[5]

Who are the Participants?

This Act provides universal access to the poor by ensuring that the fee remains at a reasonable level although in the Act itself the quantum is not specified. There is at any rate no fee for those below the poverty line. It provides for Assistant Public Information Officers at sub-district levels to facilitate filing of applications and appeals. Post offices have been so designated. There is a provision to reduce oral requests into writing together with a mandate to officials to provide all required assistance, including to disabled persons. Besides, information is to be provided in local languages.

Given the presumption that access to information is today a matter of right, there is no need to specify a reason for seeking information or personal details other than necessary for correspondence, but availing of this right is open only to citizens of India.[6]

What then are the responsibilities of a public authority? These are many.

- Appointing PIOs/Assistant PIOs within 100 days of enactment.[7]
- Maintaining, cataloguing, indexing, computerising and networking records.[8]
- Publishing within 120 days of enactment a whole set of information and updating it every year.[9]
- Publishing all relevant facts while formulating important policies or announcing the decisions which affect the public.[10]
- Providing reasons for its administrative or quasi-judicial decisions to the affected persons.[11]
- Providing information *suo moto*, including the internet.[12]
- Providing information to the Information Commission.[13]
- Raising awareness, educating and training.[14]

- Compiling in 18 months and updating regularly a local language guide to information.[15]
- Developing and organising educational programmes to advance the understanding of the public, particularly the disadvantaged, to exercise the right to information.

The government is also expected to encourage public authorities to participate in programmes, promote timely effective dissemination of accurate information on activities, train CPIOs and produce relevant training materials, user guides and related matter. The latter is being executed by the Centre for Good Governance, Hyderabad.

Does Democracy Negate Security?

With this, it should have become clear that what the Act aims at is the flowering of democracy in India. But does this flowering mean that security would be weakened? Does it imply that India must be a 'soft state'?

There have been observations primarily in the West, even by leading intellectuals, that democracy is, in fact, not compatible with security, that to cater to vested interests, with money to spend, elected representatives will inevitably bend, even yielding national interest to such pressure. In a closely argued essay "Us and Them" in the leading international journal *Foreign Affairs*[16], Jerry Z Muller, Professor of History at the Catholic University of America has argued that multi-ethnic states cannot become nations. "In short," Muller argues, "ethno-nationalism has played a more profound and lasting role in modern history than is commonly understood, and the processes that led to the dominance of the ethno-national state and the separation of ethnic groups in Europe are likely to recur elsewhere. Increased urbanization, literacy, and political mobilization; differences in the fertility rates and economic performance of various ethnic groups; and immigration will challenge the internal structure of states as well as their borders. Whether politically correct or not, ethno-nationalism will continue to shape the world in the twenty-first century."

Is India's ethno-nationalism such a challenge? Yes, indeed. Yet, any close observer of India's political evolution through the past century will have noted that despite its bewildering diversity, the practice of democracy has only strengthened India's nationhood, while neighbouring countries, with a tradition identical to ours, that had opted for authoritarian military rule and military alliances to bring political stability and hasten economic progress have, in fact, succumbed to despotism, even disintegrated. India's diversity has been its strength. It can only be a weakness if it is perceived as an instrument of dominance of one group over another, even if the dominant group is the majority, leading to what is termed as "majoritarianism." Muller's gloomy views stem from what he sees in Europe's history. On this basis, he warns the US, "A familiar and influential narrative of twentieth-century European history argues that nationalism twice led to war, in 1914 and then again in 1939. Thereafter, the story goes, Europeans concluded that nationalism was a danger and gradually abandoned it. In the postwar decades, western Europeans enmeshed themselves in a web of trans-national institutions, culminating in the European Union (EU). After the fall of the Soviet empire, that trans-national framework spread eastward to encompass most of the continent. Europeans entered a post-national era, which was not only a good thing in itself but also a model for other regions. Nationalism, in this view, had been a tragic detour on the road to a peaceful liberal democratic order." But he contradicts this rosy perception by going on to lament, "Far from having been superannuated in 1945, in many respects ethno-nationalism was at its apogee in the years immediately after World War II. European stability during the Cold War era was, in fact, due partly to the widespread fulfillment of the ethno-nationalist project. And since the end of the Cold War, ethno-nationalism has continued to reshape European borders."

But in India, there is no real majority. We are, in fact, all minorities in one way or another. In contrast to India, Europe, about the same size but in many ways less diverse, the harbinger of the concept of "nation-state," quested hopelessly for unity through war and conquest, notably the French expansion through Napoleon, Bismarck's vision of Europe under German hegemony, and Hitler's Third Reich only to find what Muller finds a precarious unity through the European Union as the 20th century, thanks to that very Europe's conflicting interests, the most violent in India's history, drew to a close. No wonder that in concluding his arguments, Prof Muller considers that "Partition may, thus, be the most humane lasting solution"! At the risk of sounding egotistic, I might say that Europe might do well to learn from us, who have borne for years the trauma of a misconceived Partition brought onto us by a European power, rather than we learn from the West, as has unfortunately been our wont

Although there will be disgruntled elements in any society, there will even be incendiaries and extremists who would love to undermine the country— but this is precisely what the RTI Act is designed to overcome, by giving each citizen a sense that he is a participant in governance through being able to hold not only the political leadership, but every section of the government from the lowest to the highest, accountable to him. Surely then, security becomes the concern not of a few, but of all.

Access to Information and Strengthened Security

First and foremost to my mind is that in a democracy, the citizen has the right to know not only the kind of person that he is electing to represent him, but also what that person has done to fulfill the mandate that he holds for the elector. Until recently, the activities of those within Parliament were shrouded in a jealously guarded secrecy. We all know of the considerable debate that accompanied the decision that all Parliamentary debate be opened to the public. The RTI Act goes further and in consequence, a citizen has the right to know every detail of his representative's comportment in Parliament. Today, you have analyses published in newspapers on the close of a Parliament session, such as on December 23, 2008, of time spent and time wasted during the session.

A major concern of all responsible Indians has been the reliability of representatives with questionable backgrounds. In a democracy, such people, if they command the support of their constituents, cannot be excluded from participating in the democratic process. But the Act now ensures that they will function under the sharp eye of the fiercest public scrutiny. No element of our political structure can now escape accountability.

We know that insidious elements thrive on secrecy. They insinuate themselves into the innards of security systems by the use of blackmail, threat and enticement. Free access to information by upright citizens will discourage such characters for they can, thus, be clearly exposed. Their activities can be kept confidential inasmuch as the security agencies may need to keep them so in order to bring them to account. Clearly, the exemptions awarded by the Act cover this. But through transparency in function, it will be impossible for public authorities to conceal realities that are better shared not only from the public but also from their own colleagues in government, a means that had been the fount both of corruption and political intrigue. And the public can now become participants in governance and, therefore, it will be in their interest, as in the interest of any public authority, to ensure that such individuals or elements of society are kept firmly under control.

Good Governance

The India Infrastructure Report, 1996, also known as the Rakesh Mohan Report, flagged the importance of infrastructure for India's policy-makers, when it stated that "Availability of adequate infrastructure facilities is vital for the acceleration of economic development of the country."

It is understood that good governance is the means adopted to deliver services to the government's clientele, in a manner acceptable to the clientele as efficient and to the provider, which is the government, as cost-effective. There is a general consensus that good governance must be participatory, transparent and accountable[17]. The present system in India, however, thanks to perceptions enshrined in the Official Secrets Act, remains firmly grounded on mistrust. If governance is to be participatory, who are the participants? It is our view that in the current political and economic environment, participants in governance include the political leadership, the bureaucracy, business, the media, financial institutions and, decidedly, the security apparatus.

The reason for this mistrust can be found in the legacy of governance in India; stemming directly at the district level from the Mughals,[18] adapted and extended with an archaic Secretariat system of the colonials. An elitist structure informed both systems and continues to subsist. The welfare state, strongly influenced by the war-time licensing legacy for distributing shortages, introduced India to independence! The economy, therefore, remained rooted in the concept of shortage.

Does this mean that the current structure is not amenable to present demands? If so, what is to replace the current structure? Assuming that it needs to be replaced, the most important step must be to develop a consensus on the objectives to be met. This must go beyond mere rhetoric like that frequently employed by intellectuals who evade basic issues other than asserting that the bureaucracy thrives by substitution of 'good ideas' with 'bad ideas'! To begin with, each participant in governance must be aware of what is expected of each. But participation in governance is too often seen as a struggle for sharing power. For example, there is the often perceived conflict of generalist vs. specialist. This brings us back to the basic proposition that governance must be distinct from the exercise of power and must comprise security of life and property, both of which are predicated on security of the nation. If this is understood, it is easier to see why perceived needs are not being met by the system even though widely understood.

Mumbai

The response to the attack on the city of Mumbai on November 26, 2008 by elements from the neighbouring country brings home this issue. The initial reaction of the security apparatus has been criticised for many reasons, several of them sound. But in a situation such as this, if the public were in full knowledge of our nation's capacities and those of every agency that participated in finally overcoming this assault and where to turn in such a situation, if it were indeed part of the security apparatus, the unwholesome exposure by the Press seeking to sensationalise issues, leading too often to unwittingly compromising security, would not have arisen.

The question does arise that given the careful planning and financing which went into the operation by the terrorists, even had the members of the public been alerted by greater access to information, would this evil onset have been foiled, or if not foiled altogether? Because of the professional nature of the operation mounted by the terrorists, once the outbreak occurred, could remedial means have not been taken more quickly, with every citizen becoming a security officer in terms of knowing where to report and how to react, thus, saving lives?

These are, of course, questions that are, and may remain, unanswered. But the right to information allows for the public authorities and the public to work together so as to make the unity of the country more apparent, deterring terrorist adventurers and bringing home to such elements wherever they may be that Indian civil society stands together and that such efforts at derailing India's unity and disrupting its economic progress are neither worth the effort nor the cost to the terrorist organisations themselves. And a cataclysm such as that which befell Mumbai would, instead of degenerating all too frequently into a blame game as it sometimes has, will have led to greater introspection, with predictably fruitful results

A good example is the US which although having effectively had freedom of information in its system since at least 1974, has evolved a government much more transparent then ours and has, after 9/11, been in a position to control such elements within that country despite a reckless foreign adventure that has cost them so dearly. This is in no small measure a result of the sense of responsibility displayed by its citizenry at large, with every institution seeking as a duty itself to do its part.

For security to be comprehensive, it is necessary that readiness to defend and react in a time of attack be a matter shared by those charged with ensuring security with, if not by every citizen then surely by most. The Right to Information Act makes that possibility real. Small wonder that police reform to transform this institution from an overbearing colonial hangover to an instrument for making citizenry secure is now an issue of high priority, with the best minds applying themselves not only from within the government but from among all sections of the public at large.

Notes

1. Section 2(f).
2. Section 4.
3. Section 2(j).
4. Sec. 19(9) (b).
5. Section 21.
6. Section 3.
7. Section 5(1).
8. Section 4(1)(a).
9. Section 4(1)(b)}.
10. Section 4(1)(c).
11. Section 4(1)(d).
12. Section 4(2).
13. Section 25(2.)
14. Section 25(1).
15. Section 26(2) & (3).
16. *Foreign Affairs*, March/April 2008, Council on Foreign Relations, NY.
17. Sebastian Morris: *The Challenge to Governance in India,* India Infrastructure Report 2002 (New Delhi: Oxford University Press 2000), ch.2, p.19.
18. The position of Collector, as the name implies was instituted by Raja Todar Mal, head of the Mughal imperial finance in the 16th century to collect land revenue, the mainstay of the Empire, under the name *Amal Guzar.*

India's Hydrocarbon Challenge

□ Shebonti Ray Dadwal

Introduction

A recent report by the Research and Information System (RIS) for developing countries projects that the Indian economy will grow by 8.5-9 per cent in 2009-10. Concomitant with a strong and growing economy, rising incomes and a vibrant market, there has been a huge surge in demand for energy, for both the power and transport sectors. In fact, the latter is the fastest growing energy-consuming sector in India, and according to the National Highways Authority of India (NHAI) statistics, the number of vehicles on Indian roads has been growing at an average pace of 10.16 per cent per annum over the last five years.[1]

According to the Integrated Energy Policy, India's energy security is premised on accessing adequate energy "in a sustainable manner and at reasonable cost" if it has to meet its goal of achieving a growth rate of 8-10 per cent. But though India has large energy resources – coal, as well as renewable energy such as wind, solar, hydro and biomass — it is deficient in hydrocarbons, that is, oil and gas resources. As a result, India has seen its level of crude oil imports as well as gas, albeit to a lesser extent, increasing incrementally, to the extent that while it currently imports around 75 per cent of its demand, this is projected to increase to over 90 per cent over the next decade.

Moreover, as the situation in the power sector continues to look grim due to inadequate generation and frequent outages, more and more oil is being diverted to the sector for use in diesel generators, as well as in the agriculture sector for irrigation purposes. Therefore, given the expected and projected rise in demand, meeting India's demand for energy is one of the most challenging tasks for the policy-makers. Nevertheless, while there are a number of options for fuelling the power sector through the use of conventional or non-renewable resources such as coal, oil and natural gas, as well as the renewable energy such as hydropower, solar, wind, nuclear, biomass, etc, the non-power sector, and particularly the transport sector, is almost entirely dependent on oil and increasingly Compressed National Gas (CNG), as no suitable substitutes have been found which can replace these as transport fuel. Hence, access to affordable and reliable supplies of hydrocarbons is crucial for India.

The Supply Problem

As India's need for oil grows incrementally, it has to perforce look increasingly to the global market for its supplies. However, the situation in the international oil market gives cause for worry – both due to the increase in the price of oil as well to decreasing reserves. In fact, part of the reason for the hike in prices is because of concerns that the world may soon be running out of conventional or 'easy' oil due to inadequate investments being made by the producing countries and oil companies as well as depleting reserves in older fields.

For countries like India, whose demand for all kinds of energy is growing, ensuring adequate supplies is of prime importance.[2] Today, India's demand for energy has seen it become the fifth largest consumer of energy, accounting for nearly 3.5 per cent of the world's energy consumption. However, it is the country's growing consumption of oil that is deemed as most critical. And the fact that it does not have sufficient indigenous resources of hydrocarbons, particularly oil, makes its oil supply policy that much more critical. India's crude oil reserves of around 5.6 billion barrels or 739 million tonnes can only sustain current levels of production for 22 years.

Till the mid-1980s, India imported only 30 per cent of its oil requirement. But with production stagnating at around 34 million tonnes (mt), the last few years have seen its oil import dependence go up to 75 per cent.[3]

The situation is slightly better in the gas sector. India's natural gas production is expected to more than double to 170 million standard cubic metres per day (mmscmd) by 2011-12 once output from Reliance Industries Ltd's (RIL's) eastern offshore KG-D6 reaches a peak. The Oil and Natural Gas Corporation (ONGC) will produce 47.06 mmscmd of gas in 2008-09, almost unchanged from 47.19 mmscmd from the last fiscal, and will rise to 51.65 mmscmd by 2011-12, while Oil India Ltd's (OIL's) production will contribute 10 mmscmd.

However, despite the optimism in the domestic gas sector, given the expected rise in demand for gas, India's Liquefied Natural Gas (LNG) imports are also slated to more than double to 23.25 mt by 2011-12 from 9 mt in 2007-08 once the Dabhol, Kochi and Mangalore terminals become operational. Petronet LNG's Dahej terminal will see capacity doubling to almost 12 mt and Shell's Hazira terminal is expected to operate at 2.5 mt; together with 81.38 mmscmd of LNG, the country's total gas availability is expected to touch 252.09 mmscmd in 2011-12 from the current 110.9 mmscmd.[4]

Therefore, as the demand for oil and gas increases, the gap between demand and domestic supply has been widening. To address this problem, India has embarked on a two-pronged strategy: one, to increase exploration and development of domestic resources – oil, natural gas as well as coal; and two, ensuring supplies from external sources.

Domestic Strategy

Exploration and Production

To accelerate the pace of hydrocarbon exploration in the country, the government launched the New Exploration Licensing Policy or NELP in January 1999, under which a total of 48 blocks (10 onshore, 26 shallow water offshore and 12 deep water offshore blocks were put on offer. Seven months later, production-sharing contracts were concluded and signed for 22 blocks. This was also the first time that deep-water acreages were offered for competitive bidding to medium and small private companies, both Indian and foreign.[5]

Since then, six more rounds under NELP have been launched, in which 162 exploration blocks have been awarded, the latest being NELP VII which was launched in December 2007, in which some 96 companies including 21 foreign companies, have submitted 181 bids for 45 exploration blocks. Apart from the two national oil companies, viz., ONGC and OIL, more than two dozen foreign companies are currently operating in the upstream sector, including International Oil Companies (IOCs) like BG, BP, ENI, Petrobras, Santos, Cairn and NIKO. Part of the reason for the renewed interest in Indian blocks is the recent rise in oil prices as well as the substantial discoveries of oil and gas off the country's eastern coat. The recent discoveries by RIL, expected to yield around 550,000 barrels of oil and oil equivalent over the next 18 months, that is, equivalent to 40 per cent of India's indigenous production of hydrocarbons, would allow India to make a saving of at least $20 billion. India's oil import bill for 2008-09 is estimated at close to $77 billion as against $67 billion in the previous year.[6] Moreover, India expects its crude oil production to increase by over 30 per cent in the next five years and its natural gas production to double from the present level of 90 mmscmd. Due to NELP, the area under exploration has increased to 44 per cent of the sedimentary basin area from 11 per cent prior to the implementation of NELP.[7]

As the response to NELPs I to VI was somewhat disappointing, NELP VII offered unprecedented concessions to attract the IOCs, hoping to attract the big ones. These included the inclusion of a consortium clause in the 57 blocks that were offered, plus an 10 extra points to foreign oil companies forming a consortium with domestic oil companies, 100 per cent Foreign Direct Investment (FDI) for Exploration and Production (E&P), freedom to market oil and gas internationally at international crude oil prices and free market-determined prices in India.

However, according to reports, the response to NELP VII too, albeit an improvement on NELP VI, has not met the expectations, with only three IOCs participating in the bidding, viz., Cairn Energy, British Petroleum and BHP Billiton. Part of the reason is the somewhat marginal role that was offered to the IOCs, with the more promising blocks being given to Indian companies, ONGC Ltd. and Reliance Industries Ltd. Moreover, there were complaints that very little data were made available on India's oil reserves, and given the high risk factor in E&P projects, many of the foreign companies were reluctant to commit to the risks. However, a major reason was the uncertainty surrounding the sudden announcement of a "retrograde" tax policy in the annual budget of 2008, a day before NELP VII bidding took place, whereby the government withdrew the income tax holiday on gas production. Given that other countries like Malaysia and Libya also opened their bidding around the same time, many of the prospective bidders preferred to place their risks in those countries.[8]

Besides the emphasis on enhancing domestic production of oil and gas, the government has also approved a comprehensive Coal Bed Methane (CBM) policy in July 1997 for exploration and production of the same. Some 23 CBM blocks have been awarded under three rounds of CBM policy. Commercial production commenced in July 2007 and so far, around 6 trillion cubic feet (tcf) of CBM reserves have been established in 4 CBM blocks.[9]

The Import Strategy

Despite several initiatives such as increase in domestic production of oil and natural gas as well as encouraging the production of biodiesel, given the growth in demand, India's dependence on the import of hydrocarbons is slated to rise over the next few years.

In a world where issue of reliability of supplies is now a major area of concern after reports that many of the oil exporting countries are reaching peak production, the issue of access to supplies is crucial for India. Traditionally, India has been dependent on the West Asian countries for its oil, and more recently gas (LNG) supplies. Since the 1980s, the country's oil imports have been increasing, so much so that today, India is the fifth largest oil consumer in the world after the US, China and Japan, and the third largest in Asia. Currently, out of India's oil imports, almost three-fourths (around 81 mt) is sourced from the Persian Gulf region. In 2007-08, India imported 89.73 mt, which is around 11 per cent more than its 2006-07 imports. Amongst its main oil sources, Saudi Arabia is India's largest supplier, followed by Iran. Some of the other West Asian states which supply India are Kuwait and the United Arab Emirates (UAE), although the former's oil supplies were less from the previous year's, while the UAE's went up considerably. Other than the West Asian states, India has been looking increasingly at Africa for oil supplies, as part of its source diversification strategy. Nigeria was the biggest exporter at 9.9 1 mt, followed by Angola, whose supplies have almost doubled in 2007-08 to 4.33 mt.[10]

In the gas sector, Petronet LNG Ltd has been scouting for more LNG deals, including from Exxon Mobil's entire stake of 3.75 million tonnes per annum (mtpa) from Australia's Gorgon project for the next 25 years for its upcoming terminal at Kochi or the existing one at Dahej. It currently gets 5 mtpa from RasGas of Qatar under a long-term LNG deal which will increase to 7.5 mtpa from 2009. The company is also in talks with Algeria, Oman, Equatorial New Guinea and Qatar for long-term contracts as it is increasing its capacity at the Dahej terminal to 10 mt (from the existing 6.5 mt) by December 2008.[11]

However, India, like other major oil consuming developing countries, in its attempt to strengthen its supply security in the face of growing uncertainties in the international oil market, is looking to increase its foreign oil and gas holdings, and to this end, the government has been encouraging the National Oil Companies (NOCs), as well as the private companies to aggressively acquire overseas oil and gas assets, and, more recently, coal assets. While

ONGC Videsh Ltd (OVL) is spearheading this strategy among the Public Sector Undertakings (PSUs), with an investment commitment of over $7 billion and an oil and gas production of 8.92 mt in the year 2007-08, private companies such as Reliance Industries Ltd, Essar and Videocon are also investing in overseas energy assets. Today, OVL has around 29 projects in 15 countries, including Brazil, Cuba, Egypt, Qatar, Myanmar, Nigeria, Iraq, Iran, Venezuela, Russia, Sudan, Colombia, Vietnam and Syria, of which the last five are producing fields while the rest are in an exploratory stage. Private oil and gas companies are not far behind. Reliance Industries Ltd has assets in Yemen, Oman, Colombia, East Timor and Australia, and may partner with OVL in Venezuela, while Essar Exploration and Production Ltd (EEPL) has assets in Vietnam, Madagascar and Nigeria.[12]

According to ONGC Chairman R.S. Sharma, "India is prognosticated to have about 0.5% of global hydrocarbon reserves (with 17 per cent of the global population). Under this scenario, overseas asset acquisition is definitely helpful."[13] Mr Sharma also believes that as ONGC has expertise in exploration and production, and the company is considering partners-in-progress in many foreign countries, it makes business sense to take up equity oil participation in those countries. Nevertheless, as political and diplomatic relations also play a vital role in foreign countries, tie-ups with global firms like the Mittals and Hindujas could further facilitate this strategy. For instance, in regions like Central Asia and Africa, an association with the Mittal group could prove beneficial. Similarly, the Hindujas have excellent networking in some oil rich Gulf countries.

However, there has been some criticism of India's asset acquisition strategy. According to Atul Chandra, former head of OVL, although acquiring overseas assets helps lock the ceiling prices for future supplies, it depends on the long-term assumption of oil prices being built into the acquisition cost. For instance, it is important that the assumed oil price for acquisition is reasonable and is lower than the market price on a long-term basis. Hence, in an environment when the international price of oil is high, there are no incentives for the countries that own the oil to sell the asset unless a very high premium is paid. The danger is that the high premium would take away the entire value from the asset, leaving almost nothing for the buyer, barring some risks. Of course, in such cases, geographical contiguity plays a major role. For instance, Chinese acquisitions in Central Asia are not only valuable for that country as they can not only receive supplies directly from these countries, thereby allowing them to reduce transportation costs, they would also ensure security of supplies in the event of a conflict. However, for a company or country that does not have a geo-strategic advantage, such a high cost of acquisition may not make sense. According to Mr Chandra, the most pragmatic acquisition strategy would be to look for good exploration blocks. But this is predicated on having a corporate team capable of evaluating possible exploration acreages in the world, which the big international oil companies have. Unfortunately, many of the Indian NOCs do not have a comprehensive set-up or data base to follow this strategy, with the result that India ends up acquiring blocks that are in the "very high risk" category. Neither are the exploration blocks acquired in the bidding round examined in depth due to shortage of time available as well as inadequate research of the area.[14]

To overcome some of these problems and to further enhance India's overall energy security, the government has set up a dedicated energy security unit within the Ministry of External Affairs in 2007 "to support India's international engagement through appropriate and sustained diplomatic interventions." The unit's charter also states that it should "support the efforts of the corporate entities, both in public and private sectors, in acquiring energy assets overseas, in transfer of new and emerging technologies to India and in building strategic partnerships with foreign companies."[15] There are plans to further set up a multi-billion dollar sovereign wealth fund to invest in energy assets overseas.[16]

Biofuels

While India has been pursuing a policy of securing overseas oil and gas fields, it is also actively pushing biomass-based fuels which can be used as a potential substitute for imported oil, and reduce the country's dependence on the same. Moreover, biofuels are also more environment friendly and can mitigate some of the emissions build-up due to the increased use of oil.

In September 2008, the government approved the biofuel policy, which states that by 2017, transport fuels in India should contain a blend of 20 per cent biofuels (a 25 per cent blend of ethanol in petrol is being used in Brazil for the last 25 years without any problem).[17]

The policy envisages two main types of biofuels – ethanol derived from plant wastes, mainly sugarcane molasses, which will be blended with petrol, and biodiesel, that is, oil produced from non-edible oilseed crops such as *jatropha curcas*, which can be blended with diesel.

The policy supports increasing biodiesel plantations on community, government-owned and forest wastelands, but not on fertile, irrigated lands. According to government estimates, the demand for high speed diesel (HSD) by the end of 11th Plan (2011-12) shall be 66.9 mt, which would require 13.38 mt of biodiesel. According to the report on biofuels, around 13.4 million hectares of land could be procured for jatropha planatations in the immediate future, which has the potential to yield around 15 mt of oil each year.[18]

However, critics of the policy state that given the fact that not more than 20 per cent can be blended with oil, biofuels cannot be a substantial substitute for oil. On the other hand, diverting large tracts of land for cultivating crops for fuel will give rise to other problems, and would exacerbate the food security issue. Moreover, there are other complications that are associated with growing jatropha. For instance, as the *Indian Express* reported on September 16, 2008, the state of Chhattisgarh, which planted 400 million jatropha saplings on more than 155,000 hectares of fallow land over three years does not have any data on the survival of the saplings or seed production. Moreover, farmers in many areas are reporting that the trees have not yet borne fruit, while in places where they have, various departments and local agencies, were waiting for guidelines on collection and sale of seeds. At the same time, the state Chief Minister, Raman Singh, who had stated that his official car would be run entirely on biofuel, has now stopped using the same.

Coal to Liquids (CTL)

In 2005, the Integrated Energy Policy report had recommended that liquefaction and underground gasification of coal be declared as one of the end uses for facilitating private sector investment in this area. Accordingly, the Coal Mines (Nationalisation) Act was amended in July 2007 to allow coal gasification and coal liquefaction as valid end-uses for captive mining thereby allowing the private sector to legitimately mine coal and use it for CTL. In July 2008, the Coal Ministry opened bidding for allocating captive coal blocks for CTL projects involving some 22 companies, including the Tata Group, Reliance Industries Ltd, Reliance Infrastructure, OIL, GAIL and SAIL.

However, the Planning Commission is believed to have expressed its opposition to CTL projects on the grounds that coal would be better used for electricity generation than for making liquid fuels as well as concerns over the benefits of converting a fuel, which is the dominant source for electricity production, into a liquid primary energy resource, given that the efficiency of the conversion process to liquid fuels is only about 42 per cent. Moreover, CTL projects make commercial sense when oil prices are high and coal prices low. In addition, there are serious environmental concerns, given that CTL technologies are highly carbon-intensive, besides the fact that they consume large amounts of water. About 12-14 barrels of water are needed to produce a barrel of liquid fuel from coal, hence the water requirement for an 80,000 b/d CTL plant would be around 125,000 cubic metres of water.[19]

Energy Pipelines

Almost all of India's oil and gas imports are transported by sea and are, therefore, vulnerable to any disruption that may occur in international waters due to conflict or acts of piracy. Therefore, to reduce its dependence on sea-based imports, the government has been planning a number of pipelines to import gas from neighbouring countries. The first such pipeline project that was taken up was the Oman-India sub-sea pipeline in the mid-1990s, but this was shelved due to lack of technical expertise and concerns regarding lack of long-term supplies. The second project that was taken up was the Iran-Pakistan-India (IPI) pipeline project, which was first proposed in 1989, to bring Iranian gas to India via Pakistan. The initial capacity of the pipeline will be 22 bcm per annum, and is expected to be later raised to 55 bcm/ per annum. Initially, around 60 mmscmd of gas would be supplied in Phase-I, to be shared equally between India and Pakistan, and later, in Phase-II, it would be increased to 90 mmscmd of gas. So far, six meetings of the trilateral Joint Working Group (JWG) of the participating countries have been held, with the last meeting being held in New Delhi on June 28-29, 2007. Though the initial cost of the project was around $4 billion, it is expected to cost $7.4 billion.[20]

However, the project has yet to be implemented despite innumerable meetings with the parties involved, due to one reason or another, ranging from differences over security, pricing, transit fees and international pressure. Hopes had been revived following the visit of Iranian President Ahmadinejad to New Delhi in April 2008, but differences over pricing, both trilaterally as well as bilaterally between India and Pakistan persist.

Pakistan and India had both wanted a fixed price for the gas imported from Iran, but following a hike in international oil prices, the Iranian government now insists on changing the gas pricing formula and would now like the gas tariff to be revised periodically, as per the trend in the international market.[21]

Moreover, besides differences among the three parties, other issues have impacted on the project, particularly opposition from the US on any energy deals between Iran and other countries. As a result, while Islamabad appears to be more willing to go ahead with the project, and has even suggested including China in the project instead of India, the possibility of piped gas from Iran to India appears unlikely, though both sides continue to state their commitment to the project.

Instead of the IPI project, Washington is now encouraging India to participate in the TAPI or Turkmenistan-Afghanistan-Pakistan-India project. However, this project appears to be even more mired in controversy than the IPI project.

First, the pipeline is expected to traverse some of the most unstable regions of both Afghanistan as well as Pakistan, and the likelihood of pipeline sabotage by militant groups in both regions is highly likely. The pipeline route is expected to pass through the west and southwest of Afghanistan through Herat, Farah, Nimroz, Helmand and Kandahar before entering Pakistan. Second, while Turkmenistan claims that it has total gas reserves of 8 trillion cubic metres, Pakistan and India want the reserves to be certified before further moving ahead.[22]

Moreover, Turkmenistan is seeking $400-450 per thousand cubic metres for the gas or $12.7 per million British thermal units (mmBtu), as against $200-230 per thousand cubic metres offered by India. The rate for gas from the IPI project is around $5.56 per mmbtu.[23] Moreover, as in the case of Iran, Turkmenistan would like to link the gas price with the international market as well as periodic revisions.

With both the IPI and TAPI projects caught in differences, the government is now mulling over reviving the undersea pipeline project from Oman's Ras al Jifan for around 8 trillion cubic feet (22.8 bcm) of gas, which would entail an investment of $2.1-3.4 billion plus transportation tariff of around $ 1.1 to $ 1.8 per mmbtu. The idea of an Oman-India sub-sea pipeline was first mooted in the late 1980s, but failed to get off the ground as it was found to be technologically too difficult as well as expensive at the time, given the then low global fuel prices and the Indian demand scenario. But with new technology now available and increasing global prices, along with growing domestic demand, the project is now being revisited.[24] The advantage of the project would be that India could be sourced from a number of suppliers, including Iran, Oman, Qatar, and possibly Iraq at a later date.[25]

Downstream Strategy

Over the last five years, though the global demand for petroleum products has increased by around 2.5 per cent, the rate of capacity addition in refineries has been around 0.7 per cent only. In fact, lack of adequate refining capacity is seen as one of the chief factors of the recent oil price hike. No new refineries have been constructed in the US and Europe in the last 20 years due to strict environmental norms. It is in this context that India can, and in fact has been, benefiting.

While India's crude oil imports have been growing incrementally since the 1990s, petroleum products exports have emerged as the single largest foreign exchange earner, accounting for 11 per cent and 15 per cent of the country's total exports in 2005-06 and 2006-07, and growing at the rate of 67 per cent and 58 per cent, respectively.

India is geographically well placed to become a major refining hub, located as it is between the producing region of West Asia and the growing markets of Asia-Pacific. Though West Asian countries had planned to construct a number of refineries, many of these have been stalled, allowing India to fill the gap. Moreover, the capital costs of Indian refineries are lower by some 25-50 per cent, and with a large reserve of trained and highly skilled manpower, India enjoys a cost advantage over other Asian countries.

Today, India has the fifth largest refining capacity, up from 19th in 1995, with a 3 per cent share of the global capacity. Hence, if as per projection made by the International Energy Agency (IEA), the world will witness a refining capacity deficit of around 112 million tones per annum (mtpa) by 2010, India can reap the benefits of its growing refinery capacity.

Indian companies plan to increase their refining capacity to 242 mtpa by 2011-12, up from 150 mtpa in 2007. Given that domestic demand is expected to be around 196 mtpa by then, the rest can be exported, earning huge revenues. Indian Oil Corporation (IOC) plans to increase its refining capacity from 60.2 mtpa to 76.7 mtpa, ONGC plans to scale up its refining capacity up to 45.5 mtpa by 2009-10 from the current 12 mtpa, Bharat Petroleum Corporation (BPCL) is setting up a 6 mtpa plant at Bina in Madhya Pradesh, Reliance Industries Ltd is constructing a 29 mtpa export-oriented refinery in Jamnagar, which is slated to be the world's sixth largest, Nagarjuna Oil Corp is planning a new refinery at Cuddalore with a capacity of 6 mtpa, Hindustan Petroleum (HPCL) is setting up a 9 mtpa refinery in collaboration with the L.N. Mittal group at Bhatinda as well as expanding its Visakhapatnam refinery capacity to 16 mtpa, and Essar is planning to upgrade its refinery to 34 mtpa capacity from the current 10.5 mtpa. Most of the new refineries will be located on the coasts, while the major centres of demand for the petroleum products are in the inland locations, particularly in the north/northwest regions.

This is impressive given that in 1947, India had only one refinery, located in Digboi with a capacity of 0.25 mtpa. Today, India has 19 operating refineries, 13 of which are in the public sector and one in the joint sector.

To facilitate India's refining ambitions, the government is encouraging joint venture cooperation in the sector, including public-private partnerships, for achieving faster growth. Under the joint venture initiative, 43 mtpa capacity will be added in the next five to six years. Out of this, IOC has tied up with Kuwait Petroleum Company for one refinery, HPCL with Oman Oil Company and Saudi Aramco for two refineries and BPCL with Oman Oil Company and Shell International for two refineries. In the private sector, Letters of Intent (LOIs) have been issued for about 41 mpta capacity. The companies to whom LOIs have been issued are Reliance (15 mpta), Essar (9 mpta), Ashok Leyland (2 mpta), Nippon Denro (9 mpta) and Soros Foud (6 mpta).

Under the EOU category, about 29 mpta capacity has been approved. In sum, additional refining capacity of about 110 mpta, excluding EOUs, is planned for implementation during the next 5-6 years.[26]

Strategic Reserves

For a while now, India has been considering setting up a Strategic Petroleum Reserve (SPR) to be accessed during supply interruptions in the event of a conflict, either in India or elsewhere. Accordingly, the government has decided to set up 5 mt crude oil storages at three locations, namely, Mangalore, Visakhapatnam and Padur (near Udupi), all of which are in coastal areas so that they are readily accessible to the refining sector through marine distribution. These SPRs would be in addition to the storages of crude and petroleum products already with the oil companies. The construction of the proposed strategic storage facilities is being managed by Indian Strategic Petroleum Reserves Limited (ISPRL), owned by the Oil Industry Development Board (OIDB). The reserves would be in underground rock caverns, considered to be the safest means of storing hydrocarbons and are expected to be operational by 2012. The estimated cost of the project is around Rs 2,400 crore at September 2005 prices, excluding the cost of crude oil for filling the caverns. ISPRL will have ownership and control of the crude oil inventories and will coordinate the release and replenishment of crude oil stock during supply disruptions through an Empowered Committee to be constituted by the central government.[27]

Conclusion

The slowdown in the global economy notwithstanding, India is expected to continue on its growth path at a rate of 7 to 8 per cent. To ensure that growth is sustained, it needs assured access to energy resources, and security of supply, therefore, is a core area of national security, second only to food security. The recent rise in oil and gas prices as well as concerns that production in several major oil fields may be nearing their peak have reiterated the importance of ensuring long-term supply security.

To this end, the government is developing a multi-pronged strategy to ensure supplies, some of which have been mentioned above. However, with growing concerns over environmental and climate change issues and the linkages that have been established between the use of fossil fuels and global warming, it is imperative that the country's energy security is not limited to ensuring supply of hydrocarbons alone. Towards this end, the government has realised the necessity of adopting a more holistic approach towards ensuring the country's energy security with linkages between the various energy sectors such as hydrocarbons, power, renewable energy, external affairs, transport, finance and other infrastructure sectors. Till the recent past, different energy sectors tended to act independently from a policy and regulatory perspective. However, it is important that cross-linkages between different energy sectors are set up to ensure that there is more integration between the same. That the government has taken cognisance of this issue is clear from the fact that the Planning Commission has brought out its report on an Integrated Energy Policy for the country to meet the common objective of energy security as a whole.

However, as per the IEA's projections, fossil fuels, and particularly oil for the transport sector, will continue to dominate the energy scenario. It would, therefore, be to India's advantage if a strategy is adopted that would allow it to use hydrocarbons more efficiently. A key factor to ensure more efficient use of energy is to ensure fair and correct pricing for various energy resources. Various committees that have been set up over the years have all recommended a graduated increase in prices to reach market determined levels. It is only when the real price of oil and gas is allowed to be reflected in the retail sector that more judicious use of resources will be seen. Hence, political imperatives governing the subsidised pricing regime that is now prevalent and which has led to large-scale adulteration and leakages in the system, should be dispensed with and market determined prices should be allowed.

Notes

1. "Indian Roads Network," National Highways Authority of India, www.nhai.org/roadnetwork.htm

2. "Supply Options", Draft Report of the Expert Committee on Integrated Energy Policy, Chap.3, p.35.
3. "Energising India for Sustainable Growth", speech by Mr Murli Deora, Minister, Petroleum and Natural Gas, Government of India, at the 19th World Petroleum Congress, Madrid ministerial session on India, July 3, 2008, http://petroleum.nic.in/speeches/03-07-2008.doc.
4. "India's Natural Gas Production to Double by 2012 – Ministry", sourced from PTI, June 19, 2008, www.zibb.com/article/3460113/indias+natural+gas+production+to+double+by+ministry
5. Press Note by Ministry of Petroleum and Natural Gas on the occasion of launch of 7th round of NELP-VII, December 13, 2007, www.petroleum.nic.in/pressrelease.pdf
6. "RIL to Pump in 120 mmscmd Gas by 2014", *Financial Express*, September 24, 2008
7 See n 2.
8. "Few Foreign Takers for India's Oil Hunt", *Alexander's Gas & OilConnections*, vol.13, issue#15, August 22, 2008, www.gasandoil.com/goc/company/cns83483.htm
9. "Augmenting Oil and Gas Production", MoP&NG, PIB Press release, August 7, 2008, www.pib.nic.in/release/release.asp?relid=41150
10. "India's Middle East Crude Imports Surge by 11%", Commodity Online, August 7, 2008, www.commodityonline.com/news/Indias-Middle-East-crude-imports-surge-by-11-10960-3-1.html
11. "Petronet Steps Up Efforts to Procure More LNG", *Hindu Business Line*, July 21, 2008
12. "RIL,OVL Buy Oil Assets in Latin America", IntelliBriefs, April 6, 2008, www.intelliBriefs.blogspot.com/2008/04/rilongc-buy-oil-assets-in-latin-america.html and "Essar Launches Largest Airborne Survey for Exploring Oil & Gas in Madagascar", Essar release, July 10, 2008, www.essar.com/oil&gas/pr2008_07_10.htm
13. "Tremendous Value to be Unlocked in Oil & Gas Biz", Interview with R.S. Sharma in BusinessLine, January 1, 2008.
14. Shebonti Ray Dadwal and Uttam Kumar Sinha, "Equity Oil and India's Energy Security", *Strategic Analysis*, vol. 29, no.3, July-September 2005.
15. Annual Report 2007-08, Ministry of External Affairs, Government of India, pp. xii.
16. "India Plans Sovereign Wealth Fund for Energy Assets Abroad," *The Economic Times*, February 19, 2008.
17. "Ethanol", Report of the Committee on Development of Biofuel, Ch 2, p.16
18. "National Mission On Biodiesel", Report of the Committee on Biofuels, Ch.5. p.113.
19. Ananth Chikkatur and Sunita Dubey , "India's Coal-to-Liquid Push Short-Sighted", July 3, 2008 , India Together, www.indiatogether.org/2008/jul/eco-ctlpolicy.htm
20. "Iran Proposes Pricing Formula for Gas Supplies to Pakistan", Fars News Agency, September 28, 2008, http://farsnews.com/newstext.php?nn=8707071269
21. See n. 20
22. "TAPI Project: Turkmenistan to Certify Gas Reserves by End of Year", *Daily Times*,

September 12, 2008, www.dailytimes.com.pk/default.asp?page=2008%5C09%5C12%5CStory_12-9-2008_pg5_5

23. "Turkmenistan Seeks $12.7 per mmbtu for TAPI Gas", *Business Standard*, September 4, 2008
24. "India Plans a Under Sea Gas Pipeline From Oman," *Payvand*, April 4, 2008, www.payvand.com/news/08/apr/1035.html.
25. "India Eyes Underwater Pipeline", *The Times of India*, May 12, 2008.
26. K. Radhakrishna, "India's Spectacular Story in Petroleum Refining", rediffnews, April 8, 2008, www.rediff.com/money/2008/apr/08petro.htm
27 "Indian Strategic Petroleum Reserves Ltd," May 15, 2008, www.isprlindia.com/aboutus.html.

N-Powering India: Opportunities and Challenges

□ **Manpreet Sethi**

Among the challenges facing an emerging India, two could particularly derail the process of its rise, and they are interconnected. The first is widespread social disharmony as a result of skewed economic development and the unmet aspirations and expectations of a growing population. The resulting discontentment, coupled with growth of sub-nationalism, opportunistically exploited by domestic vested interests or external adversaries could severely strain the country's socio-politico-economic fabric. Hence, the importance of social and human development with economic growth in order to ensure societal peace and national security.

The second challenge arises from the sheer shortage of power, especially electricity, to drive the economic processes and developmental endeavours. It is widely accepted that electricity has a direct connection with the level of development and per capita energy consumption is a parameter for calculating the human development index. Therefore, it is hardly surprising that energy and the ability of a nation to access it from reliable, secure, sustainable and safe sources tops national priorities everywhere. This is even more important for a nation like India that expects a phenomenal growth in its energy demand, estimated to be between 6-10 per cent per annum during the first quarter of the 21st century. The Power Policy of India promises electricity availability to all by 2012, and envisages electrification of all villages by 2009.[1] This appears practically impossible given that the present total power generation of about 150 giga watts (GW) is woefully short of the demand that is growing by the day. Per capita energy consumption in India at present is placed at 600 kilo watt hour (kWh) as compared to 1,300 kWh in China and 15,000 kWh in the developed economies. For this situation to substantially change, the absolute amount of energy required by India would have to at least double by 2020, double again over the next ten years, and be close to ten times the figure today by 2050. According to Dr Kakodkar, Chairman, Atomic Energy Commission, even if India's per capita energy consumption was to rise to 5000 kWh (which would still be three times less than the current consumption figures in the US), the country would suffer an energy deficit of 412 GW by 2050.[2] As is evident, the deficit itself would be nearly three times the current total power production!!

According to another estimate provided in 2006 by the government instituted Expert Committee on Energy, India's power needs would be about 960 GW by 2031-32, assuming a Gross Domestic Product (GDP) growth rate of 9 percent.[3] Since then, the global financial crisis and the consequent economic downturn have brought down the expected rate of growth of the Indian economy to 7 percent per annum. Even at this lower rate of growth, the vulnerabilities that will accompany large-scale energy import dependence are clearly evident. The government, therefore, must urgently focus on cultivating a diverse mix of energy sources that pragmatically balance considerations of cost, uninterrupted availability of fuel and environmental impact. It is important that every potential source of energy is optimally used and the menu of options is as varied as possible so as to minimise risks of disruption arising from shortages, price fluctuations or political manipulations.

What role can India's nuclear power programme play in this? What would be the advantages of investing in nuclear expansion, which is today possible with the conclusion of the Indo-US civilian nuclear cooperation agreement that has opened the prospects of India's participation in international nuclear commerce? Is India up to the challenge of making use of this opening? These are some of the questions that the paper examines. It explores the advantages, some of them unique to this country, of adopting nuclear power as a major input to the future energy mix. It also examines the challenges that lie ahead in expansion of nuclear power. While the focus of the paper is on nuclear energy, it nevertheless is premised in the belief that there is need for growth and development of *all* energy sources, existing and potential, to power India's socio-economic growth and development. The country's energy requirements are so huge that it cannot afford the luxury of banking on only one or two fuels to power its future.

Why Does India Need Nuclear Energy?

Nuclear power today accounts for 14 percent of global electricity generation[4] and the world now has more than half a century's or 11,000 reactor years of experience of handling this technology and the expertise and confidence in it has steadily grown.[5] On the other hand, the experience of very high oil prices and growing environmental concerns have led to a reconsideration of safe and sustainable energy fuels and nuclear power has surfaced as a keen contender for large-scale energy generation. Consequently, even in the US, after a period of slackness, studies such as the one conducted by the Massachusetts Institute of Technology, have emphasised the role of nuclear power for energy security.[6] In fact, the Bush Administration pushed for a return of nuclear energy as a national security concern. After nearly three decades of no new nuclear plants (though upgrades of existing plants continued), the US Nuclear Regulatory Commission (NRC) today has licence applications for 20 new plants pending before it. Even as the NRC hires more people to help cope with the application rush, new factories are also being set up to fabricate parts and components of nuclear plants.

Likewise, in Europe too, a report prepared for the European Economic & Social Committee, which advises the European Commission, emphasised that Europe needed nuclear power.[7] Consequently, some of the European Union (EU) members such as Italy and Germany that were not in favour of nuclear energy are reconsidering their phase out policies. The UK plans to build four more nuclear plants with French help, with the first one likely to be operational in 2017. Meanwhile, China is already emerging as the fastest growing nuclear power generator and if things go according to its ambitious plans, then it will be the largest producer of this energy at 130 GW by 2030.[8] In fact, of the 55 GW of additional installed nuclear generating capacity projected for Asia, 24 GW is projected for China, 12 for India and 12 for South Korea.[9] According to Organisation for Economic Cooperation and Development (OECD) estimates, if the nuclear activity planned over the coming decades remains on track, nuclear reactors would supply a fifth of total electricity generated worldwide by 2050.[10]

For India, the energy demand heartland, along with China, nuclear power holds tremendous promise. The following paragraphs identify the rationale for nuclear energy expansion in India's energy basket.

Limited Availability of Indigenous Fossil Fuels

Akin to the predominant trend worldwide (except in some countries like France and Japan), the bulk of India's existing power generation capacities exist in the thermal sector. However, unlike most big nations, India imports these traditional fossil fuels in large quantities to meet its energy demand. This obviously raises the country's vulnerabilities to unacceptable levels. For a large country like India, bulk imports of fuel are neither affordable nor strategically prudent. Moreover, with increasing worldwide competition for non-renewable hydrocarbons, viable alternatives such as nuclear energy will have to be considered. In fact, inclusion of uranium in the global energy portfolio could slow the depletion rate of fossil resources, particularly since reserves of uranium have no other major use. Nuclear power can replace the heavy dependence on hydrocarbons and the cases of France and Japan amply prove this. France today generates 80 percent of its electricity from nuclear energy. Meanwhile, Japan had managed to reduce its oil imports from 80 percent in the 1970s to 56 percent by the 1990s and today sources 30 percent of its electricity from nuclear power.[11] What is India's situation?

India has reasonable coal reserves, which according to British Petroleum estimates, comprise 8 percent of the world total. The country is the fourth largest producer of coal and lignite in the world (after the United States, China, and Australia).[12] However, India's coal reserves are of low quality (of high ash content and low calorific value) and concentrated in some parts of the country. This necessitates haulage of coal over long distances which not only raises cost but also ties down the transportation network. In fact, transport costs are three times the cost of coal when it comes out of the mine. Nevertheless, at present, coal remains the dominant fuel at 55 percent of primary energy generation and this is sustained by the import of nearly 10 million tonnes of coal annually. In fact, the Shankar Committee set up to recommend measures to meet the demand-supply gap already foresees import of 30-40 million tonnes of high-grade coal by 2011-12.[13] Meanwhile, the World Energy Outlook 2007 has calculated that India's coal imports would rise seven-fold by 2030 if the energy generation composition does not change. If the time horizon is stretched to 2050 without adding nuclear energy to the Indian energy basket, then coal imports would have to be to the tune of 1.6 billion tonnes.[14] The enormity of these figures and the gravity of the situation is self evident.

India's oil consumption in 2005 was about 2.5 million barrels per day, having doubled from the figure in 1992. As the Indian economy continues to grow even modestly at 7 percent per annum, the oil requirement of the country is expected to double again by 2030.[15] Crude oil prices are unlikely to fall below $50 per barrel in the coming few years even though they have come down from the high peak of $135 per barrel earlier in 2008. This has enormous implications not only for the strain it causes the exchequer, but also makes the country vulnerable.

The use of natural gas for energy generation is expected to increase substantially in the coming years. But, given the limited domestic availability, it will have to be sourced from outside through elaborate and long distance transportation networks of pipelines and Liquefied Natural

Gas (LNG) shipments. These will bring their own risks of terrorism, piracy and environmental spills. While "peace pipelines" are politically a laudable concept, these have enormous economic and security implications that the country must consider and be prepared to bear.

Research and development will continue towards increasing the share of renewable energy sources, including wind, solar, tidal, hydro geothermal and biofuels. However, except for hydropower in the few places where it is plentiful, none of these is likely to be found suitable, intrinsically or economically, for large scale power generation where continuous, reliable power supply is needed. In fact, the reliability and evenness of electricity supply will become even more critical for an increasingly digitised information society. Development of energy efficient technologies and measures will be explored. However, these efforts are not expected to be sufficient to meet the energy demand and can only be complementary to the addition of new generation capacities. Nuclear power, in this context, provides an important hedge for India's energy strategy given the huge deficit the country faces in domestic fuel sources.

Cost-Effective Option

Traditionally, nuclear power has been considered an expensive energy source given the high capital cost and long gestation periods required for building power plants. But recent empirical data indicates otherwise. In any case, nuclear power has long been proven a genuine economic option in terms of Long Range Marginal Cost (LRMC), or for power supply at locations far away from coal reserves, particularly if hydel sources are also not available in these areas.[16] In fact, a comparative techno-economic analysis that accounts for location of coal mines, transportation of fuel, availability of railroads, ash content and associated environmental impact and necessary mitigation measures, etc. skews the cost benefit in favour of nuclear energy. Contemporary trends such as low interest rates, high oil prices, improvements in nuclear plant capacity factors[17], reduction in construction times, etc have further rationalised the per unit cost of nuclear electricity. In fact, the construction and cost experience of the Tarapur Atomic Power Plant (TAPP) 3 and 4, India's latest nuclear plants, is illustrative of this. Not only have these plants been constructed in record time but also at a cost lower than expected. According to the Chairman of the Nuclear Power Corporation of India Ltd. (NPCIL), a Public Sector Undertaking of the Department of Atomic Energy (DAE) that is tasked with the designing, construction and operation of nuclear power reactors, the two units were built in five years at a cost of Rs 6,100 crore against an approval of Rs 6,525 crore.[18] Modern systems of construction and resource management have indeed contributed to the economics of nuclear power.

At the same time, newer methods of cost calculation that include "external costs" of health and environment favour nuclear energy since unlike thermal plants that do not account for land acquisition costs for waste disposal, etc., even though the waste generated in thermal plants is so much more, nuclear power internalises the cost of waste management and plant decommissioning. An EU study estimated that inclusion of health and environment costs would double the EU price of electricity from coal and increase by 30 per cent that from gas. The cost of nuclear power is further enhanced once carbon dioxide emissions begin to carry a significant "price." Emissions trading would provide incentives for investment in carbon free electricity technologies, and then the economics of nuclear power would improve considerably.

Increasing Environmental Consciousness

If the growing Indian economy continues to rely on traditional thermal energy sources, carbon emissions would significantly rise and environmental consequences like greenhouse effect, global warming and climate change would increasingly become a serious cause for concern. Thermal power plants pose the problem of Greenhouse Gas (GHG) emissions that cannot be wished away despite technology improvements and implementation of stringent environmental measures. Rather, the pollution is sure to increase with the upsurge in energy production from thermal plants. Table 1 shows the carbon dioxide (CO_2) emissions from different energy sources in order to illustrate that coal, oil and gas remain major sources of carbon emissions, while nuclear and other renewable energy sources figure around the lowest.

Table: Carbon Dioxide Emissions from Power Technologies in g/kWh[19]

Coal	Advanced Coal	Oil	Gas	Nuclear	Biomass	Hydro	Wind
960-1300	800-860	690-870	460-1230	9-100	37-166	2-410	11-75

As is evident, nuclear power emits the least amount of greenhouse gases. In fact, the complete nuclear power chain, from uranium mining to waste disposal, including reactor and facility construction, emits only 2-6 grams of carbon per kilowatt-hour. Given these figures, it is obvious that the strategies and technologies adopted by countries with large energy requirements will have critical implications for the local and global environment. Illustratively, France that meets 42 percent of its primary energy consumption from nuclear energy has the lowest per capita carbon dioxide emissions in Europe.

In 1997, India emitted 250 kg of carbon per person, which was approximately one-quarter of the world average and less than 0.05 percent of that in the US[20]. But with continued urbanisation, a shift from non-commercial to commercial fuels, increased use of motorised vehicles, and prolonged use of older and inefficient coal-fired plants, these values are expected to increase and nearly triple by 2020. In fact, according to the US Department of Energy, between 2001 and 2025, India's carbon emissions will grow by 3 per cent annually, twice the predicted emissions growth in the US, making India the third largest air polluter after the US and China by 2015.[21] If India is to avoid holding this dubious record, then a conscious decision to switch to more environmentally sustainable energy technologies, such as nuclear power, and commit to its rapid growth needs to be adopted at the earliest.

Availability of Mature Nuclear Technology and Expertise

With experience of half a decade in the field of nuclear technology, India, in the words of Dr Chidambaram, former Chairman, AEC, is "the only developing country that has demonstrated its capability to design, build, operate and maintain nuclear power plants, manufacture all associated equipment and components, and produce the required nuclear fuel and special materials." Indeed, India can claim to have experience in construction, operation and maintenance of a varied range of nuclear power plants – Light Water Reactors (LWR), Pressurised Heavy Water Reactors (PHWR), Fast Breeder Reactors (FBR) and Advanced Heavy Water Reactors (AHWR). It is a well known fact that having gathered several years of reactor experience in PHWR operations, India has graduated to the commercial demonstration of the fast reactor programme with the installation of first 500 MWe prototype FBR at Kalpakkam. At the same time, the country is an emerging leader in the development of reactor and associated fuel cycle technologies for thorium utilisation. A 30 KW(Th) research reactor, KAMINI, has been operational and is perhaps, one of its only kind in the world, currently operating with uranium-233 (U-233) based nuclear fuel.

The Indian nuclear plants have also achieved many international benchmarks. For instance, in 2002, the average capacity factor of Indian PHWRs was more than any reactor in the US. At the end of September 2002, Kaiga Atomic Power Plant (KAPS) recorded a capacity factor of 98.4 percent during the preceding 12 months and became the best performing PHWR among 32 reactors worldwide.[22]

It is a recognition of India's nuclear expertise that it has been invited to participate in the multinational International Thermonuclear Experimental Reactor (ITER) being built in Cadarache, France, to harness energy from nuclear fusion. Indian research in fusion had anyway been going on for the past two decades at the Institute for Plasma Research at Gandhinagar. India had planned to build an ITER scale reactor by 2030.[23] Participation in the global project will enable it to leapfrog in technology, while making a value addition to the multinational effort.

Vast Thorium Deposits

While the Indian uranium reserves at about 0.8 percent of the world's are considered to be insufficient for a power programme more than 10,000 MWe if the uranium is used on once-through basis and then disposed as waste, India has planned for spent fuel reprocessing to complement its nuclear fuel resources. The first stage of this programme involves using the indigenous uranium in PHWRs. The second stage utilises the spent fuel of PHWRs after reprocessing to extract plutonium 239 (Pu-239). This is then used in FBRs to breed additional fissile nuclear fuel, plutonium and U-233. In the third stage, thorium and U-233 based AHWRs will be able to meet the long-term Indian energy requirements. Thus, the available uranium will eventually be used to harness the energy contained in non-fissile thorium, of which India possesses about 32 percent of the world's reserves or 360,000 tonnes of high quality thorium, but which needs plutonium to kick-start fission.

How will International Nuclear Commerce Help?

Envisaging the crucial role that rapid addition of nuclear power generation could play in easing the overall energy deficit in the coming years, the government had begun exploring the option of seeking an exceptionalisation for India from the Nuclear Supplier Group (NSG) guidelines that had long prohibited any transfer of nuclear material or technology to India until it accepted the nuclear Non-Proliferation Treaty (NPT) membership as a Non-Nuclear Weapon State (NNWS) and opened all its nuclear facilities to full- scope safeguards. An opportunity to realise this presented itself in 2005 when President Bush offered to abandon the long-standing US nuclear policy towards India in order to initiate a constructive engagement in civilian nuclear cooperation. Through three years, the relevant and unprecedented agreement between India and the US suffered intense scrutiny and criticism and was nearly pronounced dead scores of times. However, braving all odds, the agreement finally cleared its last lap when President Bush signed the HR 7081 US-India Nuclear Cooperation

Approval and Non-proliferation Enhancement Act in October 2008. This has opened a range of opportunities for the Indian nuclear power programme that had been hamstrung for international cooperation since May 1974. The opportunities in particularly four dimensions of the Indian programme are worth examining.

Fuel Availability

Lack of uranium to power the Indian nuclear reactors stands out as the most serious constraint that had begun to hamper the functioning of Indian reactors over the last couple of years. It is today common knowledge that through most of 2008, the Indian power plants had to run at half their capacity levels owing to inadequate availability of nuclear fuel. This situation arose out of two factors: firstly, though the country's uranium reserves are estimated at 61,000 tonnes and have been calculated by the DAE to be enough for 10,000 MW power generation for 40 years, the uranium prospecting, mining and milling had been relatively ignored over the last few decades. Since 1968, the Uranium Corporation of India Ltd (UCIL) has been commercially producing and processing uranium ore mainly from the mines at Jaduguda, Bhatin, Narwapahar and Turamdih, all located in Singbhum district of Jharkhand. This was sufficient for the operating power plants and research reactors until about half a decade ago. However, once fast track power plant construction started from the mid-1990s onwards, a mismatch developed between uranium demand and supply. Secondly, over the decades, the uranium reserves have depleted and the ore at the Jaduguda mines is presently being obtained at much deeper levels than earlier. This pushes up the cost of recovery of uranium, which in the case of India is in any case high because of low concentration of uranium in the ore. Indian ore has uranium content as low as 0.6 per cent as compared to some Australian, Canadian and Kazakh ores containing up to 15 percent of uranium.

To meet the projected demand of the nuclear power programme, UCIL is exploring uranium deposits located in other areas of Jharkhand and in

Andhra Pradesh and Meghalaya. Progress in this direction, however, has suffered due to opposition from the local populace and non-governmental activists in the regions. Therefore, in order to tide over the domestic uranium crunch, one of the relatively immediate benefits of the recent nuclear cooperation agreement would be to allow India to access uranium from the international market at competitive prices for a programme that has planned at least five more indigenous power plants in the near future. UCIL would also be able to bid for uranium prospecting or mining in other resource rich regions of the world.

Import of Larger Reactors and Export of Smaller Ones

The Tarapur Atomic Power Plant (TAPP 3), India's 16th nuclear reactor, went critical on May 14, 2006. With this, India's indigenous nuclear programme demonstrated the capability to construct and operate 540 MWe PHWRs. TAPS 3 and 4 are today India's largest capacity reactors, with all other indigenously built plants being of 220 MWe capacity. In the future though, the NPCIL has plans to standardise on the 700 MWe plants that it today has the capability to build. Larger reactors obviously offer economies of scale and having developed a mature expertise and technological and industrial base, India has felt the need to move on to larger capacity generations.

The Koodankulam plants being acquired from Russia are of 1,000 MW. The predominant reactor capacities in many of the countries advanced in nuclear technologies average at least 1,000 MW, with France having a majority of its reactors of 1,300 MW. With the opening of international cooperation, India will have the opportunity to import larger reactors for a rapid addition to its generating capacities.

Meanwhile, given the interest in nuclear energy for peaceful purposes in many smaller countries, particularly in the Southeast Asian region, India has an opportunity to export its 220 MWe reactors that would be ideally suited for their smaller electricity grids. These reactors have proved their competitiveness in capital as well as unit energy costs and have a demonstrated record of safe operations. India also has the capability to emerge as a low cost manufacturing hub for nuclear component supplies to the resurgent nuclear industry worldwide. For instance, companies like L&T can export nuclear reactor building skills and/or operation and maintenance services.

Participation in International Projects

As has been pointed out earlier in the paper, India has already joined the prestigious ITER project. India is also member of the International Atomic Energy Agency's (IAEA's) INPRO activity and is participating in eight of the 12 collaborative projects under INPRO's Phase II programme. This programme, that seeks to build innovative energy systems with better safety, economics, waste management, and which are more proliferation resistant, is crucial for the sustained growth of nuclear power. Interestingly, research on such reactors is exploring closed fuel cycles and seriously considering reprocessing technologies as a means of extracting greater energy and reducing waste. India is among the handful of countries that have mastered the plutonium reprocessing technology and has lots to offer from its experience. Meanwhile, Indian nuclear scientists have a chance to interact with the best of their fraternity elsewhere, an exchange that was denied to them since the late 1970s.

Tide Over Delays in Moving to Thorium Cycle

India's development of the thorium cycle has now seen 37 years of work on the concept and feasibility demonstration. Of course, problems of high cost and technical complications in fuel fabrication because of high radioactivity of U-233 and reprocessing required to move to the thorium fuel cycle still persist. But then, India is among the very few countries pursuing this technology. Even the World Nuclear Association, which is dedicated to the promotion of nuclear technology, sees little scope of development of this technology as long as abundant uranium is available. However, given the peculiarities of the Indian resource base, Dr Homi

Bhabha had prescribed a three-stage programme for the country that would culminate with the exploitation of India's large thorium reserves. There is, nevertheless, a logical technical progression that is required of the PHWR and FBR stages in order to reach an optimum level of fissile material build-up that would then make the use of thorium feasible and effective. While Research and Development (R&D) continues to reach the thorium utilisation stage from multiple directions, including through the use of Accelerator Driven Systems (ADS), these are pioneering technologies that India is struggling with alone. Therefore, there can be no pre-determined dates for the advent of the third stage. Estimates vary from 2020 to 2040. In the meantime, the import of reactors from abroad would not only help India in quicker accumulation of requisite fissile material but also help narrow the widening demand-supply gap in an environmentally friendly way.

Challenges

Even though it makes eminent sense for India to not only keep the option of nuclear power expansion open, but to press for it urgently, there are certain limiting factors that must be grappled with.

Availability of Trained Manpower

The Department of Atomic Energy is estimated to have a work force of 70,000 experts today. Given the additions planned to nuclear generation capacity, it is natural that the need for more nuclear scientists, engineers, craftsmen, construction managers, plant operators and maintenance personnel would significantly swell in the coming years. The AEC has a Nuclear Training School that has until now been in charge of manpower development. However, with the need for rapid and increased numbers, it would be a challenge to recruit, educate, train and retain technical personnel, especially at a time when the private nuclear industry is also expected to be expanding worldwide. To strengthen research at universities, the DAE provides grants for projects through the Board for Research in Nuclear Sciences. A DAE Graduate Fellowship Scheme for the Indian Institutes of Technology (IITs) has existed since 2002 to promote collaborative research. IIT Kanpur already offers a course in nuclear engineering and technology, and so will Chennai from 2009 onwards. A combined thrust towards creation of a trained manpower pool will be required from academic institutions and the DAE to have enough numbers of nuclear experts available for the future expansion plans.

Limitations of Indian Manufacturing Industry

Given the high technology content, and the sensitive and precise nature of materials, equipment and processes involved in nuclear power generation, it is imperative that the Indian manufacturing industry keeps pace with the advancing nuclear science and technology and provides it with the necessary infrastructure and equipment at the required pace. In fact, this challenge could be turned into an opportunity by the industry given that the global nuclear renaissance is exposing the inability of existing manufacturers worldwide to meet the growing demand for reactor components and systems. For instance, the US does not have the capability to domestically manufacture ultra-large forgings that exceed 350 tonnes. These are necessary for making reactor vessels and its global supplier is the Japan Steel Works that has the capacity to make 5-6 such forgings every year. Given the small number of suppliers for such high end products, it is natural that their manufacturing capacities would be booked years in advance.

Given India's cost competitiveness, reasonably high engineering and technological skills supplemented by innovative techniques, the country could emerge as a hub of nuclear components and graduate slowly to more complex and high end products over the years. With the opening of international nuclear trade to India, Indian companies could enter into joint ventures or technical collaborations with known nuclear players. This could not only support the Indian nuclear expansion but also enable exports. The government could help build an enabling environment for the Indian industry by drafting necessary policies to this effect. For instance, just as there are offsets in the defence industry, a similar provision may be worked into the commercial contracts for import of nuclear reactors, making it mandatory for the seller to enhance the capability of Indian companies active in the field. It would also be of great value if the indigenous component of every new power plant is kept at a high level. This would not only enable cost benefits but also provide a fillip to the domestic industry and help provide employment to large numbers.

Necessary Legislative Processes

The existing Atomic Energy Act, 1962, does not allow private players into the field of nuclear power generation. Until now, this has been the exclusive preserve of state owned companies. With the opening of the sector to international markets, it is now necessary to amend the Act in order to allow private companies to set up and operate nuclear reactors.[24] Within India, many private companies such as Jindal Steel and Power Ltd., Tata Power Ltd., Reliance Power Ltd., and NTPC have expressed a desire to step into the field. The last,in fact, which is also the largest Indian power company, has already proposed a joint venture with imported technology to set up and make operational a 2,000 MWe nuclear power plant by 2017.[25] Several multinational companies would also be expected to bid for the multi-billion nuclear reactors market in India.

However, in order to enable this, India would also have to enact nuclear liability laws to cater for the event of an accident. This is certainly a concern for American private companies more than for the state owned enterprises of France or Russia that have sovereign immunity. Nevertheless, India will be required to sign the international legal framework for nuclear accidents, namely the Convention on Supplementary Compensation that covers claims through a global fund to pay victims.[26] Moreover, government support will also be necessary to provide risk insurance for companies building nuclear reactors which would cover events beyond the control of the owner, including regulatory and litigation delays.

Regulatory and Environmental Procedures

Given the sensitive nature of the technology and materials in use at a nuclear power plant, these have existed in a heavily regulated environment to guard against possible threats, natural and man-made, to their safety and security. For a sustainable and safe expansion of the nuclear power programme, ample attention must, therefore, be devoted to the correct and quick implementation of the necessary regulatory and environmental procedures. These are extremely essential because any accident at a nuclear site would have repercussions on the growth of the global nuclear industry. Therefore, safety performance of operating Nuclear Power Plants (NPPs) and its periodic and stringent rule-based evaluation is of vital importance in order to minimise and possibly obviate any danger to plant workers or the public. In fact, for every nuclear plant that is built and operated, the society needs assurance that the facility will be safe mainly on three accounts:

- It would not suffer an accident, leading to release of large amounts of radioactivity.
- It would not cause pollution to the environment during the conduct of its routine operations.
- It would account for the long-term storage and safe disposal of its radioactive waste.

The guarantee of these assurances requires the establishment and maintenance of effective mechanisms and the deployment of requisite measures in the design, site selection, operation and decommissioning of a nuclear plant. At the same time, relevant regulatory bodies need to be instituted to oversee and assess the implementation of safety measures against different parameters so that the individual, society and environment can be protected against radiological hazards. From the moment of site selection to the actual construction of the plant, a number of other mandatory requirements of seeking environmental clearances, rehabilitation of displaced populations from exclusion zones, development of infrastructure, etc are required to be undertaken. While India's Atomic Energy Regulatory Board (AERB) has performed this task well in the past, as the pace of nuclear activity rises, it might be necessary to expand the regulatory organisation through additional induction of trained manpower so as not to develop bottlenecks for issue of requisite licensing, while simultaneously ensuring that the most stringent standards of safety and security are maintained.

Public Perception of Nuclear Energy

Yet another significant obstacle to rapid and large-scale nuclear power expansion is public perception of this source. The Three Mile Island incident and Chernobyl have been publicised enough to create opposition to nuclear power. Unfortunately, there is very little awareness of the stringent safety regulations enforced and followed in the design, construction and operation of power plants, or of the safety record of India's power plants of nearly three and a half decades. Neither is there adequate knowledge of the fact that natural radiation exists, in some cases several times more than in the vicinity of a power plant. Fortunately, the nuclear industry is extremely conscious of the dangers involved in its activities and, hence, takes sufficient precautions to obviate chances of an accident.[27]

The other major aspect of nuclear energy that causes public concern is that of waste management and disposal of spent nuclear fuel. Fortunately, for India, this is not such a big challenge since it follows a closed fuel cycle in which the nuclear fuel after being used once is not immediately in need of disposal as waste. Rather, the spent fuel is reprocessed and the products left only after reprocessing actually constitute waste which is then vitrified and stored under water. This not only lowers the amount of waste generated, but also allows the effective use of energy potential in the spent fuel. However, the public relations department of the DAE must step up its efforts to better educate the public on the advantages and risk mitigation endeavours of the atomic establishment in order to develop the ground for greater exploitation of nuclear energy.

Conclusion

Nuclear technology in India has reached a state of self-reliance. India today has 285 reactor years of safe nuclear power generation. Kaiga 2 set a record by registering 529 days of uninterrupted run during August 2006-January 2008. Seventeen operating reactors and six more under construction indicate a high level of nuclear activity that will only pick up in the coming years as more fuel and technology is inducted into the domestic programme. The Indian nuclear power programme has also moved into the second stage of development wherein a prototype FBR is now under construction, and research and development for AHWRs is under way. India today has the capacity, technology and will to expand its nuclear power programme. International cooperation would facilitate the availability of environmentally sustainable energy to India well in time to avoid stagnation of human development.

The present historical moment has opened new vistas for the country's energy scenario. It is the bounden duty of the nation to use this to its own advantage after a careful consideration of risks and vulnerabilities. Some trade-offs will be inevitable. However, if human development, economic growth and environmental sustainability are taken as the essential parameters for these decisions, then there is a case for nuclear expansion for electricity generation, especially as part of the strategic need for as wide a diversification of the Indian energy basket as possible.

India, today, adds about 30-35 GW of power capacity every five years, which is half the planned amount. In order to add 60 GW every five years for the next 25 years, the right choices must be made now. For a sustained progress to usher in a resurgence in civilian nuclear power, realistic action will be necessary on several fronts: a supportive policy environment, including through legislative changes, commensurate industrial investments, help from universities and training institutes for manpower requirements and support from the academic and strategic community to monitor trends and identify limitations to forewarn against possible dangers. A comprehensive policy on its expansion must be urgently drafted and implemented if India is not to let unavailability of power stand in the way of its economic growth and development.

The IAEA Director General El Baradei once rightly pointed out, "Disparity in energy supply, and the corresponding disparity in standards of living, in turn, creates a disparity of opportunity, and gives rise to the insecurity and tensions…." India cannot afford such fissures. It is imperative that the necessary large-scale energy generation is achieved through environmentally friendly resources and technologies, because otherwise the financial and human costs of coping with environmental disasters could severely undercut the benefits of economic growth.

Notes

1. M.R. Srinivasan, "The World's Energy Resources and Needs", *Nuclear India*, vol.39, no. 1-2, July-August 2005.
2. "Uranium Import can Stave Off Looming Energy Crisis: Kakodkar," *Hindu Business Line*, July 5, 2008.
3. Swaminathan Anklesaria Aiyar, "Nuclear Power Gives Energy Security," *The Times of India*, July 20, 2008.
4. This has dropped from 16 percent in 2005 as a result of six of Japan's nuclear power plants remaining shut as a consequence of the earthquake early in 2008 and several of France's reactors going in for simultaneous repairs.
5. World Nuclear Association, http://www.world-nuclear.org
6. MIT Study, *The Role of Nuclear Power* (Massachusetts: MIT, 2004).
7. "EU Report Supports Nuclear Role," *Nucleonics Week,* January 29, 2004.
8. At present, the US is the largest producer of nuclear power at 98 GW. China has 11 operational nuclear power plants that provide 1.3 percent of its total generating capacity.
9. World Nuclear Association, http://www.world-nuclear.org
10. Tara Patel, "Nuclear Reactors May Supply a Fifth of Power by 2050", http://www.bloomberg.com, October 16, 2008.
11. Jasjit Singh, "Growing South Asian Interests in the Persian Gulf Region: Problems and Opportunities," *Strategic Analysis*, vol. 23, no. 9, December 1999.

12. Tata Energy Research Institute (TERI), "Indian Energy Sector," March 2003, http://www.teriin.org/energy/Indian energy sector.htm.
13. "Capacity Build-up in Coal Essential," *The Hindu*, May 22, 2006.
14. As estimated by Dr Kakodkar, n.2.
15. World Energy Outlook, 2007.
16. Yoginder K. Alagh, "Economics of Nuclear Power in India," *Nu-Power International*, vol. 11, no. 1-3, 1997, p. 22.
17. Power plant reliability is measured by capacity factor, or the percentage of electricity actually produced compared to the total potential electricity that the plant is capable of producing. On this parameter, nuclear power plants have shown out as the most reliable sources of electricity production, with average capacity factors exceeding even 90 per cent in many countries, compared to only 68 percent for coal, 35 percent for natural gas and 34 percent for oil. Capacity factors for renewable energy sources are also low at about 30 percent.
18. IBNLive, May 22, 2006
19. Rangan Bannerjee, "Assessment of Role of Renewable Energy Technologies", Greenhouse Gas Pollution Prevention Project - Climate Change Supplement, The Louis Berger Group Inc. Global Environment Team available at http://www.climatechangeindia.com/gep_ccs/.
20. Ibid.
21. Figures as cited by Condoleezza Rice, Remarks at the Senate Foreign Relations Committee on the US India Civil Nuclear Cooperation Initiative, April 5, 2006. Available at US State Department website.
22. K.S. Parthasarthy, "Nuclear Growth Despite Fuel Resource Constraints," *Navhind Times*, June 6, 2008.
23. "India will Join ITER Next Month in Brussels," *The Hindu*, April 26, 2006.
24. Interestingly, the US Atomic Energy Act, 1946, was revised in 1954 to permit private sector involvment in reactor development, though the Congress retained ownership over nuclear fuel. The 1954 Act also established the AEC as the agency to oversee reactor construction and use. This distinguishes the manner in which the nuclear power generation is regulated as compared to conventional electricity production.
25. "India to Broaden Corporate Base for Nuclear Power," *World Nuclear News Overview*, October 31, 2008.
26. This convention will come into force once five or more nations collectively having 400,000 MWe of installed capacity ratify it with the IAEA. Four have already done so – the US, Morocco, Argentina and Romania, totalling a capacity of 3,19,256 MWe.
27. For more on the safety aspects of the Indian power programme, see Manpreet Sethi, "Nuclear Safety: Critical for Future Nuclear Expansion," unpublished paper in a series of papers for a DAE Project on "Nuclear Energy for India's Energy Security."

Renewable Energy : A Major Instrument for Sustainable Living in The 21st Century

□ **Ashok Parthasarathi**

By the year 2100, our only one earth is expected to be populated by 9 billion inhabitants compared to the present approximately 6.5 billion, if nations, institutions and individuals continue to follow the "business as usual" values, mindsets, policies, institutional perceptions and images of "the good life" in the crucial areas of demography, production and consumption of goods and services, and due to the environmental impact of the 1.5 billion rich inhabitants of today. As the distinguished eco-environmentalist Lester Brown of the Brookings Institution, Washington D.C., has been arguing in his 20 books and countless articles over the last 35 years, "Business as usual is no longer a viable option for the human race." What is that "business"? A state of affairs in which the 1.5 billion consume 60 per cent of the world's natural resources in pursuit of a life-style which causes 70 per cent of the world's commons to be irretrievably polluted, including emitting so much greenhouse gases as to lead to global climate change, making the planet uninhabitable. But the tragedy is that the remaining 5 billion of our planet are madly copying the "business" of the 1.5 billion's life-styles, led by the 1.2 billion Chinese as the "vanguard", with a 1.1 billion Indians dangerously closely following suit. Meanwhile, at least 20 island nations will be totally submerged due to sea level rise, tornados, "quixotic" rainfall patterns, serious food shortages in the poorest countries and many other global scale disasters occurring at an accelerating and ever expanding rate.

There are many villains in such a horrible "destination disaster" staring the human race in the face – indeed, have been so doing for the last 20 years – but the central one is what drives our global system – carbon-based energy sources. There is widespread agreement among scientists, both natural and social, Non-Governmental Organisations (NGOs) academics, thinkers, futurologists and, reluctantly, even governments and some industrialists that: (a) we have to, as the whole human family, de-carbonise our economics and societies as the topmost priority, with maximum speed; (b) rich countries have to put a lid on their carbon-led economic growth in 10 years from now; (c) the poor countries of today have to make a sharp and fundamental shift in their development strategies from those based on "copy- catting" the so-called rich countries and economic growth mania, to strategies based on meeting the basic needs of all their peoples through basic institutional, structural, policy and value changes across the board so as to at least start achieving near zero growth population in the very next generation.

Role of Renewable Energy in a New Development Paradigm

It is against the above background that we have to conceptualise and methodologise the future of an emergent India. Fortunately, we had in Indira Gandhi a visionary leader who saw the crucial importance of renewable or so-called "non-conventional energy" as far back as 1973. She saw what was then called the "oil shock" of August 1973, under which the Organisation of Petroleum Exporting Countries (OPEC) hiked crude oil prices from around US$ 10 a barrel to a whopping US$ 40 a barrel, as portending a huge signal to the world about the need to move swiftly to a massive shift to renewable energy sources from fossil-fuels. She sent me, then her Science and Technology (S&T) Adviser, a Note on August 28, 1973 which read as follows;

> I have always been keenly interested in harnessing solar energy for our development, particularly of our rural areas. Every time I asked why we

were not doing more work in that direction, I was told the costs of solar electricity systems were too high in comparison with fossil fuels to be economically viable. Now that oil has hit US$ 40 a barrel, may be the position has changed. Please investigate and give me a report

I worked for two months in that no fax, no-internet world, using all my contacts with the S&T community worldwide and at home to give her a detailed status report and Action Plan. Based on that report, she directed the setting up of a new Central Public Sector Company, Central Electronics Ltd. (CEL), to be the pilot plant builder and our first commercial Solar Photo Voltaic (SPV) manufacturer, with generous Government of India funding and coupled to Research and Development (R&D) programmes on solar cells, panels and systems at the National Physical Laboratory and Indian Institute of Technology (IIT) (Delhi). Action was taken immediately to comply with her directive and the embryo of SPV i.e. Solar Electricity Programme, was launched in 1976.

The world SPV scene and its evolution from 1973 to date is set out in Milestones in **Annexure-1.**

The Indian Scene

As mentioned earlier, we started R&D on solar energy systems as far back as 1976 when a pioneering group of physicists and electronic engineers was nucleated in the public sector company, Central Electronics Limited (CEL). Laboratory work to produce solar cells and panels was completed in 1980. Based on that experience, a five-year National Solar Photo Voltaic Demonstration (NASPED) programme was funded by the Department of Science & Technology of the Government of India for execution by CEL. Designing and building an SPV Pilot Plant of 1 megawatt (MW)/ year production capacity at CEL was the core of NASPED. Using that plant and the solar cells and modules produced by it, a number of complete rural oriented SPV power systems, ranging from domestic and village street lighting to solar water pumps for irrigation and drinking, SPV-powered radio and TV sets, SPV-powered energy systems for powering static and mobile wireless communication sets, etc. were prototyped and produced in small lots for extensive field and user evaluation. Based on the success of NASPED, the DST financed the setting up at CEL of developing world's first commercial plant of 5 MW/ year capacity in 1985.

Meanwhile, the Government of India set up in 1982 a dedicated new nodal department to promote all types of non-conventional energy sources called the Department of Non-Conventional Energy Sources (DNES) which was upgraded to the world's first full-fledged Ministry of Non-Conventional Energy Sources (MNES) in 1991, recently renamed Ministry of New and Renewable Energy (MNRE). Its vision is to develop new and renewable energy technologies, processes, materials, components, sub-systems, complete systems and products and services at par with international specifications, standards and performance parameters, and deploy and commission such indigenously developed and manufactured products and services in furtherance of the national goals of energy security and energy independence. A large, broad-spectrum, programme covering the entire range of new and renewable energy technologies launched then is being implemented, led by MNRE, today.

Present Situation

India is among the top rankers in both grid interactive and stand- alone renewable energy systems. The country has a total estimated renewable energy potential of about 1,20,000 MW—80,000* MW from wind turbines, about 2,500 MW from small hydropower plants of capacity of 25 MW or less, and 2,500 MW from biomass/bio-energy. In addition, the country

---

* Includes removal of old wind turbines of unit rating of 225 KW and "re-powering" the sites on which such turbines are located with new current 2MW and upcoming 5MW turbines.

receives solar radiation on an average of 6.6KWH/day, the highest in the world. Currently, wind, biomass and small hydropower contribute around 11,300 MW which accounts for around 8 per cent of the installed national power generation capacity from all sources, with a share of around 4 per cent in the electricity mix. In addition, 200 MW is generated as captive/ distributed renewable power. Wind energy is one of the fastest growing renewable energy resources with an installed capacity of over 9,800 MW as of end March 2008, which is the *fourth* largest in the world after the US, Germany and Spain. Biogas and solar lighting have reached 4.0 million and one million households respectively for meeting cooking, lighting and water pumping requirements, mostly in rural areas. As of March 2008, about 2.5 million SPV systems of some 28 different types for rural, remote area and industrial applications were in operation. All of these were designed, engineered and manufactured within the country. In addition, around 2.6 million sq m collector area of solar *thermal* systems have been designed and deployed to meet domestic and institutional hot water needs and industrial process heat requirements. A major Village Electrification Programme is being implemented to electrify 7,000 villages through renewable energy resources by 2012.

Government policy initiatives have facilitated the Design and Development (D&D), manufacture and use of renewable energy systems based on an array of technologies. Many technologies are currently at the threshold of economic viability. Strong R&D capacity and manufacturing base for a number of renewable energy systems and devices, namely, solar water heating systems, solar cookers, solar photovoltaic systems, wind turbines and biomass power generation systems has been created in the country. Domestic R&D has enabled wind electric turbines of capacities each of 1.5 MW and 2 MW (the latter of which is at the cutting edge of world manufacturing levels) to be designed, developed and manufactured in the country on a large scale and even exported to some 10 countries.

Major Renewable Energy Projects

Wind Power

We have today some 7 major wind turbine manufacturers operating in the country. These range from BHEL to Suzlon Energy to Vestas of Denmark, Enercon of Germany, Tata Wind Power, Renewable Energy Systems and Wind Power, Netherlands. They all manufacture complete turbines with high indigenous content and, therefore at lowest cost. They are all internationally certified. Furthermore, wind-solar and wind-diesel hybrid power plants have also been developed, produced and deployed widely across the country for a variety of applications. The 9,800 MW of wind power installed and operational includes several wind farms of 50-80 MW capacity each. Wind power has the intrinsic advantage of having a capital cost of Rs. 40 million per MW, the same as that of coal-fired power plants, but with the huge advantage of an erection and commissioning time of only three months vis-à-vis at least forty months in the case of the most efficiently built and commissioned coal-fired thermal power plants. Then, of course, there is the zero carbon footprint and the huge carbon credits that can be secured from wind power plants. Furthermore, there is the phenomenal Compound Average Growth Rate (CAGR) increase from 2 per cent in 2002 to 21 per cent in 2008.

However, the wind energy sector also has a number of problems. The first is an intrinsically technical one — the Plant Load Factor (PLF) cannot be, even theoretically, in excess of 40 per cent, a fact, that has to be lived with. Secondly, it is a structurally closed sector, in which the only people who know anything about it are in its circle of influence – consultants who get business or the business itself. Thirdly, all the turbines manufactured in the country today are based on foreign technology and so are unable to extract energy from winds of speed less than 40 km/hr. However, large areas of our country have annual average wind speeds of only 18-20 km/hr. So we need a joint R&D programme of the Centre for Wind Energy Technology (C-WET) and the wind turbine manufacturers to develop a family of such India-specific wind turbines which can give 25–30 per cent PLFs in 18–20 km/hr winds – which has not been done by MNRE even after all these years. Fourthly, the wind power industry has been driven by the 100 per cent tax concession in the very first year of operation. What is needed is a new policy which provides incentives for maximum power generation and reduces the aforementioned large capital subsidy. It took so much time to overcome the resistance of many so-called wind project/farm developers to such a long overdue change that it was only in July 2008 that the earlier disastrous policy was amended by the Government of India to a generation-based incentive scheme for wind power but limited to wind farm projects of total capacity upto to 49 MW. Fifthly, in several states e.g. Maharashtra, wind energy turbines function at a PLF as low as 11.8 to 12 per cent compared to global PLF averages of 25–30 per cent. Sixthly, there is a sizeable percentage of non-performing wind power plants among existing wind farm projects. These have to be made operational at high PLFs on a crash basis by both technological upgradation and better after sales service by wind turbine manufacturers. Seventhly, there is an urgent need for MNRE to fund R&D jointly with the wind turbine industry to develop larger capacity turbines of at least upto 5 MW capacity and much small turbines of 300–500 KW capacity optimised for Indian wind profiles, locations and loads, as the Chinese have done in Inner Mongolia at the scale of 300,000 in number.

The largest wind power generating capacity today (800 MW) comprises the wind farms in Tamil Nadu. Our leading and largest wind turbine manufacturer M/s Suzlon Energy, a wholly Indian owned company, has not only installed and operationalised several such wind farms at home (Suzlon has sold 52 per cent of all installed wind turbines in the country) but also undertaken major wind farm projects worldwide. These include a 750 MW wind farm in Turkey (in the final stages of operationalising), a 330 MW wind farm in France, and a 600 MW wind farm in China – all against stiff international competition. It has also operationalised many other such wind farm projects in 22 countries. Most recently, the author facilitated them in obtaining a 100 MW order from Pakistan. Suzlon's production and global sales in 2007-08 were around Rs.15,000 crore! It has also bought foreign wind manufacturing companies in Germany and Spain. So, in as short a period as seven years since inception, Suzlon has become a very large and truly global company, based entirely on indigenous technologies through a massive R&D manufacturing and marketing effort.

Solar Photo Voltaic (SPV) Power

India *as an entire nation* has the highest solar insolation of 1,700- 1,900 watts/sq metre. It was, thus, only natural for us to give priority to build up almost totally indigenous technology-based SPV power sources and solar thermal power sources right from the start of our overall renewable energy programme in the mid-1970s, as briefly recounted earlier.

The largest SPV system in the country today is the 25 MW composite power plant for home and street lighting and a wind-SPV hybrid energy plant to power a large cold store for fish. This project, which is the largest in Asia, is located on the large Sagar Island in the Sundarbans in the Ganga-Brahmaputra delta, 150 km from Kolkata. The beneficiaries are some 12,000 fisher-folk. The households are charged Rs.7 per KWh which they are willing to pay as there is no other mode of electrification for the island because it is 10 km from shore. The 25 MW SPV power plant is built up from 25 KW, 50 KW and 100 KW sub-units. The whole project, which was started in mid-1997, was completed in mid-2003. The modular power plants were manufactured, installed and maintained by India's five top SPV manufacturers, four in the public sector and one in the private sector. Over the last 15 years, the cost of solar electric power has come down from Rs. 15/unit to around Rs. 9 today. In the next 3–5 years, this cost will come down to Rs. 5/unit. The recent increase in crude oil prices, should enable the Rs.5/unit of solar–based power to be fully competitive with small, medium and large diesel generating sets in the next two to three years! The recent decisions of nine states and four Union Territories to provide generous rebates, VAT exemption, etc. is making solar power fully competitive with the domestic grid electric power tariff in Delhi of Rs. 4.5/unit.

This has led the Cabinet Committee on Energy chaired by the Prime Minister to very recently launch a massive "Solar Lantern Mission" to produce 10 million such lanterns in the next *two* years. This will totally eliminate use of kerosene lanterns, with the consequent advantages of saving kerosene, eliminating the annual kerosene subsidy of around Rs.12,000 crore; eliminating kerosene-based polluting fumes; providing the poor with vastly superior lighting quality from solar lanterns; and, finally, reducing carbon contributions of 1.14 tons, thereby contributing to a reduction in global warming. This Solar Lantern Mission is actually part of a much larger and more comprehensive "Solar Energy Mission."

Our largest public sector SPV company, Central Electronics Ltd. (CEL) makes 28 different types of SPV systems, all based on in-house R&D, in rural, remote area and industrial categories. These range from portable solar lanterns to SPV water pumps for both rural drinking water and irrigation systems for whole villages and agriculture respectively to special vaccine storage refrigerators at village Primary Health Centres, to SPV power sources for charging the batteries of wireless communication sets of military pickets on the world's highest battlefield on the Siachen Glacier in the Himalayas at 16,000-18,000 ft and all along in the northeast and in Rajasthan's desert districts of Jodhpur, Jaisalmer and Barmer, particularly along the Indo-Pak border. SPV power sources also power the communication control and instrumentation systems on offshore oil and gas platforms.

The SPV cell, panel and systems production using crystalline silicon technology is the overwhelmingly dominant solar cell technology in use worldwide today – 2/3rd of total world capacity of around 8,000 MW in 2007, with an investment of US $ 100 billion in that year and US$

60 billion in 2006. This is the case in the US, Europe, Japan, India and China, with individual plant capacities at the maximum level of 750 MW/year achieved by Sharp Solar of Japan. However, Sharp itself has commissioned some months ago a giant plant of 1,000 MW/year capacity plant based on double junction amorphous silicon solar panels with a multi-crystalline top layer. Sharp is already going further to set up another 1,000 MW/year capacity based on (non-silicon) thin film Copper Indium Gallium Selenide (CIGS) though it has not yet decided which CIGS technology to use. The attractiveness of these efforts is driven by small capacity thin film CIGS solar cell plants being able to produce solar panels at much lower cost of raw materials and even some production equipment, compared to the hitherto standard crystalline silicon-based solar cell and panel technologies which still use 300–350 micron "thick" raw silicon wafers.

The issue has been complicated by the silicon ingot and wafering production industry worldwide being unable to keep pace with the explosive growth of the SPV cell and module production industry worldwide. This shortage has hit our SPV industry seriously as this has made it unable to purchase raw silicon wafers and even cells since 2006. All the wafer and cell production capacities worldwide are booked full with orders up to not only 2010 but, in some cases, up to 2012! So MNRE is promoting on a super priority basis the establishment of at least one large polysilicon production plant of around 3,000 TPA capacity by 2010/11 backed by two silicon ingot and wafering plants in the same time-frame. But MNRE has taken this crucial decision at least two years too late. The decision should have been taken in 2006/07. Why the delay? Mainly because of generalist IAS civil servants being appointed secretaries of MNRE rather than renewable energy specialists and that too with tenures of 8–10 months on average!! This has been one crucial failure of MNRE which has held back our total national SPV production capacity to a mere 100 MW. This is because both existing public and private sector companies and new entrepreneurs in both sectors have been hesitating to undertake large capacity expansions and set up large new capacity solar cell and panel plants. It is tragic that this should be the case despite there being a huge domestic demand — a demand that is crying out to be met with solid and large funds, both governmental and private, readily available with users/customers of SPV products to make purchases and undertake deployments. What is more, as already indicated, this tragedy is being allowed to occur in a country with the highest solar energy availability all year around, and our conventional grid electricity supply – even in urban areas let alone on a gigantic basis in semi-urban, semi-rural and entirely rural areas – so uncertain, and the huge benefit of avoiding greenhouse gases and the availability of large carbon credits to prevent further global warming!!

Biomass Power/Cogeneration Programme

The biomass power/cogeneration programme is implemented with the main objective of promoting technologies for optimum use of the country's biomass resources for grid and off grid power generation. Biomass materials successfully used for power generation include bagasse, rice husk, straw, cotton stalk, coconut shells, soya husk, deoiled cakes, coffee waste, jute waste, groundnut shells, sawdust, etc. The technologies being promoted include combustion/ co-generation and gasification either for power in captive or grid connected modes or for heat applications.

The current availability of biomass in India is estimated at about 500 million metric tonnes per year. Studies sponsored by MNRE have estimated surplus biomass availability at about 120-150 million metric tonnes per annum, covering agricultural and forestry residues corresponding to a potential of about 16,000 MW. This apart, about 5,000 MW additional power could be generated through bagasse-based co-generation in the country's 550 sugar mills, if these sugar mills were to adopt technically and economically optimal levels of co-generation for extracting power from the bagasse produced by them.

Technologies

Combustion

The thermo-chemical processes for conversion of biomass to useful products involve combustion, gasification or pyrolysis. The most commonly used route is combustion. The advantage is that the technology used is similar to that of a thermal plant based on coal, except for the boiler. The cycle used is the conventional Rankine Cycle, with biomass being burnt in a high pressure boiler to generate steam and operating a turbine with such generated steam. The net power cycle efficiencies that can be achieved are about 23-25 per cent. The exhaust of the steam turbine can either be fully condensed to produce power, or used partly or fully for another useful heating activity. The latter mode is called co-generation. In India, the co-generation route finds application mainly in industries.

Gasification

Instead of combustion, it is possible to convert the biomass into producer gas by gasification (partial combustion). Thermo-chemical gasification involves burning the biomass with insufficient air so that complete combustion doesn't occur, but a gaseous product is obtained. That "producer gas" is a mixture of carbon monoxide and hydrogen. Gasifiers are classified as updraft or downdraft depending on the direction of flow of the biomass and producer gas. India has significant experience in atmospheric gasifiers.

Co-generation in Sugar Mills

The sugar industry has been traditionally practising co-generation by using bagasse as a fuel. With the advancement in the technology for generation and utilisation of steam at high temperature and pressure, the sugar industry can produce electricity and steam for their own requirements. It can also produce significant surplus electricity for sale to the grid using the same quantity of bagasse. The sale of surplus power generated through optimum cogeneration would help a sugar mill to improve its viability, apart from adding to the power generation capacity of the country.

Deployment

The ministry has been implementing a biomass power co-generation programme since the mid-Nineties. A total of 194 biomass power and co-generation projects aggregating to 1590 MW capacity have been installed for feeding power to the grid. In addition, around 172 biomass power and co-generation projects aggregating to 2,050 MW of electricity are under various stages of implementation. States which have taken leadership positions in the implementation of co-generation projects are: Andhra Pradesh, Tamil Nadu, Karnataka and Uttar Pradesh, The leading states for biomass power projects are Andhra Pradesh, Karnataka, Chattisgarh, Maharashtra and Tamil Nadu a manufacturing capability exists in the country for the equipment/machinery required for setting up biomass projects. Except for some critical control equipment, and high efficiency turbines, most of the equipment can be procured from indigenous sources. As to promotional policies besides, the central financial assistance (mentioned subsequently), fiscal incentives such as 80 per cent accelerated depreciation, concessional import duty, excise duty, tax holiday for 10 years, etc., are available for biomass power projects. State Electricity Regulatory Commissions have determined preferential tariffs and Renewable Purchase Standards (RPS). The Indian Renewable Energy Development Agency (IREDA) provides loans for setting up biomass power and bagasse co-generation projects.

Table

	Special Category States (NE Region, Sikkim, J&K, HP & Uttaranchal)	Other States
Bagasse Co-generation (private sugar mills)	Rs 18 lakh X (C MW) □0.646	Rs 15lakh X (C MW)□0.646
Bagasse Co-generation projects by cooperative/ public/joint sector 40 bar & above 60 bar & above 80 bar & above	 Rs. 40 lakh * Rs. 50 lakh Rs. 60 lakh Per MW (maximum support Rs. 8.0 crore per project)	 Rs. 40 lakh * Rs.501akh Rs. 60 lakh Per MW (maximum support Rs. 8.0 crore per project)

**For new sugar mills (which are yet to start production and sugar mills employing backpressure route/seasonal/incidental cogeneration) subsidies shall be one-half of the level mentioned above.*

Biomass Gasification

During the XIth Plan period, a target of 1,700 MW has been fixed for biomass grid connected projects. Biomass gasification systems ranging from 5 kW to 250 kW have been developed indigenously for electrical, thermal and mechanical applications using various types of woody and non-woody biomass, including agro residues, forest waste, etc. In addition, 100 per cent producer gas engines are also being manufactured in the country. Presently, there are about 15 biomass gasifier manufacturers in India.

Major beneficiaries of this programme are mainly individual households and village communities for meeting the unmet demand of electricity in electrified or yet to be electrified villages. In addition, cottage industries, small-scale industries, public utilities, etc. are also availing the benefits from distributed power, generated through biomass gasifiers. Gasifier systems are now being used for decentralised / distributed power generation, captive power and for meeting thermal applications use in industries. So far, about 100 MW capacities of biomass gasifier systems have been installed by industries for captive and thermal application. In addition, about 8 MW equivalent systems of small capacity have been deployed for meeting the unmet demand of electricity in rural areas for energising water pumps, flour mills, agricultural farm implements, etc. MNRE provides subsidy @Rs.15 lakh per 100 kw biomass gasifier systems for rural areas and Rs.l0 lakh per 100 kw for captive power system in industries.

Bio-fuels

India's energy security would remain vulnerable until alternative fuels to substitute or supplement petro-based fuels are developed based on indigenously produced renewable feed stocks. India now recognises biofuels as potential future fuels that assume special importance, particularly from energy security point of view. The consumption of diesel in India is enormous, compared to petrol consumption. Therefore, India needs to focus more on bio- diesel than ethanol for supplementing the petro-based fuels. The rationale for taking up a major programme for the production of biofuels for blending with gasoline and diesel in India emanates from ensuring energy security, reducing imports, generating employment for the poor, meeting the global environmental concern on climate change and to achieve a number of other national objectives.

The Government of India has taken up a number of initiatives. The subject of biofuels being inter-disciplinary and multi-institutional, specific responsibilities has been given to the ministries/departments concerned.

Many developments have taken place in the area of biofuels in recent years in India. The government has approved 5 per cent mandatory blending of ethanol with petrol across the country, except in few smaller states and 10 per cent mandatory blending from October 2008. The government has also approved to permit sugar factories to produce ethanol directly from sugarcane juice to augment availability of ethanol and reduce oversupply of sugar. Five per cent ethanol blended petrol is already being marketed.

There are problems of state taxes and restrictions imposed by the states for production of ethanol, and on its sale, purchase, movement, storage, transportation both within and outside the states, which need to be critically examined and solutions found for uninterrupted production and supply of fuel ethanol. Many states in India have taken up cultivation of jatropha and pongamia on wastelands. More than six lakh hectares of wastelands have been planted with jatropha and pongamia for production of non-edible oils for use as biodiesel feedstock. R&D work has been initiated by several institutions to improve productivity of primary material and processing techniques, but blending of bio-diesel with diesel has not yet been commercially started.

The Planning Commission's Report on development of bio-fuels (2003) estimated that to meet the 20 per cent blending requirement of biodiesel with diesel, in 2011-12, 11.2 million hectares of land would require to be planted with jatropha. Further, the Planning Commission has identified 2.0 million hectares of land available under joint forest management (JFM) and forests in different states. MNRE has prepared a national policy on bio-fuels. The policy framework outlines the strategy to achieve energy security in the country through sustainable production, conversion and applications of bio-fuels. Ministry of Science & Technology and Ministry of Agriculture have initiated R&D and demonstration projects on various aspects of plantations.

Of late, the issue of food versus fuel has been under serious deliberation the world over. Rise in the prices of foodgrains has been attributed, by some, to the increasingly better standards of living in the developing countries. Others have held biofuels as one of the major factors in pushing up the prices of foodgrains. Diversion of land, which grew cereals for human consumption, into production of biofuels would wipe out any advantage that might accrue to the world as a whole. In the Indian context, the focus is on the production of indigenous biofuels from non-food feedstocks by raising these on waste and marginal lands only.

With a favourable bio-fuels policy, India may be able to supplement the petro-based fuels by about 20 per cent with bio-fuels in 10 years from now. The second generation biofuel technologies based on non-food feedstocks, are going to play an important and decisive role for increasing the blending capacities to considerable levels.

Fuel Cells

World Status

Current estimates indicate that there would be more than 1.2 billion vehicles in the world by 2020 – an increase by about 70 per cent compared with today's numbers. The issues relating to carbon dioxide emissions, energy security and urban air quality are the drivers for policy-makers and automobile companies to look at green alternatives to fossil fuels for transport applications. In the short-term, biofuels and Internal Combustion Engine (ICE) battery hybrids are expected to be the technical solutions, but in the medium term, the choice would be restricted to plug-in electric vehicles and fuel cell-based vehicles. Therefore, worldwide, there has, for the last decade, been a growing interest in fuel cell technologies for automotive and also stationary power generation applications. Fuel cells generate electricity through an electric chemical conversion process.

The development of hybrid electric drive trains may prove to be beneficial for fuel cell vehicles also as there are many common components. Currently, the emphasis is on the development and use of rechargeable lithium in batteries for hybrids and also pure electric plug-in electric vehicles. The plug-in electric vehicles may be good for driving within cities. But they have problems of range and durability – constraints which fuel cell powered vehicles do not have. Therefore, fuel cells are expected to play a significant role either as a primary source of power or a range extender or as an auxiliary power unit in automobiles in coming years.

In terms of technology choice for automotive applications, especially for cars, Polymer Electrolyte Membrane (PEM) fuel cells are predominantly the technology used by major automobile manufacturers such as General Motors, Honda, Toyota, Daimler, Chrysler, Hyundai and Volkswagen. The low operating temperature combined with good durability and range makes such fuel cells ideal for use in cars. For auxiliary power units in automobiles, some companies e.g. BMW of Germany plan to use even Solid Oxide Fuel Cells (SOFC). It is estimated that, world-wide, about 1,500 fuel cell vehicles were on the roads in 2007. Some companies have recently introduced the concept of fuel cell cars, whereas others have introduced fuel cell truck fleets as demonstration projects. These projects in the USA, Japan and Europe have already fully established the technical feasibility – reliability, petrol/diesel comparable ranges and speeds as also the superb environmentally benign nature of fuel cell-based vehicles. However, cost-wise it would still take some more years before fuel cell vehicles are commercially available.

For stationary power generation, Molten Carbonate Fuel Cell (MCFC), PAFC, PEM and SOFC technologies are those of choice at present. The estimated worldwide cumulative fuel cell installations for such applications are about 500 MW presently. The major share of installations is based on MCFC (about 40 per cent). The contribution of PAFCs and PEMFC is about 20 per cent each and that of SOFC, less than 20 per cent. PAFC was one of the earliest technologies to be commercialised and some experts who were writing off PAFC a year ago, have started to reevaluate their views about the future of PAFC technology not just for stationary power generation but also for vehicular applications. US companies like Hydrogen Inc. and UTC Inc are active in the development, prototype production and marketing of PAFC units of 200 KW and 400 KW capacities. UTC Fuel Cells' 400 capacity fuel cell vehicles currently cost US$ 3,500–15,000. But the company expects to bring down the cost substantially as their aggressive marketing of such vehicles gathers momentum and crude prices remain at the present high levels. The choice of fuel cell technology type to be used, however, is often linked with availability of fuel. If high purity hydrogen is available from the chlor-alkali industry, then PEMFC is a preferred choice of fuel cell for stationary application. Electricell, Hydrogenics, Nuvera Fuel Cells are some of the companies engaged in development and marketing of PEMFC. In the case of PEMFC, most of the companies are restricting their unit size upto 50 KW. However, some companies like Nuvera Fuel Cells plan to develop and market units of up to 500 KW. If fuel comes from sources such as biogas, then MCFC seems to be the preferred fuel cell technology because MCFC involves a high temperature auto reformation process which is lightweight and of high efficiency. Ansaldo, Fuel Cell Energy, GenCell and CFC Solutions are some of the companies that are engaged in development and marketing of MCFCs. Fuel Cell Energy plans to have multi MW MCFC systems. In the area of SOFCs, General Electric, Mitsubishi Heavy Industry and Siemens Power are some of the companies that are active. General Electric has a plan to develop and market multi MW SOFC systems with coal gasification units. For domestic applications, both PEMFC and SOFC-based systems are being deployed in Europe and Japan.

Status in India

Efforts for development of small scale fuel cells (typically 5 MW–10 MW ratings) in the country were initiated in the early 1990s by the Corporate R&D unit of Bharat Heavy Electricals Limited (BHEL). However, by 2001, 50 KW capacity PAFC power packs were developed and successfully demonstrated in a chlor-alkali plant in Andhra Pradesh. A 200 KW PAFC unit was also imported by BHEL from Toshiba of Japan and field tested. With the emergence of PEMFC as the most versatile among different fuel cells technologies, the focus of BHEL's R&D shifted to this technology from PAFC, though the Naval Materials Research Laboratory (NMRL), Ambernath, is still devoting efforts for development and commercialising of PAFC technology. NMRL has developed and demonstrated PAFC units of up to 10 KW capacity, along with suitable reformers for production of hydrogen from methanol.

Like elsewhere in the world, major efforts in India are devoted to development, demonstration and commercialisation of PEMFC technology. A number of organisations that include SPIC Science Foundation, Chennai; BHEL, Centre for Fuel Cell Technology (CFCT) , Chennai; National Chemical Laboratory, Pune; and the Indian Institute of Technology, Chennai, are currently actively engaged in research, development and demonstration of PEMFC technology. These organisations are working on development of catalysts, gas diffusion layers, bipolar plates, membranes, etc. for PEMFC and also development of stacks and system integration.

Efforts made in India have led to the development of PEMFC stacks of capacity upto 5 KW, which have been demonstrated in decentralised power packs and in a fuel cell–battery hybrid vehicle by SPIC Science Foundation. This includes five 3 KW capacity PEMFC-based uninterrupted power supply (UPS) units. PEMFC has been demonstrated in vehicles e.g. vehicles of 1.2 KW as an add-on to their basic electric vehicles made by the electric vehicle pioneers at the world level, REVA Electric Car Company, Bangalore. Most of the present efforts are focussed on development of PEMFC stacks of 80–100 KW capacity suitable for vehicular applications by SPIC Science Foundation. BHEL and CFCT have developed PEMFC power packs up to 5 KW capacity.

In addition, research and development efforts are also being pursued for development of other types of fuel cells like SOFC, alkaline fuel cells, direct alcohol fuel cells, by a number of research groups, namely, Central Glass and Ceramic Research Institute, Kolkata, BHEL, Indian Institutes of Technology, Delhi and Kanpur, and NMRL, Ambernath. However, commercial production of fuel cells in the country has not yet started. The University of Petroleum and Energy Studies, New Delhi, is currently undertaking a study on estimation of by-product hydrogen available in the country in different sectors, namely, fertilisers, petroleum refineries and the chlor-alkali industry. The excess by-product hydrogen from these sectors could be utilised in hydrogen-fuelled automobiles and power generation applications.

Electric Vehicles

Electric Cars

This is an area in which a major exciting development has taken place in the last 15 years. This is the emergence and rapid growth to maturity of the electric car called REVA, designed, developed and manufactured on a growing scale by the REVA Electric Car Company, Bangalore. The company is a joint venture between the Maini Group of India and AEV LLC of California. A brief history of the Reva electric car project is as follows:

- 1994-2001: REVA invests in 7 years for core EV technologies, acquiring 10 patents.
- 2001: Launch in Bangalore.
- 2002-2003: Enters Greece and Cyprus.
- July 2004: Launch in London. REVA develops India's first fuel cell car prototype.
- April 2005: Reva NXG concept car showcased at EVS21 in Monaco. Designed in collaboration with Dilip Chhabria and Simputer developer Encore Software, the NXG is powered by sodium nickel chloride batteries, with an extended range of 200 km per charge and a top speed of 120 km per hour.
- December 2006: Draper Fisher Jurvetson and Global Environment Fund invest $ 20 million.
- 2007: REVA featuring superior AC drive train technology and increased performance and enters Norway, Spain and Ireland. Sales pass 2,000 units, becoming the world's leading EV manufacturer.
- 2008: Launch in Delhi and enters Belgium and Chile. Passes milestone of 50 million customer driven kilometeres, a world record for Electric Vehicles (EVs).

Reva is an electric vehicle that is best suited for inner city mobility. It is an automatic, easy to drive, park, has a range of 80 km and a maximum speed of 80 km/h. One of its key advantages is that it has zero emissions and is very quiet. REVA will be produced at the 2,000 No. level in 2008-09 and at the 6,000 No. level in 2009-10.

Reva seeks to launch in five to ten new markets per year with the goal of being sold in 40 to 50 countries by 2015 and to establish it as a global EV brand. In India, it aims to be present in 50 cities by 2015. Another automobile manufacturer, Hero Honda plans to launch its first electric passenger car by 2013. It will not be based on the conventional battery mode, "to avoid high running costs" (which, however, REVA has not faced). Hero is developing the passenger car based on capacitors which can be charged instantly and give a longer range.

The company has built a new state-of-the-art and state-of-the-environment assembly plant in Bangalore with a capacity of 30,000 units per year, due for completion, April 2009. The plant is built to LEEDS guidelines, with rain water harvesting, solar charging, solar heating and natural ventilation in order to minimise carbon footprint and hopes to reach this capacity soon.

A strong new product development programme in place to introduce one new model and one variant every year, incorporating new technologies as these evolve. More than 100 employees in the R&D team are focussed on developing new EV solutions. Advancement in electronics and battery technologies will allow greater driving distances and shorter charging times. Increase in volumes will, in turn, lower product pricing and significantly enhance demand over the next 3 to 5 years.

The electric car in India segment faces regulatory, fiscal/legislative and infrastructure challenges. During the initial stages, there were no regulations for electric vehicles in place and Reva worked with agencies, State and Central government departments in India and abroad to enable EVs to be introduced. In the European Union (EU), for example, the Reva is classified as a quadricycle and is limited by weight and power parameters. Several changes in government policies at the state and central levels over the years have had negative impact due to excise duties, taxation, sales and road tax, import duties on technology and components without any thought to subsidies. In contrast, we have seen central, regional and city authorities introduce emissions-based incentives, and privileges for owners of EVs, particularly in Europe.

Likewise charging infrastructure support from some governments in Europe has helped Reva but engagement with several players and governments in India for public-private partnerships for charging infrastructure support has not yet materialised. Therefore, clear, consistent signals and coordinated regulatory and fiscal policies, to both push and pull new technologies and behaviour, are all essential to encourage the switch to emission free motoring.

Electric Motorcycles and Scooters

Reva car apart, eight electric motorcycle and scooter manufacturers have also launched domestic manufacture in recent years. The leaders are: the Yo Bykes manufacturer Electrotherm, the leading conventional motor cycle manufacturer Hero Honda and the E-scooters of the public sector company, Scooters, India. The first claims it has sold around 50,000 e-bikes since it began production, the latter has sold 20,000 or more e-scooters since it launched its product in the market. Meanwhile, Hero Honda is planning to expand its product portfolio in e-bikes by launching two more such bikes in the high speed category. Their top management has announced that the company wants to be a major player in electric transportation in which they are investing heavily in R&D. In the next five years, Hero plans to launch electric three and four wheelers also. Their investment in 2008-09 itself was about Rs. 800 million for expansion of storage capacities, R&D and indigenisation.

Research, Design and Development (RD&D)

Research and Development activities of the MNRE aim at resource assessment, technology development, demonstration and commercialisation. A comprehensive policy on RD&D is in place to support R&D in the area of new and renewable energy, including associating and supporting RD&D carried out by industry.

Specialised technical institutions, namely, the Solar Energy Centre (SEC), Gwalphari, Gurgaon, Haryana; Centre for Wind Energy Technology (C-WET), Chennai, Tamil Nadu; and Sardar Swaran Singh National Institute of Renewable Energy (SSSNIRE), Jalandhar, Punjab, have been established by MNRE to play a major role in providing technical back-up and R&D on renewable energy technologies/systems. In addition, research, development, demonstration and testing activity in the area of small hydropower and biomass power are supported at centres established in the Indian Institute of Technology, Roorkee, and Indian Institute of Science, Bangalore. The complete technology of a locally designed, engineered, field proven and mass manufactured biomass power plant has also been internationally patented and the technology exported to Switzerland. SEC provides testing and standardisation facilities for our renewable energy industry, helping to provide customer confidence in solar energy systems/devices and new product development. C-WET has been playing a vital role in the wind resource assessment programme of the country. It is also the national testing and certification agency for indigenously produced wind turbines. This has greatly helped technology upgradation of locally-made wind turbines and also facilitated the export of wind energy equipment.

However, despite having been set up as far back as 1999 and with a very competent wind energy technologist with 25 years of professional life in wind turbines, inducted as full time Director of C- WET in 2002, the Centre has not yet taken up a joint technology development programme with the wind turbines manufacturing industry to realise a uniquely Indian family of wind turbines capable of achieving high PLF in our widespread very moderate –18 km / hour average wind speeds and high survival conditions in areas prone to cyclonic storms. C-WET has also likewise, not developed higher power wind turbines in the 3 MW –5 MW range, capable of such high PLF in moderate wind speed areas. This is not surprising considering that the annual budget of C-WET even for 2008-09 was only Rs. 90 million and its professional staff complement was only 20 persons. Fortunately, many of the wind turbine manufacturers are undertaking in-house R&D at levels ranging from 3-5 per cent of sales covering both products and production processes.

To return to the overall R&D effort on all types of renewable energy systems, MNRE also operates an R&D support programme in government laboratories (other than C-WET) and IITs and universities. However, this programme, on which only Rs 1,000 million was invested by MNRE on an array of small R&D projects in some 45 institutions during the 10th Five-Year Plan period as a whole (2001-07), is targeted to increase to only Rs 3,000 million in the 11th Plan now under way. The figure should really be Rs 30,000 million or Rs 3,000 crore per year. The absorptive capacity for such a steep step-up exists in government laboratories, IITs, universities and, above all, in industry. That sum is the kind being invested not only by the USA, Germany and Japan but even by China, a recent entrant into the field.

The underlying purpose of such RD&D effort is to make our renewable energy industry technically and commercially competitive at the world level. In addition, the share of indigenously designed, developed and manufactured new and be renewable energy systems / devices has to continue to increase and be tightly monitored for its eventual growth to a dominant position. Research, design and development efforts are focussed on manufacture of complete systems, even if these efforts are required to be shared among different institutions. The focus of such efforts is on system design, engineering and manufacturing at lower costs of systems/ products falling in the following categories :

- Solar thermal energy (high temperature) power generation systems.
- Solar thermal energy for domestic and industrial applications.
- Energy efficient buildings utilising renewable energy concepts.
- MW scale solar power generating systems.
- Increasing the number, scale and efficiency of major SPV cells, modules and systems manufactured to attain world levels.
- MW scale wind turbine electric generators for low wind regimes.
- Biomass integrated gasification as opposed to the present combustion-

based combined cycle systems.

- Simulators for simulation and modelling of renewable energy based grid-interactive power stations.
- Alternative biofuel, synthetic and hydrogen systems.
- Hybrid renewable energy systems of all types
- Geothermal and tidal energy systems.
- Energy intensive storage devices, including those for renewable energy-based grid power.
- Fuel cell-based systems for industry and transport.
- Electric vehicles which have longer range, higher speed and lower cost.

The Future

The country has set a goal of 14,000 MW of renewable power by the end of the 11th Plan (2012) (from 11,300 MW at the end of the 10th Plan) apart from a growing programme for solar water heating and remote village lighting. A substantial financial outlay of Rs. 1,600 crore has been made in the Plan budget of MNRE for a major step-up in R&D, manufacturing and new systems design. An encouraging development of the last two years is that major private sector companies like Reliance and Moser Baer have announced plans to set up large grid connected SPV power plants of 10-15 MW capacity. Moser Baer has recently announced that it would be making an investment of US$ 1.8 billion over the next three years to expand the capacity of its 40 MW thin film solar cell manufacturing plant in a suburb of Delhi to 600 MW/year. But we have to take such grandiose plans with a pinch of salt, particularly because those companies are coming into the solar photo volataic world on a completely "green" basis. With these initiatives and smaller plant capacity expansion by the five public sector SPV manufacturers, the goal of achieving a 10 per cent share of grid interactive solar power by 2012 and 20 per cent by 2020 looks entirely feasible. However, as pointed out earlier, we could achieve the cost competitive target price for solar power in two years if we had an interested government which alone can provide the five-fold increase in the plan budget of MNRE that is urgently needed *and which will pay itself back in three years!*

In alternate fuels such as biofuels, synthetic fuels and hydrogen, the current aim is substitution of petroleum and diesel by up to 5 per cent by 2017 and 10 per cent by 2032!. These are far too pessimistic a set of goals. There is no need to wait till 2017 and 2032. The aim is to facilitate development of indigenously designed, developed and manufactured new and renewable energy products and services that are reliable, convenient, safe, efficient and affordable. Emphasis is being placed on reducing the cost of renewable power projects and products and increasing their efficiency. In emerging technologies, the country has been pursuing research and development in hydrogen and fuel cell technologies for transportation and power generation. A National Hydrogen Development Board has been established, chaired by Ratan Tata, to oversee efforts for development of hydrogen as a renewable energy source for large-scale commercial use. However, for four years, the Board appears to have achieved little. Now Indian Oil Corporation (IOC), our oil and gas behemoth (the largest company in the country with sales of Rs. 220,000 crores and holding 41st position in the Fortune 500 companies) has on its own come out with a programme of distributing hydrogen and CNG-hydrogen on a commercial basis! With this announcement by IOC sometime ago, many small car manufacturers are in a scramble to start commercial production of the engines for this clean fuel. How has it been possible so quickly? Because the top managements of leading auto companies like Maruti, Tata Motors and Mahindras, and bus and truck manufacturing companies, Tata Motors and Ashok Leyland had the vision and foresight to develop full prototypes of those special engines sometime ago shows how our top companies have now fully come of age! The sky is the limit for non-conventional energy with the cost of oil being US$ 180. But the government must act now in a massive way and across the board.

Annexure-I

The World's SPV Scene and Its Evolution

1973 & 74	The very first 100 per cent PV solar company, Solar Technology International, was established in the USA in 1973 and in 1974, the Japanese "Sunshine Project" was launched
1975	The big American oil company Atlantic Richfield Co (ARCO) bought Solar Technology International and, thus, ARCO Solar Inc. was created.
1976	The US Department of Energy (DoE), along with its Photovoltaics Programme, was created. DoE, as well as many other international organisations began funding SPV R&D at appreciable levels, and a terrestrial solar cell industry quickly evolved.
1977	World production of SPV cells and panels in California was around 500 KW – ARCO Solar was the largest manufacturer and had an aggressive R&D programme with more than 140 PhDs working on all aspects of solar cell and panel technologies.
1979	ARCO Solar built the biggest solar cell, panel, and photovoltaic systems production plant of I MW/year capacity at that time.
1980	ARCO Solar was the first to build a SPV power plant with peak power of over 1 MW.
1981	ARCO Solar built another 1 MW SPV power plant over 108 double-axis trackers in Hesperia, California. This was the world's first MW-Scale tracking power generation plant.
1980-82	Meanwhile, following the pioneering initiative of ARCO Solar, other major international oil companies like Exxon of the US, BP of the UK, Total of France came to set up SPV companies over 1980-82.

1984 ARCO Solar Silicon, the first silicon technology based solar SPV panels produced.

1985 ARCO Solar designed and installed a 6 MW solar power plant with its own prime panels of that era. This system was tied to the Southern California Edison's Grid Power System.

1986 ARCO Solar introduced a G-4000, the first commercial thin film photovoltaic panel.

1990 ARCO Solar increased the thin film system production capacities in Camarillo, California to 7 MW per year. Siemens bought ARCO Solar and established Siemens Solar Industry which at that point in time became the world's biggest photovoltaic company.

1992 ARCO Solar opened PV cell and panel production plants in Japan and Germany.

2000 Shell Corporation bought Siemens Solar worldwide and re-named it Shell Solar.

2002 Sharp Solar of Japan has been the biggest producer of solar cells and panels

2004 Total global PV cell production increased from less than 10 MW per year in 1980 to about 1,200 MW per year. The total global PV installed capacity panels reached about 3,000 MW.

2005 Solar World of Germany bought Shell Solar. By that time, Sharp Solar of Japan had already become the largest producer of solar cells panels.

2007 The cumulative worldwide production of solar panels stood at 8,000 MW. Sharp Solar Corporation reached a new solar cell production milestone with over 2,000 MW in cumulative production (equivalent to one quarter of the 8000 MW produced worldwide).

2008 Sharp Solar has a 720 MW/annum capacity and is building

1,000 MW double junction amorphous silicon solar cells with a top layer of multi-crystalline silicon manufacturing plant (the world's first) about to go into full commercial production by end 2009. Current worldwide annual production capacity of SPV has exceeded 4,000 MW.

Food Security for a Growing India

☐ M.S. Swaminathan

Food security involves physical, economic and social access to every child, woman and man in our country. For achieving sustainable food security, we need to enhance farm productivity. About twenty years ago, I coined the term "Evergreen Revolution" to emphasise the importance of improving productivity in perpetuity without ecological harm. Fortunately, we have a large untapped production reservoir which we can harness for ensuring Food for All under conditions of diminishing per capita availability of arable land and irrigation water as well as expanding biotic and abiotic stresses. I would like to deal with some of the opportunities in this paper.

The highest wheat production so far was 76.7 million tonnes in the year 1999-2000. Since then, production and productivity have been declining. How do we turn the tide and how can we step up the production of wheat and rice during the coming years? First, it must be emphasised that seed reserves are important for crop security, just as grain reserves are important for food security.

Second, in wheat, there is a vast untapped production reservoir available in Uttar Pradesh (UP), Madhya Pradesh (MP), Bihar and Rajasthan. The Indian Council of Agricultural Research (ICAR) Wheat Directorate in Karnal has calculated that we can produce an additional quantity of about 24 million tonnes of wheat immediately by bridging the gap between potential and actual yields, with technologies and varieties now on the shelf (Table 1).

Table: Achievable Targets by Bridging Yield Gaps Through Available Technologies Under Irrigated Conditions (based on National Demonstrations)

State	Current Area 2003-04 (000 ha)	Current Yield gap t/ha	Additional production possible (000 t)
UP	8,418.0	1.346	11,330.5
MP	2,831.8	2.071	5,864.7
Rajasthan	2,103.1	1.646	3,461.7
Bihar	1,483.0	1.196	1,773.6
Haryana	2,303.0	0.581	1,338.0
Gujarat	660.7	0.714	471.7
Maharashtra	581.1	0.656	380.0
Karnataka	97.0	0.998	96.8
Punjab	3,444.0	0.241	82.9
			24,800.0

It will be prudent to launch a well-designed farmer-centric compensatory production programme in Bihar, Rajasthan, MP and UP, with priority attention to soil health enhancement and varietal choice.

Similarly, there is vast scope for increasing rice production in West Bengal, Assam, Orissa, Andhra Pradesh, Tamil Nadu, Karnataka and even Kerala during the rabi season. The yield of boro rice is high in Assam and West Bengal. Over 27 high yielding rice hybrids are now available to suit different agro-climatic and growing conditions, as well as grain quality requirements. They are from both the public and private sectors. Pusa RH-10 is a superfine, aromatic grain hybrid suitable for cultivation in northwest India. KRH2 is a high yielding and widely adapted hybrid, while DRRH 2 is an early hybrid with a good yield potential. States with an unutilised yield reserve in their Agricultural Production Bank should be encouraged immediately to initiate action with the guidance of experienced farmers and scientists to utilise the yield reserve wisely to improve production and productivity. The precise agronomic package will have to be developed on a location specific basis with the help of agricultural universities.

Rabi and boro rice production can be enhanced considerably by giving attention to balanced fertilisation, particularly to the supply of the needed micronutrients like zinc, boron and sulphur. Together with plant protection, the enhancement of soil health will help to improve productivity, at least by an additional tonne per hectare. There are nearly 5 million hectares under rabi and boro rice in the country and improved varieties are available for all the states where rice is cultivated between November and May. Striking progress in improving the yield of rainfed maize, soybean, sorghum, green gram, blackgram, pigeon pea, chickpea, finger millet (ragi), pearl millet (bajra), castor, etc., can be achieved through balanced fertilisation (NPK and the needed micronutrients). Seeds of improved varieties should be maintained in Village Seed Banks in rainfed areas, so that alternative cropping strategies can be introduced, depending upon monsoon behaviour. Improved cultivars alone can enhance productivity by 10 to 50 percent. Varietal choice should be based on the likely moisture availability. The short duration chickpea variety Shwetha (ICCV2) has revolutionised chickpea production in Andhra Pradesh. The productivity increased from 470 kg per hectares in 1993 to 1,084 kg per hectare in 2004. The area under cultivation also increased seven-fold. There are nearly 12 million hectares of rice fallows in Madhya Pradesh, Orissa, Jharkhand, Chhattisgarh and West Bengal. In such rice fallow areas, chickpea can be grown by using residual soil moisture. Simple seed priming technologies like soaking seeds in water and micronutrient solution for 6 hours and drying in the shade will help in establishing a good chickpea crop in rice fallows. In Madhya Pradesh, 2 million hectares remain fallow during the kharif season. Using broad bed and furrow, balanced nutrient management and short duration soybean cultivars like Samrat, farmers in the Vidisha district were able to take a crop of chickpea or wheat during rabi and thereby double their income. Many such simple steps in soil-water-crop management can lead to major advances in both crop output and farmers' income. This is the pathway to making farming economically viable.

The timely and adequate supply of credit, seeds and electricity, together with addressing the micronutrient deficiencies in the soil will help to offset the loss in production during kharif. States which had heeded the National Commission on Farmers (NCF) appeal made in December 2005 that 2006-07 may be observed as the Year of the Farmer and Agricultural Renewal were in a much better position to improve rabi production. The five-pronged strategy recommended then consisted of soil health enhancement, water harvesting and management, credit and insurance, technology and inputs and remunerative marketing.

Adaptation to climate change is an urgent task. The Climate Management Unit of the National Rainfed Area Authority should develop computer simulation models of weather behaviour coupled with the public policy and agronomic responses needed to meet diverse possibilities.

Extending the Gains

Eastern India [eastern UP, Bihar, Chattisgarh, Orissa, West Bengal, Assam and the Northeast (NE) states] have a large untapped production reservoir even with the technologies now available. In these areas, poor water management, rather than water availability, is the major constraint. The Indo-Gangetic plains offer scope for becoming the major bread basket of India through an appropriate mix of technology, services and public policies. In many of these areas, the aquifers should be enriched during the southwest monsoon period, and extensive ground water use should be promoted during the October-April period. Given the right strategy, the Ganges Water Machine could become the main anchor for our food security system. Bihar, in particular, is a sleeping giant in the field of agriculture. The work of the Indian Agriculture Research Institute (IARI) in the Dharbhanga district and Sone Command area has shown that the wheat yield can be increased substantially with good seeds and improved agronomic practices. The major bottleneck is, however, the absence of a grain purchasing machinery which will provide the Minimum Support Price (MSP) to farmers.

Action to extend the gains of higher productivity and profitability should cover all rainfed areas. This should be a priority task of the National Rainfed Area Authority. The recommendations of the Swaminathan Committee on

"More Income Per Drop of Water (2006)" should be converted into action plans on a location and farming systems basis.

Making New Gains

The immediate prospect for making new gains lies in the areas of post-harvest technology, agro-processing and value addition to primary produce. The National Commission on Farmers (NCF) has made several recommendations in this area in its first four reports. In the longer term, there is need for new yield and quality breakthroughs in major crops through genomics and gene pyramiding. For example, super wheats capable of yielding about 8 tonnes per hectare are now in the breeders' assembly line. Such wheats have a complex pedigree and illustrate the importance of genetic resources conservation and exchange.

Super wheats are semi-dwarf with robust stem, broad leaves, large spikes, with more grains per panicle and more grain weight. The super wheat architecture in the breeders' assembly line, both at the International Maize and Wheat Improvement Centre (CIMMYT) and All India Coordinated Wheat Improvement Programme (AICWIP), is derived from a blend of Tetrastichon (Yugoslavia), Agrotriticum (Canada), Tetraploid Polonicum (Poland) Gigas (Israel), Morocco wheat (Morocco) and semi-dwarf wheats currently grown in India

We can produce 100 million tonnes of wheat by 2015, by the following two steps:

- Average yield of 4 tonnes per hectare from 25 million hectares.
- Harnessing the large untapped yield reservoir in eastern, central and western India.

Every state should develop a detailed agricultural strategy for its major farming zones and systems based on the three-pronged approach, defending the gains already made, extending the gains to dry farming areas and making new gains through value addition.

Surface water is going to be a serious constraint since, as pointed out earlier, ground water is being overexploited. Fortunately, the Union Ministry of Water Resources has recently initiated a "More Crop and Income per Drop of Water" Movement. During this year, starting with the kharif, 5,000 Farmer-Participatory-Action- Research-Programmes have been initiated throughout the country by the ICAR through the agricultural universities, ICAR Research Institutes, the International Crop Research Institute for Semi-Arid Tropics (ICRISAT) and the Water and Law Management Institutes (WALMIs). 50 such institutions have been entrusted with the responsibility of organising jointly with farm families 100 Action Research Programmes each for demonstrating that it is now possible to increase yield and income per drop of water by creating synergy among water, crop variety choice, agronomic practices, particularly relating to soil macro and micronutrients, and agricultural implements. Each programme will cover a minimum of one hectare and will be implemented in a participatory mode, with the farm family having a sense of ownership of the programme.

The emphasis will be on rainfed areas where catalytic technological and management interventions will be introduced to make a striking impact. The programme will be so designed that a small government project leads to a mass movement in the area of water conservation and use efficiency, as happened in the case of National Demonstrations in Wheat during 1964-65. The economic benefit to the farmer as a result of this programme should be measured. Each Action Research Programme will need about Rs. 50,000. Thus, the total cost of 5,000 Farmer Participatory Action Research Programmes will come to Rs. 25 crore. A well-planned water literacy drive, together with the revitalisation of traditional systems of water conservation will also be undertaken as a part of this programme. Also, Action Research Projects in irrigated areas will aim at phasing out flood irrigation by the end of the 11th Plan.

Integrated Asset Reform

There is need to complete the unfinished agenda in land reforms. Kerala, West Bengal and now Tamil Nadu have set good examples on the distribution of both ceiling surplus land and appropriate government land to the landless poor. Tamil Nadu's recent step in providing two acres of land to the landless labour families is commendable. This should be emulated by all states. We should revive the spirit of Acharya Vinobha Bhave. In addition, there is need for aquarian reform for the equitable and efficient utilisation of all community and government water bodies. Aquarian reform is also needed in respect of marine fisheries and coastal aquaculture. This should be high on the agenda of the National Fisheries Development Board. Because of population pressure, land and aquarian reforms alone may not be adequate to provide productive assets. Land and aquarian reforms could form part of an integrated asset reform system designed to provide some productive asset to everyone in the village. Livestock rearing, training in market driven skills or any other form of income security could all form part of an Integrated Asset Reform Policy. Livestock provide good opportunities for strengthening both income and nutrition security. A Livestock Development Council would help to promote integrated attention to all aspects of livestock care and sustainable use. Every veterinary and animal sciences university should establish a Vidya Dairy on the model of the one at Anand, and a Vidya Abattoir to promote the efficient use of the entire animal biomass like skin, bones and blood.

From Suicide Relief to Suicide Prevention

In the agrarian distress hotspots, there is need for a paradigm shift from "Suicide Relief" to "Life-saving Support." While immediate relief measures are important, prevention should be the goal. This is the pathway to providing every farm and landless labour family with an opportunity for a productive and healthy life.

The occurrence of farmers' suicides in several states of the country (particularly Andhra Pradesh, Maharashtra, Karnataka, Kerala and the Punjab), since the year 2000 marks a sad chapter in India's agricultural history. The extreme step of taking one's life marks the loss of hope in the prospect for leading a productive and satisfying life. The farmers' suicide tragedy has several dimensions – economic distress and despair, breakdown of social and state support systems, and psychological nightmare. The response to this situation has also to be multi-dimensional, with priority going to mitigating economic distress. The various steps taken by central and state governments, the Prime Minister's visit to Vidarbha and the different relief packages announced so far have helped to instil hope in the minds of farm families. Nevertheless, farmers' suicides are continuing. We need to develop strategies for concurrently increasing on-farm productivity and off-farm employment. Multiple livelihood opportunities are essential for strengthening the farmers' coping capacity to withstand crop losses caused by drought, floods and biotic stresses.

The agrarian crisis sweeping the country underlines the need for attending to the problems of farmers and farming with the same seriousness as we did in the early Sixties. According to most estimates, farming is no longer remunerative and over 40 percent of farmers would prefer to quit farming if they have an alternative option. Farming is both a way of life and the principal means of livelihood for nearly two-thirds of our population. It is clear that we will not be able to make progress in improving our per capita income or human development indicators, if agriculture continues to stagnate. This is why our Prime Minister is calling for a second green revolution. Green revolution is another term for improving production through productivity enhancement. The first green revolution was based on the development and spread of new genetic strains of wheat, rice, maize and other crops characterised by their ability to utilise irrigation water, sunlight and plant nutrients effectively and convert them into grains. This revolution was confined to areas with assured irrigation. However, even in these areas like the Punjab, farming is becoming unremunerative due to adverse ecological and economic factors, and farmers are getting heavily indebted. The challenge now is to fight and overcome the fatigue of the green revolution in its heartland.

About 15 years ago, I stressed the need for developing technology and public policy for an evergreen revolution designed to improve the productivity of crops in perpetuity without associated ecological harm. The evergreen revolution is based on an appropriate blend of different approaches to sustainable agriculture such as organic farming, green agriculture, eco-agriculture and agriculture based on effective microorganisms. While organic farming excludes the use of mineral fertilisers, chemical pesticides and genetically modified crops, the green agriculture practised widely in China is based on the principles of integrated pest and nutrient management and integrated natural resources conservation and enhancement. We need to promote green agriculture, which allows the use of the minimum essential mineral fertilisers and chemical pesticides in both rainfed and irrigated areas.

Our agriculture is becoming a gamble in both the monsoon and in the market. Public investment in irrigation has been going down. Fortunately, the Bharat Nirman programme will help to reverse this trend. We should restrict the use of the term "second green revolution" to improving productivity in dry-farming areas. In such areas, pulses, oilseeds and other high value but low water requiring crops can be grown. At the moment, we are importing pulses and oilseeds from abroad and keeping our dryland farmers poor. The first and foremost requirement in such areas is the operation of a minimum support price and an efficient farmer centred marketing system. Also we should include jowar, bajra, ragi and pulses in the Public Distribution System. "Coarse grains" should be redesignated as "nutritious grains" since they are rich in nutrients. At the moment, our food security system is being developed on the basis of imported grains. This will destroy the prospect for a second green revolution or even an evergreen revolution. This is why the National Commission on Farmers has recommended the establishment of an Indian Trade Organisation, to protect the interest of our farmers. The recent collapse of the World Trade Organisation (WTO) negotiations in agriculture has clearly underlined the fact that the industrialised countries have no intention of reducing subsidies to farmers in their countries. Most of these farmers are corporate agribusiness companies. It is well known that these subsidies run to hundreds of billions dollars each year.

In the rainfed region of Vidharbha, it is too risky to adopt expensive technologies like Bt cotton. Small farmers who take loans for cultivation expenses have no coping capacity to meet the calamity of crop failure. Traditional crops like jowar should once again be revived. The funds allotted under the Prime Minister's package for seed replacement should be used to promote jowar, pulses, fodder jowar and legumes. Also, organic farming and crop-livestock integration should be promoted on both ecological and economic grounds. Vidharbha can be declared as the organic farming zone of Maharashtra, so that its oranges, jowar, cotton and other crops become known as organic products and thereby gain in market value. Fortunately, considerable technical knowledge is available on scientific organic farming in Vidharbha.

Farmers are crying for life-saving support today and not just schemes which may fructify a few years later. India's future lies in becoming a strong, vibrant and prosperous agricultural nation. This can be achieved through a convergence of technology, public policy and services and farmers' enthusiasm, leading to an ever-green revolution.

Sustainable Agriculture

Sustainability science is a multi-disciplinary field. It requires a holistic approach while analysing problems like I did in my analysis of high yield agriculture in 1968, before the term green revolution was coined by Mr William Gaud of the USA in September 1968. The kinds of problems that exploitative agriculture can create were described by me at the Indian Science Congress held in Varanasi in January 1968 in the following words:

> Exploitative agriculture offers great dangers if carried out with only an immediate profit or production motive. The emerging exploitative farming community in India should become aware of this. Intensive cultivation of land without conservation of soil fertility and soil

structure would lead, ultimately, to the springing up of deserts. Irrigation without arrangements for drainage would result in soils getting alkaline or saline. Indiscriminate use of pesticides, fungicides and herbicides could cause adverse changes in biological balance as well as lead to an increase in the incidence of cancer and other diseases, through the toxic residues present in the grains or other edible parts. Unscientific tapping of underground water will lead to the rapid exhaustion of this wonderful capital resource left to us through ages of natural farming. The rapid replacement of numerous locally adapted varieties, with one or two high-yielding strains in large contiguous areas, would result in the spread of serious diseases capable of wiping out entire crops, as happened prior to the Irish potato famine of 1854 and the Bengal rice famine in 1942. Therefore, the initiation of exploitive agriculture without a proper understanding of the various consequences of every one of the changes introduced into traditional agriculture, and without first building up a proper scientific and training base to sustain it, may only lead us, in the long run, into an era of agricultural disaster rather than one of agricultural prosperity.

Such an inter-disciplinary science has to be built on the following foundations:

- **Ethics**. Ethical considerations will have to guide human behaviour in relation to natural resources exploitation. Bioethics and environmental ethics are now developing into well-defined scientific areas. The ethical responsibility of safeguarding the environment rests on professionals, political leaders and the public. In the past, by investing conservation with spiritual significance, every individual was made to integrate ethics in day-to-day life.
- **Economics.** Ecological economics does not permit depreciation of natural assets. Thus, it has a time dimension of infinity. Ecological economics is also a fast developing science and it will help to measure the benefit-risk structure of development projects from the point of

view of their long-term impact. Ecological economics should become part of the curriculum in technological and management institutions. All dependents on natural resources for their enterprises should understand that good ecology is the pathway to good and enduring business.

- **Equity.** The concept of equity is now discussed in terms of both intra-generational equity and inter-generational equity (i.e. safeguarding the interests of the future generations). For example, over-exploitation and pollution of the aquifer will deny opportunities for ground water availability to future generations. Similarly, the melting of ice and glaciers, resulting in water shortage in cold desert areas like Ladakh, will force the future generations to migrate from the area. Climate change leading to the melting of ice will not only cause floods in the plains but also a rise in sea level over a period of time. Another important component of equity relates to the gender dimension of sustainability science. Women have been the great conservers of biodiversity and natural resources. Their role should be acknowledged and strengthened.
- **Energy.** Energy is a key factor in terms of both economic development and climate change. Integrated energy supply systems involving the optimum use of all renewable forms of energy like solar, wind, biomass, biogas, geothermal, etc., have to be developed. Other opportunities like hydrogen and nuclear energy will have to be integrated into an overall sustainable energy security system.
- **Employment.** Many of the livelihood opportunities in developing countries are based on the use of natural resources like land, water, forest and biodiversity. Emerging technologies tend to promote jobless economic growth. In population rich but land and water hungry countries, there is need for job-led economic growth. Therefore, development experts and technology developers should take into account the impact of new technologies and management procedures on job and livelihood security. Jobless growth is joyless growth in population rich countries.

- **Education.** Education is a cross-cutting theme and has to take into account all the above-mentioned factors. Environmental literacy should be based on the principle of "do ecology." For example, in the case of biodiversity, there is need to create an economic stake in conservation. Orphan crops can be saved only if there are markets for them. Similarly, in the case of nature tourism, those who operate houseboats or hotels in eco-sensitive areas should be aware that good ecology is good business. Environmental education should also be based on practical examples, which can drive home the message which is to be conveyed. Therefore, it should be based on field projects which can demonstrate how to organise ecotourism, conduct green audit, manage rain forests sustainably, etc. Just as action research programmes help to gather data on the economics and ecology of development projects, action education will derive its roots from field experience.

Countries like ours require "do ecology" and not just "don't ecology." Education should go to the grassroot level and, in this respect, India is fortunate to have grassroot democratic institutions like Panchayats and Nagarpalikas. Elected members of these bodies should become environmentally literate. This is where modern information, communication technology involving the integrated use of the internet, cable TV, community radio and the cell phone will help to achieve last mile and last person connectivity in terms of knowledge empowerment. Distance education methods as promoted by the Indira Gandhi National Open University will be very important for reaching the unreached and voicing the voiceless.

Speaking on "Agriculture in our Spaceship Earth" in 1973, I proposed a twin strategy to deal with the growing damage to our life support systems. These were, "do ecology" for developing countries, and "don't ecology" for industrialised countries. The first revolves around activities, which will generate an economic stake in conservation and help to reduce poverty. The "don't ecology", in contrast, largely relates to regulations and restrictions in areas such as carbon emissions and the unsustainable consumption of natural resources. Two examples of "do ecology" given below have a large potential for extrapolation.

First, the tsunami of December 26, 2004, resulted in severe loss of life and property along coastal Tamil Nadu, where I now live. For over 15 years now, we have been trying to persuade coastal communities not to destroy the mangrove forests along the coast. But their livelihood preoccupations did not allow them to pay heed to that request. The tsunami miraculously changed their outlook. Villages adjoining the thick mangrove forests were saved from the fury of the tsunami, because of the speed-breaker role played by the mangroves. In villages, where mangroves had been destroyed either for fuel wood or aquaculture ponds, several hundred fisher people died. This area is near the temple town of Chidambaram, where centuries ago, the temple builders had chosen a mangrove species as the temple tree. Following the tsunami, there was a sudden awareness of the reason for this choice, and local people now refer to mangroves as "life-savers." What we could not achieve in 15 years by arguing that mangroves would serve as a bioshield in the event of sea level rise, was achieved in a day.

The same tsunami brought home to farmers near the shoreline the importance of conserving local land races of rice. Several thousand hectares of rice fields along the coast got inundated with sea water. Most varieties perished, but a few salt-resistant ones withstood the inundation. Conservation of local biodiversity got a shot in the arm, and now every farmer wishes to maintain a field gene bank (i.e. *in situ* on-farm conservation) and a seed bank. The calamity became an opportunity to prepare both fisher and farm communities to meet challenges linked to a rise in sea level. The bioshield and agro-biodiversity conservation movements in this area have now become community-driven.

A second example relates to the revitalisation of the conservation traditions of tribal communities in the Eastern Ghats region. Fifty years ago, the tribal communities in the Koraput region of Orissa were familiar with more than 1,000 land races of rice, but at the turn of the century, this figure had come down drastically. "Dying wisdom" became linked to vanishing crops.

It became clear that the only way tribal families would once again start conserving agro-biodiversity would be by creating an economic stake in conservation. A dynamic programme of participatory conservation and breeding, coupled with agronomic improvement, soon led to a big spurt in the production of Kalajeera, an aromatic local variety, which is being snapped up by the market almost as soon as it is harvested. The same has started happening in Kerala with medicinal rices like Navara used in traditional ayurvedic practice, and with under-utilised millets in the Kolli Hills region of Tamil Nadu.

To cut a long story short, "do ecology" is triggered either by an ecological disaster or an economic opportunity. Preaching does not help. Enlightened self-interest, however, motivates people and leads to harmony with nature. This is happening in the green revolution areas of the Punjab too. Thirty years ago, when I pointed out to Punjab farmers that the "green revolution" was becoming a "greed revolution" because of the excessive use of mineral fertilisers and the over-exploitation of ground water, they listened politely, but did not change course. Now, in a despairing mood, they are ready to change. The economics of unsustainable farming has become adverse, leading to indebtedness, and occasional suicides. The "climate" has become opportune for farmers to take to conservation farming.

Developing countries with pervasive poverty and expanding populations should spread a "do-ecology" methodology, which can confer tangible ecological and economic benefits to the people. The industrialised countries with high standards of living and a highly educated population should press ahead with "don't" regulations.

The Brundtland Commission Report was appropriately titled "Our Common Future", to emphasise that, irrespective of political frontiers, our future is ecologically intertwined. I would like to add that without a better common present, the hope for a better common future may remain elusive.

Both unsustainable life-styles and unacceptable poverty must vanish, if humankind is to have a better common present and future.

Sustainability science is, thus, multi-disciplinary and multi-dimensional. For each area of human activity, there is need to develop technologies which can help to achieve the desired goal without associated ecological harm. For example, in the case of agriculture, which occupies the largest land area and utilises over 75 per cent of water resources, there is need for developing methodologies for achieving an evergreen revolution which alone can ensure enhancement of productivity in perpetuity. Conservation farming and green agriculture, which involve the use of integrated natural resources and pest management techniques, are the pathways to an evergreen revolution. This will call for anticipatory research as, for example, in the case of meeting the challenges of climate change, as well as participatory research and knowledge management in order to ensure that the recommended practices are socially compatible and feasible. Also, education has to be derived from the adoption of an agro-climatic and agro-ecosystem approach, taking into consideration the specific needs and opportunities prevailing in arid, semi-arid, hill, coastal, irrigated and island ecosystems. Harmony with nature should become a non-negotiable ethic. The rise and fall of great civilisations in the past have been related to the use and abuse of land, water and other natural resources. Therefore, sustainability science should hereafter guide all technology development and dissemination programmes.

Finally, population growth should not exceed the population supporting capacity of ecosystems. The human ecological footprint should be reduced through limiting wants and avoiding waste. Today, over a billion women, men and children of the human population are living in absolute poverty and destitution. Another one billion are leading unsustainable life-styles. Therefore, the ethical principles propagated by sustainability science should aim to curtail both poverty and unsustainable consumption of natural resources. This is the challenge before us from the point of view of ensuring the well-being of both the present and future generations. To meet this challenge, we must integrate the best in traditional wisdom and frontier science like biotechnology and information and communication technology. I would, therefore, like to deal briefly with methods of achieving such a blend.

Knowledge is a continuum. Present-day discoveries often have their roots in prior knowledge. Unfortunately, the Intellectual Property Rights (IPR) regime tends to ignore the contributions of traditional knowledge in the creation of new knowledge. This has led to accusations like biopiracy, plagiarism, knowledge piracy, etc. The World Intellectual Property Rights Organisation (WIPRO) has, hence, emphasised the need for recognising the role of traditional knowledge in the growth of contemporary science and technology. Fortunately, the Global Biodiversity Convention adopted at Rio de Janeiro in 1992 and the Food and Agriculture Organisation (FAO) International Treaty on Plant Genetic Resources for Food and Agriculture (2001) have both stressed the importance of recognising and rewarding traditional knowledge as well as the contributions of rural and tribal families to genetic resources conservation and enhancement through knowledge addition on their practical value. Our national legislations, the Plant Variety Protection and Farmers' Rights Act (2001) and Biodiversity Act (2002), have both emphasised the importance of recognising and rewarding traditional knowledge and local agro-biodiversity, which often constitute the backbone of our food and livelihood security systems.

Traditional Knowledge and Modern Science

Traditional knowledge has led to the growth of indigenous systems of medicine like *ayurveda, unani, siddha,* etc. There is a growing awareness of the importance of traditional systems of medicine. Saving plants for saving lives and livelihoods has become a global goal. Unfortunately, however, there is still no methodology for rewarding traditional knowledge, since it involves community recognition, although there are systems in place for providing financial recognition in the field of genetic resources conservation and sustainable use. For example, the Gene Fund provided for in the Plant Variety Protection and Farmers' Rights Act and the Biodiversity Fund provided under the Biodiversity Act can be used for rewarding and strengthening the *in-situ* on-farm conservation traditions of local communities. It should be emphasised that while cryogenic ex-situ conservation leads only to the preservation of specific genotypes, on-farm conservation results in both preservation and evolution. New genotypes through mutation and recombination can occur under conditions of *in-situ* conservation, while *ex-situ* methods involving cryogenic storage can only lead to preservation without loss of viability. Therefore, we should do everything possible to promote *in-situ* conservation by recognising and rewarding traditional knowledge and conservation techniques.

Anil Agarwal and Sunita Narain (1997) have chronicled the dying wisdom in relation to water harvesting and conservation techniques developed over the ages. The US National Academy of Sciences has published a series of books on the Lost Crops of the Incas, Lost Crops of Africa, etc. The World Health Organisation (WHO) has been appealing to save plants to save lives, with reference to medicinal plants. Therefore, no further time should be lost in preventing the erosion of traditional knowledge and local biodiversity. Saving plants and traditional wisdom are particularly important to face the challenges arising from global warming and climate change.

There are around 1,500 gene banks in operation today in different parts of the world, providing facilities for *ex situ* conservation for an estimated six million species and varieties. More recently, the Scandinavian countries have established a long-term seed storage facility under perma frost conditions known as the Svalbard Global Seed Vault, which can hold over six million seed samples. This will serve like a Noah's Arc in order to preserve for posterity a sample of genetic diversity currently occurring on our planet. However, as already emphasised, cryogenic preservation will not give us the benefit of natural evolution and the further development of new genes and genotypes. This is why recognition of traditional knowledge and traditional conservation ethos is exceedingly important. Sacred groves and sacred trees constituted important methods of conserving economic, ecological and spiritual keystone species. These are also tending to get neglected.

Several steps have been taken in India to recognise and preserve traditional knowledge. A database on indigenous innovations is being kept at the Institute of Management, Ahmedabad, under the leadership of Prof Anil Gupta. The Foundation for the Revitalisation of Indigenous Health Traditions (FRLHT) is also maintaining a database on our heritage of both medicinal plants and traditional medicine and health practices. There are many other initiatives including the Community Gene Bank of the M.S. Swaminathan Research Foundation (MSSRF). We are yet to start a similar programme with animal genetic resources. India is very rich in animal wealth but, unfortunately, many important breeds, including the Vetchur cow of Kerala, are now endangered animals. We should institute a Breed Saviour Award to accord recognition to those who are conserving local breeds of cattle, buffalo, sheep, goat, poultry, etc. In the case of poultry, there is indiscriminate killing of native breeds of birds in order to prevent the spread of the H5N1 strain of the avian influenza. In this process, we may lose genes for resistance, in case any of the local breeds possess such genes. Therefore, we should establish an offshore quarantine island in one of the unmanned Lakshwadeep group of islands, where, in a high security greenhouse testing of local poultry, breeds for resistance to the H5N1 strain could be conducted. We must strengthen our infrastructure for searching for, and saving, genes, which can help us to overcome emerging challenges caused by both climate change and transboundary pests.

There is need for national and international financing instruments for promoting the conservation of traditional knowledge and endemic bioresources. At the international level, the Global Environment Facility (GEF) is financing measures to implement the Convention on Biological Diversity, the Framework Convention on Climate Change and the Convention to Combat Desertification. More recently, a Global Crop Diversity Trust was set up in 2004 with an initial capital of US$ 260 million.

The Trust supports information systems for the concerns of agricultural biodiversity, including databases, documentation of collections and the exchange of information through networks. These international initiatives are important but what is more important is the spread of genetic literacy among our population. Every child, woman and man should become aware of the value and significance of traditional wisdom and local biodiversity. This will become easier if there is an economic stake in conservation. We should establish biovalleys in areas rich in bioresources. The aim of the biovalley is to promote an era of biohappiness arising from the conservation and sustainable and equitable use of biodiversity, leading to more jobs and income for the local population. Otherwise, both traditional knowledge and native biodiversity may tend to disappear. The power of "the seeing eye and understanding heart" will be evident from the outstanding contribution of the farmer-breeder, Mr Joseph Sebastian, whose cardamom variety "Njallani Green" is the ruling variety in the Idukki district. Njallani has helped to improve the productivity and profitability of cardamom and illustrates the power of indigenous knowledge and observation power.

Food Security

Food security is likely to be a major casualty in an era of climate change. Adverse changes in precipitation, temperature and sea level will harm present and potential food production. This will be a disaster, particularly in countries like ours where population is still growing and per capita land and water resources are shrinking. The most urgent task, therefore, is to strengthen our agricultural production systems under conditions of uncertain weather patterns.

Fortunately, short and medium term weather forecasting techniques are improving with reference to reliability. The Indian Meteorological Department has predicted a normal southwest monsoon this year. What should we do to maximise the benefits of a normal monsoon using environmentally benign technologies? In my view, we should launch immediately a "bridging the yield gap movement" using clean technologies associated with conservation farming and green agriculture. The idea of the movement is to bridge the prevailing gap between potential and actual yields.

Our food security must be based on home grown food, so that it is both reliable and affordable. Food and fuel will be the most expensive commodities in the coming years. Home grown food-based food security will also help to strengthen rural livelihoods and achieve the goal of food for all, surely and speedily.

Climate Change and the Global Security Environment

□ **Chandrashekhar Dasgupta**

Climate change has ascended to the top of the international agenda within a short period of two decades. It figures prominently in the annual UN General Assembly debates, in the deliberations of specialised agencies such as the World Trade Organisation (WTO), International Labour Organisation (ILO), World Health Organisation (WHO), International Meteorological Organisation (IMO) and International Civil Aviation Organisation (ICAO), in international financial institutions, including the World Bank and the regional development banks, and in such plurilateral forums as the G-8 or G-77 Summits. Successive reports of the Intergovernmental Panel on Climate Change (IPCC), which includes leading scientists from around the globe, have shown with increasing degrees of certainty the expected impacts of climate change induced by human activities.

Side by side with the increasing certainty of the advent of climate change and its consequences in diverse areas of human activity, there has developed a growing awareness of the possible long-term impacts of climate change on the global security environment. A seminal paper on the subject appeared as early as in 1991[1] and, in recent years, we have witnessed an outpouring of officially sponsored, as well as independent academic studies on the potential security implications of climate change. These include an official European Union (EU) paper, *Climate Change and International Security* [2]; a scholarly volume by the German Advisory Council on Global

Change (WBGU) entitled *Climate Change as a Security Risk*[3]; a current UK Ministry of Defence project to identify regions where global warming might spark off conflict or security threats[4]; the US Council on Foreign Relations Special Report, *Climate Change and National Security*[5]; the Centre for Studies and International Security (CSIS)/Centre for a New American Security publication, *The Age of Consequences: The Foreign Policy and National Security Implications of Global Climate Change*[6]; and the US Centre for Naval Analysis study, *National Security and the Threat of Climate Change* (April 2007).

Climate change impacts will modify the physical geography of our planet and will also alter the distribution and accessibility of natural resources such as freshwater, arable land and seabed mineral reserves. These changes can potentially affect the international security environment.

Geographical changes induced by global warming are already in evidence in the Arctic region. The polar ice cap is shrinking, opening up new navigable sea lanes and raising expectations that the vast oil and natural gas reserves concealed under the ice cap might become accessible in the foreseeable future. As a result, regional states have revived conflicting claims to territories, territorial waters and Exclusive Economic Zones (EEZs). Rising sea levels will result in the submergence of low-lying islands and coastal areas, altering the configuration of coastlines. These changes could lead to disputes over maritime borders or EEZs.

Climate change will also alter the distribution of natural resources such as freshwater and arable land. This could lead to intensified competition for resources – within and between states – and might even lead, in extreme cases, to violent civil strife or international conflict. As a result of intensified domestic conflict, weak states might become ungovernable "failed states." Deprivation and new inequalities in the distribution of scarce resources might also create fertile breeding grounds for terrorism. Environmental degradation could provide an impetus to massive migration flows. Contests over scarce resources, if unresolved through diplomacy, could conceivably even lead to international armed conflict.

Changes in rainfall and glacial melt patterns will affect freshwater availability. Water scarcity and drought conditions may lead to a reduction in the extent of arable land while, on the other hand, higher temperatures may open up new cultivable areas in regions (e.g. Siberia) that currently suffer from extreme cold. Sea level rise will result in submergence of low-lying coastal and island territories and will lead to domestic population shifts or migration across international borders. Marine resources – more specifically, fisheries – will be affected by changes in temperature and shifts in oceanic currents. Competition and conflict may be stimulated by a reduction in the availability of a natural resource (e.g. freshwater) as well as increased availability. For example, the retreat of the Arctic ice cap may make it possible to exploit currently inaccessible seabed oil and natural gas resources. This has already led to more assertive postures in regard to territorial claims in the Arctic region. These are some examples of the ways in which climate change might alter the distribution of natural resources, sharpening domestic or international contest and, possibly, contributing to conflict.

In general terms, therefore, climate change will have a significant impact on the global security environment over the course of this century. However, we must enter a couple of caveats.

First, the present state of scientific knowledge does not enable us to predict with any degree of certainty the precise extent or timing of climate change impacts in a specific geographical area. Few detailed micro studies of climate change impacts are currently available[7]. Moreover, the degree of temperature increases and associated phenomena such as sea level rise or more frequent extreme weather events can only be projected within a fairly broad range. We can draw up scenarios that indicate possible or plausible impacts but these should not be misconstrued as *predictions*. A large measure of uncertainty is associated with these scenarios.

Second, climate change will occur in progressive stages over an extended period of time stretching at least into the next century. Thus, the security implications of climate change will unfold incrementally, almost imperceptibly. Its impacts will not be clearly distinguishable from other factors. Climate change will generally aggravate *existing* tensions and strengthen trends that are already in evidence. It is unlikely to be identified as the primary cause of any particular conflict.

With these words of caution, we may now proceed to sketch some plausible scenarios concerning possible impacts of climate change on global security.

Arctic Dispute Scenarios

Will the *Arctic region* become an international "hot spot" as a result of the impacts of climate change? Temperatures in this region are rising twice as fast as the average for the rest of the planet and the Arctic ice cap is shrinking at a relatively rapid rate. The US Geological Survey estimates that 25 per cent of the world's undiscovered oil and gas reserves lie under the ice cap. The prospect of these resources becoming exploitable has already sparked off a new territorial contest in the Arctic region. Russia, Denmark and Canada are vigorously pursuing overlapping border and EEZ claims – claims that had lain dormant for many decades.

The melting of the polar ice cap is also opening up new shipping lanes in the Arctic Ocean. The Northern Sea route (along the Russian coast) and the Northwest Passage (linking the Atlantic and Pacific Oceans) will become navigable for at least part of the year. These are likely to become major shipping lanes, providing much shorter shipping routes between Europe and the Far East. The opening of the Northwest Passage would reduce the route by as much as 4,000 nautical miles, compared to the present route through the Panama Canal. The new Arctic navigation routes have already brought into focus a dispute between Canada and the United States over the Northwest Passage, which Ottawa claims as its internal waters, while, according to Washington, these are international waters through which all countries have the right of free passage.

Russian claims in the Arctic are based on the contention that the undersea Lomonosov Ridge, which runs from its coast to the North Pole, should be regarded as part of its continental shelf. In a spectacular assertion of Moscow's claim, in August 2007, a Russian submarine planted a national flag, fashioned from titanium, directly below the North Pole. Since the Lomonosov ridge extends also to Greenland (territory of Denmark), the Danes have advanced a similar claim in the Arctic region. Canada, on its part, has forcefully asserted its claim to territorial rights over an Arctic archipelago consisting of some 19,000 islands and all adjacent waters. In a major policy statement in July 2007, Prime Minister Harper said, "More and more, as global commerce routes chart a path to Canada's North and as the oil, gas and minerals of this frontier become valuable, northern resource development will grow ever more critical to our country." Harper added, "Canada has a choice when it comes to defending our sovereignty over the Arctic, we either use it or lose it. And make no mistake, this government means to use it."[8] Indicating that his government meant business, the Prime Minister announced that Canada would build an Arctic deep-water port and induct into service eight new patrol ships equipped with ice-cutters.

The vigorous revival of conflicting claims in the Arctic is the most telling example, to date, of the impact of climate change on international security. The multiple conflicting claims in the Arctic region have a potential for generating tensions between major claimants. It should be added, however, that an outbreak of armed conflict is very unlikely, in view of the state of diplomatic relations between these countries.

Water Stress Scenarios

Freshwater resources are expected to shrink significantly as a result of global warming in regions that are dependent on glacial melt water during the dry season. As glacier melt increases, the runoff will increase initially but will subsequently decline sharply as glaciers disappear. The resultant water stress will sharpen international competition and could even result in armed conflict in extreme cases.

In *West Asia,* the flows of the Jordan and Yarmuk rivers, upon which Israel, Jordan and Palestine are dependent, is expected to decline significantly.

Freshwater availability in Israel is expected to fall by as much as 60 per cent during this century. It has been estimated that by 2025, per capita water availability in Israel would be less than 500 cubic metres, as against the minimum requirement of 1,000 cubic metres for an industrialised country[9]. This is likely to exacerbate current tensions over access to water, particularly between Israel and its Arab neighbours, Jordan and Palestine. It has even been speculated that reduction in river flows might actually precipitate armed conflicts in this water stressed region. However, as we shall see below, "water wars" are a rather unlikely prospect.

Central Asia (comprising Kazakhstan, Kyrghyzstan, Tajikistan, Turkmenistan and Uzbekistan) is heavily dependent on glacier melt water. Glaciers in the region are retreating quite rapidly. According to a United Nations Development Programme (UNDP) estimate, the Tajikistan glaciers lost a third of their area during the second half of the 20th century[10]. The major upstream states controlling the headwaters are the mountainous countries of Kyrgyzstan and Tajikistan, while the major downstream consumers are the agricultural areas of Turkmenistan and Uzbekistan, particularly areas under water-intensive cotton cultivation. Some analysts have sketched possible conflict scenarios between downstream and upstream states over shrinking water supplies[11].

Reduced rainfall could aggravate instability in the *Horn of Africa*. It has been asserted that the almost 40 per cent decline in rainfall in northern Darfur during the 20th century led to intra-tribal tensions and that freshwater stress is one of the complex factors underlying the current conflict in the area[12]. Increased water stress in Somalia and Ethiopia can exacerbate ongoing internal conflicts and further weaken the state structure.[13]

It has been asserted that water stresses induced by climate change are likely to lead to "water wars" in regions such as West Asia. A careful examination suggests that this is an unlikely, though not impossible, outcome. Water stress is commonly resolved through upgradation of technology, adoption of superior water management practices and closer international cooperation, rather than by recourse to arms. In any case, from a military viewpoint, an armed conflict scenario seems implausible unless the aggrieved party (which can only be the downstream state) is militarily superior to the upstream state.

Migration Scenarios

Climate change will result in sea level rise because of the melting of the Arctic ice cap as well as thermal expansion of the oceans. Sea level rise threatens to submerge low-lying coastal areas and islands. The Ganges-Brahmaputra basin in South Asia, the Mekong delta in Southeast Asia, the Yangtse basin in China and the Nile delta in North Africa are among the densely populated territories facing a potential threat of submergence and population displacement.

The implications of sea level rise for *Bangladesh* have attracted a great deal of attention[14]. Sea level rise could lead to the submergence of vast areas of this densely populated, low-lying, riverine country. During this century, climate change impacts will lead, unless checked, to a massive increase in illegal migration across international borders, leading to economic, social and political instability in adjoining regions, in particular, India's north-eastern states. In the words of an American analyst, "Devastating floods in Bangladesh could send tens of thousands of refugees across the border to India, potentially leading to tension between the refugees and recipient communities in India."[15]

Africa is highly vulnerable to climate change. It has been estimated that the extent of arable, rainfed land in North Africa and the Sahel might shrink by as much as 75 per cent on account of water stress and soil degradation. Because of sea level rise and ingression of saline water in the Nile delta, Egypt might lose as much as 12-15 percent of its arable land by 2050, affecting 5 million people. We have already noted the impacts of climate change in the Horn of Africa. In Southern Africa, too, droughts may lead to acute food shortages. Europe is concerned that environmental stresses could stimulate large-scale emigration from the African continent to more prosperous areas across the Mediterranean.[16]

Similarly, concern has been voiced in the United States over the possibility that climate change might raise the tempo of migration from *Caribbean* islands such as Haiti and Cuba, or from neighbouring *Mexico*. The Caribbeans will be severely affected by extreme weather events related to climate change. Northern Mexico is likely to suffer from severe water shortages. An exodus could generate political tensions between the United States and these countries.[17]

In the words of the scholarly WBGU analysis, "If global temperatures continue to rise unabated, migration could become one of the major fields of conflict in international politics in the future."[18]

Destabilisation, "Failed States" and Radicalisation Scenarios

Quite a few states may be unable to cope with the adverse impacts of climate change because they lack effective governance structures and adequate financial resources. These poor and weak states are likely to suffer acutely from food shortages, rampant disease and frequent humanitarian emergencies. In these circumstances, law and order might collapse altogether and civil wars might break out. Analysts have argued that climate change poses the potential risk of destabilising state structures or even causing their total collapse ("state failure"). Moreover, deprivation and violence may create a fertile ground for extremist ideologues propagating terrorist creeds. Concerns have been expressed over the possibility that severe climate change might provide an environment conducive to the spread of terrorism. Thus, an American study on the security implications of climate change draws attention to possible state failure, creating "ungoverned spaces" where terrorists could organise their operations with ease.[19]

It may be noted that a number of states in India's neighbourhood face major challenges to their stability. Thus, the German Advisory Council for Climate Change (WBGU) includes Afghanistan, Pakistan, Myanmar and Sri Lanka in its list of weak and fragile states[20]. State failures on our periphery would obviously affect our security environment. It must be emphasised, however, that the impacts of climate change will be felt incrementally and are unlikely to be severe for the next three decades at any rate. Fragile or weak states have to contend with many more immediate and pressing challenges. Climate change will not be the primary cause of political instability or terrorism.

Should climate change be viewed as a "threat to international security" or a "threat multiplier"? This is not a purely academic question. Britain succeeded in placing climate change on the agenda of the UN Security Council in 2007. It must be questioned whether the move was appropriate.

The recent EU paper, "Climate Change and International Security" suggests that "climate change is best viewed as a threat multiplier which exacerbates existing trends, tensions and instability." However, Roy Anderson, the Chief Scientific Adviser to the UK Ministry of Defence, has offered a more precise definition. He describes climate change as "a key strategic factor affecting societal stresses and the responses of communities and nations to those stresses."[21] Climate change is, thus, best viewed as a "societal stress multiplier", not a "threat multiplier." Societal stresses can and should be peacefully resolved before they develop into "threats." On the basis of a careful analysis, the WBGU study comes to the conclusion that "climate-induced interstate wars are unlikely to occur. However, climate change could well trigger national and international distributional conflicts and intensify problems already hard to manage such as state failure, the erosion of social order, and rising violence."[22]

In the case of climate change, the need of the hour is for international cooperation to mitigate the phenomenon and to adapt to its impacts. Its worst stresses can be avoided altogether if highly industrialised countries reduce their greenhouse gas emissions and if developing countries are financially and technologically enabled to cope with its impacts. This would prevent aggravation of existing tensions and instability.

The proper UN forum to address climate change is the General Assembly and its subsidiary bodies, not the Security Council. The Security

Council was created to address urgent "classic" security issues resulting from aggression or the use of force. Climate change lies outside its competence. Addressing the potential impacts of climate change requires international cooperation to reduce global greenhouse gas emissions on an equitable basis and to promote economic and social development in developing countries in order to enable them to cope more successfully with the climate change impacts. These are matters that lie within the mandate of the UN General Assembly and its organs. All countries must have an equal say in these matters and no country, or group of countries, can claim an exclusive right of veto. The inscription of climate change on the Security Council agenda was an ill-advised move.

Apart from the more distant global implications of climate change, future planning for the armed forces must take into account the direct domestic impacts of climate change, such as severe and frequent storms, coastal erosion, floods and droughts.

Climate change will have implications for the survivability of a country's military as well as civil infrastructure – roads, railways, airports, bridges, port facilities, buildings, etc. Increased frequency and intensity of floods, storms, typhoons and other extreme weather events require enhanced standards for infrastructure construction. Sea level rise will have additional implications for coastal infrastructure. As a recent US study notes, "Climate change could, through extreme weather events, have a more direct impact on national security by severely damaging critical military bases."[23] The study notes that Miami, the site of the US Central Command (CENTCOM), has been identified as most vulnerable to hurricane storm damage. US military assets located in Miami are also vulnerable to extreme weather events.

Since defence-related infrastructure is usually expected to last for 50 years or more, the likelihood over a period of time of more frequent and intense extreme weather events as well as sea level rise needs to be taken into account. Building standards will have to be raised in many cases. Moreover, proneness to such risks must be considered in selecting sites for new installations. In the case of coastal installations, the effects of sea level rise have to be fully considered.

Climate change is expected to precipitate increasingly frequent humanitarian crises associated with storms, floods, droughts, etc. The armed forces are likely to be called upon to assist in disaster relief operations with much greater frequency in the future. Apart from the domestic context, appeals for assistance might also be received from severely affected neighbouring countries. A capacity to respond to such appeals for assistance would enhance a country's soft power. Cooperation with other states present in the region could meet urgent humanitarian needs, promote goodwill and assist in promoting stability. In drawing up their future plans, it would be prudent for the armed forces to take into account domestic as well as regional disaster relief contingencies.

Conclusion

From an Indian perspective, the long-term impacts of climate change will tend to aggravate two major existing national security concerns. First, the presence of relatively weak and unstable states on our periphery poses a potential concern because of the risk of a possible spillover of problems across our borders (as was the case, for example, in the former East Pakistan in 1971). Weak and unstable states are often also breeding grounds and launching sites of international terrorism. Climate change is expected to exacerbate the economic, social and political stresses on fragile states. Second, sea level rise in low-lying Bangladesh is expected to lead to the submergence of a significant part of its territory during the course of this century. This will tend to further aggravate the long-standing problem of illegal immigration into India, with all its consequences for social stability in the northeastern states.

As a result of climate change impacts, the armed forces are likely to be called upon to assist in domestic humanitarian contingencies with increasing frequency. They may also be required to undertake disaster relief operations in India's neighbourhood in cooperation with friendly countries. This may require advanced planning and preparation.

Finally, the direct impacts of climate change should be taken into account in all infrastructure planning. This may require upgrading current construction codes and standards. Site selection of coastal installations should also take into account the effects of expected sea level rise and proneness to extreme weather events.

Notes

1. Thomas F. Homer-Dixon, "On the Threshold: Environmental Changes as Causes of Acute Conflict", *International Security,* 16, Autumn, 1991.
2. *Paper from the High Representatitive and the European Commission to the European Council, S113/08.* Hereafter, Solana (EU High Representative).
3. R.Schubert, H.J. Schellnhuber, A.Epinay, R.Grieszhammer, M.Kulessa, D.Messner, S.Rahmstorf, J.Schmid, German Advisory Council on Global Change (WBGU), *Climate Change as a Security Risk* (London & Sterling, VA: Earthscan, 2008).
4. See report in *The Guardian,* September 11, 2007.
5. Joshua W. Busby, CSR No. 32.
6. Kurt M. Campbell, Jay Gulledge, J.R. McNeill, John Podesta, Peter Ogden, Leon Furth, R. James Woolsey, Alexander T.J. Lennon, Julianne Smith, Richard Weitz, and Derek Mix, *The Age of Consequences: The Foreign Policy and National Security Implications of Global Climate Change,* CSIS/ Centre for a New American Security, November 2007.
7. India's N*ational Action Plan for Climate Change* (2008) calls for such studies in the Indian context.
8. http:pm.gc.ca/eng/media.asp? id=1742
9. Campbell, et al, n.6, p.60.
10. UNDP, *Human Development Report 2006. Beyond Scarcity: Power, Poverty and the Global Water Crisis.* Cited in Schubert, et al, n.3, p.88.
11. See, for example, Schubert, Ibid., pp. 89-90.
12. See, for example, Solana, n.2, p.6.
13. See, for example, Busby, n.5, p. 8 and Solana, n.2, p.6.
14. See, for example, Campbell et al, n.6, p.57 and Busby, Ibid., p.8.
15. Ibid., p.8.
16. Solana, n.2, p.6.
17. See, for instance, Busby, n.5, p.6, Campbell, et al, n.6, p56, and CNS, p.7.
18. Schubert, et al, n.3, p.174.
19. Centre of Naval Analysis (US) report, cited in Busby, n.5, p.9.
20. Schubert, et al, n.3, pp.45-46.
21. n.4.
22. Schubert, et al, n.3, Summary for Policy-Makers, p.1.
23. Busby, n.5, p.6.

Water Security: Indian Concerns

□ Ramaswamy R. Iyer

Preliminary

This paper, based on the author's experience in water policy planning, needs, rights, conflicts, social justice, ecological sustainability, etc is a contribution within the overarching theme of comprehensive secutiry. The paper would seek to explore the idea of water security. In that context, some preliminary cautions seem desirable.

First, there is a difference between 'security' in the police and military senses and 'security' in other usages such as social security, food security, water security, etc. Under certain circumstances the non-military and military senses of 'security' may tend to overlap, but it is useful to maintain the distinction between the two.

Secondly, when we deal with water conflicts, environmental concerns, and so on, the guiding spirit has to be one of constructive cooperation and harmony. In using the language of security in that context, we have to take care not to introduce and adversarial orientation into the discussion.

Thirdly, if we wish to talk about water in the language of security, we must distinguish between 'water security' and 'water and security'. The first is a concern about water: its availability, adequacy, reliability and quality.

The second conveys a sense of insecurity in the conventional military sense, but arising from control over water. This paper shall be concerned mainly with the former.

Predictions of a Crisis

When we talk about water security, the first question that comes to mind is whether India is confronted with the prospect of severe water scarcity or a water crisis in the not too distant future. There is a widespread view that a water crisis is looming on the horizon. The demand for freshwater is expected to increase sharply and rapidly because of the growth of population, the pace of urbanisation and the processes of economic 'development', increasing the already severe pressure on the available (finite) supply. In a report a couple of years ago, the World Bank spoke of the Indian water economy as "bracing for turbulence." Many writers also refer to declining per capita availability of water resources. Following the World Bank, it has also become the practice to talk in terms of the inadequacy of "water storage per capita." All this involves fallacious thinking, but it is proposed to enter into that analysis here. Those interested may refer to the article "Water: A Critique of Three Concepts" in the *Economic and Political Weekly* of January 5, 2008. However, it is necessary to deal with the broad question of a looming crisis, taking note of the basic numbers of India as a whole:

- precipitation over the Indian landmass: 4,000 billion cubic metres (BCM);
- available surface water resources (as measured near the terminal points of river systems): 1,953 BCM;
- available groundwater resources: 432 BCM;
- *usable* surface water resources: 690 BCM;
- *usable* groundwater resources: 396 BCM;
- total usable water resources: 1,086 BCM.

There is some difference of opinion as to whether the river-flow figure of 1953 BCM mentioned above includes a part or the whole of the groundwater availability of 432 BCM, but we need not go into that question here. The above numbers are those of the National Commission on Integrated Water Resources Development Plan, hereafter NCIWRDP, in its Report of September 1999. The Central Water Commission's numbers are slightly different, but we may ignore those differences for our present purposes.

It must be mentioned that the accepted projections of water availability have been questioned by N. K. Garg and Q. Hasan, in an article in *Current Science* of October 10, 2007, and by Prof. T. N. Narasimhan of the University of California Berkeley, in an article in the June 2008 issue of the *Journal of Earth System Science,* published by the Indian Academy of Sciences, Bangalore, from two different points of view. These criticisms need to be examined carefully. However, as the official estimate of 'usable' water resources is much lower than that of 'available' water resources, there seems to be a safety margin, and perhaps the over-estimation has not done too much harm. It may be added that the Central Water Commission's final and authoritative reply to the criticisms is not yet available.

As against the above numbers of availability, the NCIWRDP gives a low estimate of 970 BCM and a high estimate of 1,180 BCM as the country's total water requirements by 2050. This is what gives rise to predictions of scarcity or crisis.

Examining Demand Projections

That seems plausible, but it must be noted that 'demand' is a crucial factor here, and that this will, in turn, depend crucially on how we use water. 'Demand' is, therefore, what we should look at first, and very carefully, before we even begin to think of supply-side answers.

Before proceeding further, let me make what may seem to be a semantic point. In relation to water, a basic life-support substance, the very term 'demand' seems to me questionable. We can talk about the need for water; or about the fundamental *right* to drinking water; but 'demand' seems the wrong word. This is not a quibble about words: the terminology assimilates water to the general run of consumer and industrial goods and reduces it to a commodity subject to the market forces of supply and demand. Without entering into an elaborate discussion of that issue, let me say merely this: the usual approach prevalent in the case of consumer or industrial goods, of projecting a future demand and bringing about a supply-side response to meet that demand, will be inappropriate in the case of water; instead, reversing that approach, we must start from the fact that the availability of fresh- water in nature is finite, and learn to manage our water needs within that availability.

However, leaving that point aside and accepting the term 'demand', let us look at the different components in the overall projection. Taking agriculture first, the benefits of irrigation are evident, but as a water-user, it has much to answer for. It is the largest user of water (around 80 percent). It is also an extremely inefficient water-user (35 to 40 percent as estimated by NCIWRDP; and yields in irrigated agriculture in India are quite low, and projected at only 4 tonnes per hectare even in 2050 (NCIWRDP). Substantial improvements in efficiency in water-use in agriculture (in conveyance systems, crop-water requirements, irrigation techniques, yields) are needed, and if achieved, could sharply cut down the agricultural demand for water.

An even more important point is that supply creates demand and necessitates more supply. The availability of irrigation water leads to the adoption of water-intensive cropping patterns. More water is needed even to *continue* with this kind of agriculture; and of course, there is a desire to *expand* that agriculture, creating a demand for still more water, until the demand becomes unsustainable. There is always a demand for more water and still more water. We have to get away from this kind of competitive, unsustainable demand for water.

In rural and urban water supply, the tendency is to project future needs on the basis of fairly high per capita norms, and the thinking is in the direction of enhancing the norms. However, is that necessary? In Delhi, for instance, the actual supply by the Delhi Jal Board is upwards of 200 lpcd, which is higher than the current norm and higher than the supply in other cities. The problem is that it is unevenly and inequitably distributed. There are areas where people – poor people – have to manage with 30 lpcd or less, and other areas where people – the middle classes and the rich – use 400 to 500 lpcd or more. What we need to do is to enforce economies on those (whether in rural or urban areas) who use too much water, and improve availability to groups or areas that receive too little. If this were done, it might not be necessary to raise the average.

In industrial use of water, multiple recycling and re-use needs to be insisted upon, allowing minimal make-up water: we must move towards a situation in which 90 per cent of the requirement of water for industry would be met through recycling. That might be very difficult today, but it must be our goal.

Strenuous efforts need to be made to maximise what we get out of each drop of water in every kind of water-use. Further, the amount of waste that is taking place in every use needs to be tackled: the waste must be reduced, and a part of it must be recovered for certain uses.

If we do all those things, the figure of future requirement of water may become significantly smaller than current projections, and we may be able to avert a crisis or at least minimise it. It is not my intention to promote complacency by discounting the predictions of a crisis. I am only pointing out that a crisis can be minimised if we do certain things. The actions that I have outlined will be undoubtedly difficult. However, they are doable.

Supply Side Actions

Even with strenuous efforts of 'demand management', some supply-side action may still be needed. There are only three ways in which water available for use can be augmented: rainwater-harvesting, groundwater drilling and large projects (for storage, i.e., dams and reservoirs, or for long-distance water transfers such as the ILR Project). Each of these would have its impacts and consequences. The impacts and consequences of large dams are by now fairly well known. In recent years, the reckless exploitation of groundwater and the consequent depletion and/or contamination of aquifers have begun to cause serious concern. Rainwater-harvesting has barely begun to be promoted, but some critics have already started cautioning against extensive recourse to this; in their view, this will have implications for overall basin hydrology.

Obviously, none of these possibilities on the supply side can be ruled out altogether. It seems evident that all three supply side actions might need to be adopted, but with great care and in a wise and prudent combination. In the past, our planners and engineers have treated big projects as the first choice, and relegated local rainwater harvesting initiatives to a minor, secondary role. My own recommendation would be to reverse that approach and treat local, community-led augmentation as the first choice, with big dams and long-distance water transfers as projects of the last resort, to be adopted only where they are the unique option or the best of available options; and the imposition of severe restraints on the exploitation of groundwater.

The justification for that reversal of approach is clear enough. First, if the approach of local augmentation holds promise of some addition to availability, then special emphasis needs to be laid on it; that will not happen so long as the orientation continues to be towards big projects. A change in that orientation is, therefore, necessary. Secondly, the option of local augmentation, where available, seems preferable to bringing in water from large and distant storages, except where the latter is the only course open or the best of available options. Thirdly, big projects have formidable impacts and consequences (ecological, social, human); those of small local interventions are far more manageable.

If we forget for a moment the questionable calculus of supply and demand and look at 'security' from the point of view of protecting the ecological system and planet earth, we begin to realise that by building a series of large 'water resource development' projects we may not be *ensuring* security but *endangering* it. Besides, such projects, while perhaps conferring greater water or energy security on some groups of people, may inflict severe insecurities on others.

Rivers Shared with Neighbours

A second aspect of water security relates to inter-country relations. India shares river systems with Pakistan, Nepal, Bhutan and Bangladesh. In all these cases, except that of Bhutan, the water relations are beset with problems.

From the point of view of water security, the question for this paper would be: is India's water under threat from her neighbours? The important point is that it is usually the lower riparian country that feels a sense of insecurity, not the upper riparian. The lower riparian tends, rightly or wrongly, to worry about upper riparian control over waters, and to entertain apprehensions about the upper riparian depriving it of water or harming it through the release of flood-waters. It follows that the national security implications (if any) of water conflicts in South Asia are likely to worry Pakistan and Bangladesh (the lower riparians) more than India (the upper riparian). India is, of course, a lower riparian in relation to Nepal and Bhutan, but those countries as upper riparians need not cause us much anxiety: they do not constitute serious threats to India's water.

However, three points need to be noted (i) there is a degree of dissatisfaction with the water-sharing under the Indus Waters Treaty 1960 and a series of differences over Indian projects on the western rivers; (ii) talks over several decades with Nepal on a number of large projects (mainly hydroelectric, but with irrigation components) have made no headway, and the Mahakali Treaty of 1996 has been virtually inoperative; (iii) despite the Ganges Treaty of 1996, prickly water relations persist between India and Bangladesh. These are elaborated in the following paragraphs.

India-Pakistan

The Indus Waters Treaty 1960 certainly settled the water-sharing dispute between India and Pakistan, and it has managed to survive four wars. In that sense, it must be regarded as a success. However, the allocation of waters under the treaty is regarded as unfair in both countries. Many in India feel that the allocation of 80 per cent of the waters to Pakistan and 20 per cent to India was a very unfair settlement; and many in Pakistan argue that the territories that went to India under partition were historically using less than 10 percent of the Indus waters, and that the treaty was generous to India in giving it 20 percent of the waters. Each side thinks (or many on each side think) that its negotiators did a bad job. The arguments on both sides are fallacious, but there is no need to go into them. When prolonged inter-country negotiations by teams acting under governmental briefings lead to a treaty, and the treaty is approved and signed at the highest levels, it must be presumed that it was the best outcome that could have been negotiated under the given circumstances; either side is then precluded from saying that it was unfair, unequal, poorly negotiated, etc. If a degree of dissatisfaction with the treaty arises in the course of operation of the treaty, that would be a matter for inter-country discussions within the ambit of the treaty. If a renegotiation of the treaty were now undertaken, each side will try to improve its position, and the outcome cannot be predicted. Good sense would lie in working the existing treaty in a spirit of constructive cooperation.

A more important criticism of the treaty is that it carried out a surgery on the river system, dividing it into two segments, one for Pakistan and one for India. It can be argued that dividing the river-system into two segments was not the best thing to do, and that the better course would have been for the two countries jointly to manage the entire system in an integrated and holistic manner. However, given the circumstances of Partition and the difficult relationship between the two newly formed countries, it would have been naïve to expect that such a joint integrated cooperative approach would work. Even now, with some improvement in the political climate, one is not sure that the vision of more enlightened cooperation on the river system will be easy to realise. If the best course is unavailable, then we have to settle for the second best; that is what the treaty represents.

While the water-sharing on the Indus system by India and Pakistan stands settled through the Indus Waters Treaty, its operation has been characterised by sharp differences between the two countries over the design and engineering features of several hydroelectric projects. In one case, namely that of the Baglihar project, the differences were settled through a reference to a neutral expert, but a fresh controversy has now arisen over the brief reduction of flows during the initial filling of the newly constructed reservoir. One must hope that this will get resolved soon. Differences over the Tulbul and Kishenganga projects still remain. These matters do not fall within the ambit of the present paper which is about water security and not about difficulties in the construction of hydroelectric projects. However, it must be noted that the persistent and seemingly intractable differences over the Indian projects on the western rivers are illustrative of a lower riparian's deep visceral anxiety (whether warranted or not) about perceived threats from upper riparian control.

India-Nepal

Turning to the prolonged and infructuous talks between India and Nepal over several projects, there are two questions here: the wisdom of undertaking such projects in the Himalayan region, and the difficult India-Nepal relations.

The postulation of several large projects in Nepal arises from the idea of the huge hydroelectric potential in the Himalayan rivers. There is, in fact, no such natural potential in a running river; it exists only in a falling river i.e. in a waterfall. In a running river, the hydroelectric potential is not natural but man-made: it is created by a dam. The statement that "there is a hydropower potential in the Himalayas" can, therefore, be translated as "there is a technical possibility of building dams". However, the potential for building dams means also a potential for ecological damage, human misery and possible disaster in the event of heavy floods. The dangers are particularly acute in the Himalayan region, given the friability and proneness to mass-wasting of the mountains, the huge load of sediment that the rivers carry and the added danger of seismic activity. While the project planners might claim that they have answers for all these problems, the precautionary principle would suggest that we leave the Himalayan rivers alone.

As for India-Nepal relations, these have been badly mismanaged on both sides. Confining ourselves to water, let us take a brief look at the past. The Kosi/Gandak agreements of the 1950s were not inspired by any large visions of 'regional cooperation'; they were essentially projects conceived by India to meet its requirements or solve its problems, with some benefits to Nepal included. That was the way (myopic, in hindsight) the projects were designed, with Nepal's agreement, but they were subsequently criticised in Nepal for conferring substantially more benefits on India than on Nepal, though this was inevitable, given the relative magnitude of cultivable areas in the two countries. The projects also suffered from poor design, inefficient implementation and bad maintenance (not to mention corruption); even what was promised was not delivered either in Nepal or in India. The Kosi/Gandak agreements, initially signed in 1954/1959, were amended in 1966/1964 to take care of Nepalese concerns, but the sense of grievance was not wholly removed. The bitterness generated by these experiences coloured all subsequent dealings between India and Nepal. Suspicion and mistrust grew and became a massive impediment to good relations between the two countries. The Indian handling of that difficult and complex situation can hardly be said to have been wise or sensitive. The Tanakpur episode made things worse. Eventually, a new chapter in Indo-Nepal relations seemed to open with the Mahakali Treaty of February 1996. Unfortunately, that treaty, signed after extensive consultations with a view to avoiding the mistakes of the past, has remained a dead letter, contributing to a worsening of India-Nepal relations rather than a dramatic improvement as had been hoped. That old acrimony has been revived by the Kosi floods of 2008. There were some brief mutual recriminations over the responsibility for the failure of an embankment in Nepal, but the exchange of blame seems to have given way to talk of cooperation. Once again, one hears talk of a high dam on the Kosi.

However, there has been a political change in Nepal, and a new government is in position. There is, thus, an opportunity for India and Nepal to forget the past and build a new relationship. It is for the two governments — in consultation with the civil societies of the two countries — to spell out the elements of that relationship. Let me merely say that constructive cooperation between the two countries can do enormous good to both, and that such cooperation can take other forms than large dam projects in the Himalaya.

India-Bangladesh

In the relationship between India and Bangladesh, the dispute over Ganga waters was for two decades an important component, perhaps the most important one; and though it now stands resolved by the Treaty of December 1996, its potential for re-surfacing should not be under-estimated. The dispute began with the planning and eventual construction by India of a barrage across the Ganga at Farakka for diverting a part of the waters of the river towards its Hooghly arm with the objective of keeping the Kolkata Port flushed and operational and protecting the water-supply of the city from the incursion of salinity. This caused serious concern in what was then East Pakistan and later became Bangladesh. Bangladesh regarded this as a "unilateral diversion" of the waters of the Ganga by India with severe adverse effects on it (Bangladesh); and felt that this was a case of a larger and more powerful country disregarding the legitimate interests of a smaller and weaker neighbour. That view became widely prevalent in Bangladesh, cutting across all kinds of divisions. A national sense of grievance grew and became a significant factor in electoral politics. In its extreme form, the nationalistic position became a myth, with India being cast in the role of a demon: whether Bangladesh was afflicted by drought or by floods, the responsibility was laid at India's door. Farakka was blamed for all kinds of ills. The dispute now stands resolved through the Ganges Treaty of 1996, but twelve years after the treaty, feelings still tend to run high over Farakka in Bangladesh, as if it continues to be a live grievance!

After a shaky start because of low flows in the river in the 1997 lean period, the Ganges Treaty seems to have worked to the satisfaction of the two governments in the ensuing years. The treaty provided for a review on request by either country at the end of five years, i.e., in December 2001, but neither country moved for a review at that stage, presumably because neither had any serious problem with the treaty. More recently, there have been some media reports about a review but the present official position is not known.

Agreements or treaties are needed on some of the other rivers (Teesta, Muhuri, Manu, Gumti, Khowai, Brahmaputra, Dharla and Dudh Kumar) that have been identified as important. Water sharing talks are proceeding on the Teesta river on which both countries have built barrages, but no breakthrough seems to have been achieved. Meanwhile, Bangladesh has developed an acute sense of anxiety over the implications for itself of the Indian River-Linking Project.

Whether the Ganges Treaty will continue to work well, whether treaties or agreements will be reached on other rivers, and whether an understanding will be reached on the River-Linking Project, are political questions and not water-related questions. It is common knowledge that the political relations between the two countries had been far from good for some years now. In recent years, attempts to improve those relations, appear to be succeeding. We must hope that India-Bangladesh political relations will continue to improve in the next few years so that water and other issues can be resolved.

India-Bhutan

With Bhutan, India's water relations, if that description fits the case, consist essentially in cooperation in the construction and operation of hydro-electric projects in Bhutan for the sale of the generated power to India. The price of electricity initially negotiated was rather low but that has been corrected by subsequent revisions. The revenues to Bhutan have been very large and have made that country richer than other South Asian countries in terms of income per capita. The India-Bhutan relationship, unlike the India-Nepal relationship, has been largely unproblematic. One phase of a project has led to another, one project to a second, and so on. The only question here is how this succession of projects in the Himalayan region, of larger and larger capacities, and the concomitant developments in terms of growing wealth and the inevitable intrusion of the outside world and modernity, are reconcilable with Bhutan's deep concern for ecology and heritage, its concept of Gross National Happiness and its pursuit of the middle path. However, that is a matter for Bhutan to reflect on, and I am sure that reflection and introspection are going on. This is not a question of India-Bhutan relations, and we need not spend more time on it.

China and the Brahmaputra

Until some years ago, water did not figure in the talks between India and China, but during the last few years, it has become part of the agenda. Unfortunately, there is not much to say. From time to time, there are media reports and articles about Chinese plans to divert the waters of the Brahmaputra (or Tsangpo) northwards, worrying people in India and Bangladesh, but there is hardly any actual and reliable information.

It must be noted that two different kinds of intervention are being talked about. One is the idea of a massive project, the world's largest, for the generation of hydroelectric power at the point on the river where it takes a U-bend before entering India. This may be feasible or not; it may have horrendous ecological consequences; but if the waters are returned to the river after they pass through the turbines, it may not affect the flows to India and Bangladesh. The other is the idea of a diversion project. If Brahmaputra waters do get partly diverted northwards, this will definitely reduce the flows correspondingly, and is, therefore, a matter of concern to downstream countries. We are told that there are no such official plans, but we cannot be too sure of this for several reasons.

Firstly, certain parts of China are indeed desperately short of water and the Chinese authorities have to provide water to those areas from some source. Secondly, there is plenty of water in the Tsangpo, and it may seem natural to the Chinese to consider diverting some of it to water-short areas. Does not some such thinking lie behind our own river-linking project?

Thirdly, while there may be no official decision and no 'project', the media reports cannot be dismissed as entirely without foundation. The idea of a diversion has indeed been mooted in academic papers. It must have been discussed in government circles too. Fourthly, the project may be gigantic and may present horrendous problems, but that is unlikely to deter the Chinese; it may even enthuse them. Fifthly, if they have decided that the national interest demands the south-north diversion of waters, they are unlikely to be unduly worried about international law or lower riparian concerns. However, and this is the sixth point, what *may* deter them would be political and foreign policy considerations; if they feel that good relations with neighbours are desirable for political reasons, then they may indeed take their concerns seriously and try not to upset them.

From the Indian point of view, the point to examine would be the quantum of possible diversion and the impact it would have on the flows to India. In the absence of hard information, we may have to work out possible scenarios. There is a view that any diversion by China will not affect us badly because precipitation further down contributes a good part of the river's waters. It seems to me that complacency on that ground would be dangerous. We shall have to keep questioning the Chinese constantly on their plans and expressing our concerns. This will have to be as important a part of the agenda for India-China talks as the border issue. We have to do our best to ensure that the Chinese do not undertake any major intervention in the river, or that in doing so, they keep us in the picture and take our concerns into account fully in the planning, construction and operation of the intervention.

It appears that the Ministry of External Affairs are seized of this matter and that the subject does figure in the talks with the Chinese. I can only say that it is necessary to be extremely watchful and take timely action, as there is not much point in complaining about reductions in flows after a dam has been built.

It is necessary to make common cause with the Bangladeshis on this matter. That may not be easy, as they may turn round and say that China is only doing to India what India has been doing to Bangladesh. We have to find ways of overcoming that awkwardness. A joint India-Bangladesh approach to China on this matter would be far more effective than separate approaches.

Corporate Control over Water

We now come to another kind of 'security' concern which does not often figure in security debates, namely, that of corporate control over water, a scarce natural resource. This arises in the context of the current trends of privatisation and globalisation. In international circles there has been a strong advocacy of private sector participation in water projects (reformulated as "Public-Private Partnership" or PPP to make it more acceptable), and of the privatisation of water services. In India, a major change from the National Water Policy of 1987 to that of 2002 was the introduction of a provision to facilitate private sector participation. The idea of privatisation has been spreading through the water sector loans that many state governments have negotiated with the World Bank.

Against that background, some important questions arise. For instance, can the state divest itself of the responsibility for the provision of water, leave it to market forces, and retain only a regulatory role? Moreover, what exactly does privatisation mean in relation to water? Can we distinguish between the privatisation of a *service* (i.e. the distribution of an allocated quantum of water) and the privatisation of the *resource* itself (i.e. allowing private entities to acquire a measure of control over some part of the water resource endowment through the construction of storage or diversion structures on rivers, or the installation of tubewells or borewells on groundwater aquifers)?

There are some difficulties even in relation to mere water supply, i.e. distribution. First, if the right to water as a basic essential for life is a fundamental right (or even a human right, as is widely believed), then the state presumably has the responsibility of ensuring that right. Is that responsibility adequately discharged by licensing and regulating private providers? In the event of a failure on their part, will not the responsibility revert to the state? Secondly, while the fact that water is a basic right does not imply that it should be supplied free to the middle and affluent classes, it may be necessary to subsidise supplies to the poorer sections, and perhaps some free supplies may be called for. No one should be denied this basic life-support requisite because he or she is unable to pay for it. How will the corporate private sector deal with this? It must necessarily operate on the basis of profit, and the pricing may put the water beyond the reach of the poor. Disconnections for non-payment of bills may result in the denial of this prime need. This is not a hypothetical question. Steep price increases following privatisation and large numbers of disconnections have actually occurred in some countries. What kind of 'water security' will the less affluent have under those circumstances?

Moreover, it is difficult to confine privatisation to the distribution service. Sooner or later, this will lead to the entrustment of the water-source itself, i.e. a part of the river or of an aquifer, to the distributor. In Chattisgarh, 20 km of the Sheonath river were leased to a private corporate entity for a water supply scheme. There was a public outcry against this, and the state government tried to cancel the arrangement, but that effort ran into legal difficulties. This is an unresolved issue. The Plachimada (Coca Cola) case is not one of privatisation, but involves a question of private corporate control over water resources. The villagers complained that their water resources were adversely affected by the powerful borewells of the Coca Cola company, and got a favourable judgment from the court; this was overturned by a Division Bench on appeal by the company; and the case is now before the Supreme Court. When the judgment comes, it may turn out to be a landmark one.

An important question that will need consideration in this context is whether allowing the domestic private sector to exploit national natural resources, particularly water, will not make it difficult to deny a similar right to foreign investors in terms of the World Trade Organisation (WTO) regime and the principle of "national treatment" of foreign investors, and if so, whether there is not a danger of our losing control over our own natural resources. This is not an alarmist view. In many cases of privatisation of water services, behind the national bidders stand major international corporations. There are some half a dozen giant corporations seeking control over significant proportions of the world's water resources. Vandana Shiva in this country and authors abroad such as Maude Barlowe, Tony Clarke, and Ricardo Petrella, have been writing about the dangers of such corporate control.

The present paper will not enter into the ideological aspects of this debate; it is concerned only with the security aspects. The 'security' concern here is three-fold: the life-security of the individual; the security of the community's traditional control over a common pool resource; and the security of the country's (or the state's) control over a vital natural resource. These security concerns need careful attention.

Water and Climate Change

Finally, water security needs to be considered from the perspective of climate change. The report on Climate Change and Water by the Technical Support Unit of Working Group II of the Intergovernmental Panel on Climate Change (June 2008) says, among other things, the following: "Areas in which runoff is projected to decline, face a clear reduction in the value of services provided by water resources. Increased annual runoff in some areas is projected to lead to increased total water supply. However, in many regions, this benefit is likely to be counterbalanced by the negative effects of increased precipitation variability and seasonal runoff shifts in water supply, water quality and flood risks." The implications of this for India are not quite clear. But one point needs to be made here, namely, that the impact that climate change will have on water resources is a subject on which all the South Asian countries will have to think, study, plan and act jointly in consultation and coordination, not nationally, in a fragmented fashion.

Comprehensive Security Challenges for an Emerging India

□ **Kapil Kak**

Comprehensive Security

In a conceptual approach to India's comprehensive security, the questions that often arise are: What is being secured? Whose security is at stake? What is the nature of the security objectives outlined by India's leadership before and after independence? In addition, today, over six decades later, what could be the comprehensive security determinants for India in the years ahead? These questions arise because security has come to acquire different meanings to the state, its people and social groups. Looking back, the concept of peace, and its linkage with comprehensive security, did not escape the attention of the visionary Jawaharlal Nehru, who, along with the leaders of the freedom movement led by Mahatma Gandhi, conceptualised the idea of India as a democratic, liberal, secular, pluralistic, inclusive and federal republic. Addressing a Youth Conference in Poona (now Pune) on December 12, 1928, Nehru said, "We shall have real security and stability… only when it has come to signify the well-being of the vast majority of the people, if not all, and not of small groups only."[1]

Unsurprisingly, the concept was to form a key component of the (National) Objectives Resolution moved by Nehru in the Constituent Assembly even before India framed and eventually adopted the Constitution in 1949. It included freedom, justice, equality, minority protection, territorial integrity of sovereign rights over land, sea and air, and India's rightful and honoured place in the world for it to make full and willing contribution to the promotion of world peace and the welfare of humankind. In the introduction to this volume, K. Subrahmanyam has justifiably defined comprehensive security to include "avoidance of shortage of basic requirements of a country's population: clean air, water, healthy surroundings, environmental security, food, healthcare, education, employment, old age care and good governance".

Hence, we see that while the state has, for centuries, been the sole referent of security, today, the individual and family, and civil society, have joined it in a matrix in which human security is perceived to becoming a co-equal of national security. Our approach, thus, has to take a broad and holistic view to incorporate the dimensions of socio-economic security and stability. This should augment traditional elements of diplomacy and the armed forces to seek partnerships in strategic, economic and technological dimensions, and territorial protection of the nation's frontiers, air spaces and maritime domains as also safeguard and project national interests through 'hard' and 'soft' power capacities.

One must not miss the linkage of comprehensive security with cooperative security in that one country alone cannot obtain and sustain the latter. In an interdependent world, today's crises — be it international terrorism, nuclear proliferation, financial sector meltdown, energy security or climate change — can only be resolved through significant cooperation of the bulk of the international community. The Tamil poet, Subramanian Bharati, was perhaps spot on when he enunciated the concept of *Vasudeva Kuttumbukan* (the whole world is a family), anticipating today's globalisation. A shift from a "state-centred, competitive approach to national security to a global cooperative framework linking peace, democratization, development and the environment"[2] would make for security acquiring both a comprehensive as well as a cooperative character. Creative forces of competition ought to be moderated by "cooperation and harmonization to derive the bounty of sharing enunciated in the *Rig Veda*."[3] Given India's post-independence record of seeking global peace as a primary security objective, at the present juncture of its rising international profile, India may perhaps be in the best position to lead the way.[4]

India and the World

India's Emergence

At the threshold of the second decade of the 21st century, a vibrant democratic India has a medium-size one trillion dollar economy, economic growth rate of 7-8 per cent, and the fastest growing market of a 300-500 million English-speaking middle class, with a scientific-technological orientation. These expanding economic and knowledge capacities, with a supportive military muscle, accompanied by a palpable power shift from North Atlantic-Europe to continental Asia, have facilitated the denouement of the industrialised world perceiving India's emergence as being benign. Not surprisingly, a RAND Corporation Study (1997), in recognition of this strategic transition, inferred: "For India, survival means survival as a great power, and security has become synonymous with the safety that enables India to develop, maintain and prosper in political eminence".[5] But as Henry Kissinger once said: "Each success only buys an admission ticket to a more difficult problem." For India, the 'difficult problem' does not arise in its dealings with the world, but in the increasingly iniquitous distributive mechanisms of the socio-economic growth and development process, with attendant governance and internal security dimensions. As we shall analyse later, these challenges, in the decades ahead, would be awesome and staggering.

Last Two Decades: Ordeal by Fire

To analyse the external security challenges that could confront India over the next two decades (2010-2029), a brief recapitulation of how it very successfully handled such challenges during the preceding two decades (1990-2009) could provide useful insights. Firstly, the seminal initiation and sustenance of economic reforms and liberalisation – which are now irreversible – yielded consistently impressive Gross Domestic Product (GDP) growth rates of near 7.5 per cent. Secondly, the conduct of the Pokhran II nuclear tests was a bold stratagem in the face of imminent widespread international opprobrium and economic sanctions. Thirdly, restrained yet successful military force application during the Kargil conflict, under a nuclear overhang, served to demonstrate India's determined yet fully responsible conduct as an emerging power.

Fourthly, India's effective coercive diplomacy through mobilisation of the armed forces, and unmanageable international pressure, in the wake of the Pakistan-sponsored attack on India's Parliament in 2001 compelled Pakistan to drastically reduce terrorist attacks in Jammu and Kashmir. This enabled the conduct of free and fair elections in that state and initiation of the India-Pakistan peace process, which, though currently stalled, could, and must, resume. Lastly, India successfully pursued hard-nosed national-interest driven negotiations with the US for the strategically seminal bilateral nuclear cooperation agreement of 2008. It not only exceptionalised India in the global nuclear order, despite not being a signatory to the nuclear Non-Proliferation Treaty (NPT), but also constitutes a major initiative towards diversifying energy sources. Throughout the aforementioned two decades, 1990-2010, despite different ruling parties at the Centre, India demonstrated resoluteness in pursuing and safeguarding its external strategies in alternate cycles of restraint and boldness – classical attributes of a rising power.

Global Configuration

Globally, an essentially 'polycentric' power configuration comprising the US, a rising China and India, and the European Union (EU), Russia and Japan would dominate. The United States National Intelligence Council (November, 2008) has identified the US, China, India and Japan as the world's four largest economies in 2025. Arvind Virmani in his essay (Chapter 2) as also Goldman Sachs, predict that the world's three largest economies in 2050 will be China, the US and India to constitute a "tripolar global structure." Others rightly aver that this "depends a great deal upon India pressing on with economic and structural reforms."[6]

The 'silver bullet' for acquiring huge advantages in the international system and enhancing strategic autonomy is for India to sustain an economic growth rate of 8-9 per cent over the next two decades. India would also need to stabilise its strategic neighbourhood, sustain military modernisation, induct and absorb strategic technologies and actively seek to influence multilateral groupings like the UN, G-20 (that produces 80 per cent of global GDP) and World Trade Organisation (WTO), etc in formulating global rules of the game that serve its interests.

Rivalry between the USA, the extant superpower, and China, the aspiring superpower, constitutes the greatest global security challenge. In Asia, the expected China-India contestation could also be a reality. However, India must continue to desist joining a US containment of China. Mutual concerns of the US and Russia in the Caucasus would persist. Despite its many limitations, Russia appears committed to reconfigure the post-Cold War shift in the balance of power in favour of the US. Given Europe's dependence on energy supplies, it is disinclined towards a Cold War against Russia, a perspective that India – with its multi-vector foreign, security and economic policy compulsions – would require to strongly support.

Regional Dimensions

As Sanjaya Baru argues so unquestionably in Chapter 3, "India's relations with major powers, neighbours and with other powers, especially in Asia, are shaped by the single-minded focus on improving the well-being and livelihood security of our people and in creating the external environment for India's long-term development." India's primary security arena is the geo-strategic space between the equator and 40 degrees north latitude and 30 degrees to 110 degrees east longitude. This includes "three concentric circles...Hindu Kush to Irrawadi, Aden to Singapore and Suez to Shanghai."[7] India must seek to contain, if not prevent, the growth of forces that could operate from these concentric arcs to its detriment.

In the aforementioned area, expeditionary military capacities, especially in the maritime and aerospace power dimensions, would require to be built up and sustained for force application when our national interests so dictate and India's expertise is sought by the UN and/ or willing coalitions. Needless to mention, such security contingency management may need to be worked out with regional organisations like the Gulf Cooperation Council (GCC), Association of Southeast Asian Nations (ASEAN), and Shanghai Cooperation Organisation (SCO), etc. The countries in this region of crucial security importance would need to work together to promote an economic and cooperative security architecture that is expandable for a larger polycentric Asia.

"Security of West Asia", as G. Parthasarathy rightly avers in Chapter 4, "is critical to India's energy security for uninterrupted supplies." The safety of four million Indians there who remit over $ 40 billion every, year is another dimension: 1,15,000 of them required airlift back home during the Gulf War in the largest airlift in world history after Berlin. Staying away from the Arab-Israeli dispute, India must engage the Arab world, and Israel and Iran in balance, and sustain the same. As to the East Asian region, the vision of a East Asian Community, run on the basis of ASEAN plus 6 (including India), as against the plus three formulation, offers creative opportunities. India must engage China, Japan, South Korea and other heavyweights to work together for a peaceful and secure Asia that is in tune with their aspirations. Uninterrupted trade and energy flow is their common objective; they could coordinate counter-terrorism, anti-piracy and disaster relief operations in the Indian Ocean region where India would obviously need to be in the lead.

Central Asia, the heart of the Eurasian continent, with substantial reserves of oil and gas, has begun to energise trade, stimulate regional cooperation and promote all round prosperity. The future security of these states hinges on the interplay in the strategies and actions primarily of the

US, Russia and China, and how astutely these states are able to balance the same. A moderate religious orientation of the regimes and governing elites remains India's key security objective for this region: their leadership perceives India as a role model and has often publicly advocated the need for it to secure the states and society from the threats of religious radicalism and terrorism. India's political, economic and military capabilities may require to be brought to the table to safeguard its vital stakes here. Given Pakistan's unwillingness to provide land transit to us, access through Iran and Afghanistan needs to remain a high priority, regardless of US reservations.

In Southern Asia, India would continue to be at the centre of a ring of fire with Pakistan-Afghanistan at one periphery and a potential strategic challenge from a rapidly rising China at the other. The Sino-Pak nuclear-conventional arms nexus, and a worrisome dyad of religious extremism-cum-international terrorism, with possible intersection of the latter with Weapons of Mass Destruction (WMD), poses formidable challenges to India, the region, and the world. Hence, military defeat of the virulent and resurgent Al Qaeda-Taliban-LeT (Lashkar-e-Tayyeba) combine alone can prevent exacerbation of India's security dilemmas. Shared India-US interests in the detoxification of the Pak-Af Indus-Hindu Kush faultline require to be balanced against Pakistan's obsessive and unwarranted apprehensions on this count.

A paradigm shift in India's approach to Pakistan, at a grave juncture in its history, may be an idea whose time has come. Beyond a point, coercion has its limits. This is not to suggest a rollback on the insistence that Pakistan prosecute the 26/11 conspirators and accomplices, or lowering of guard on India's nuclear and conventional deterrence, including punitive force application and covert capabilities, but a transition from an establishment-centric to an elected government-cum-civil society focussed approach. In such a dispensation, the latter would be progressively assisted to reconfigure Pakistan's internal DNA. Such a grand strategic tack change could help move Pakistan on to an entirely new plane. Reengagement both on the back-channels as also the composite dialogue track would serve to reduce tensions, enhance India's international heft, restart the stalled peace process, help move towards resolution of the Jammu and Kashmir issue (with rich internal security dividends), generate greater economic cooperation, foster people-to-people exchanges and mutual confidence that could result in eventual normalisation of bilateral relations.[8] Did the 'badly' (but craftily?) drafted Sharm el-Sheikh July 2009 joint statement envisage such a creative leap that ended in a thud because Indian public opinion had not been prepared for a strategic shift? Perhaps the time has come for Prime Minister Manmohan Singh to "Churchill-like, summon the Indian people to support a new policy that reflects the merger between Indian interests and Pakistani survival."[9] It would certainly be a long haul, but India must think, act boldly and persevere like a classical great power.

For the near term, Afghanistan's stabilisation remains the key to India's security, and the pursuit of its strategic interests. A moderate, neutral and independent Afghanistan—without interference from neighbours—should be India's objective. Hence, India cannot countenance the extremist-religious democracy-threatening Taliban dominance and linkages with Al Qaeda and LeT. Nor must it succumb to calls for shouldering any military burdens. "The need", as Satinder Lambah avers "is to isolate the Taliban and build the security forces of Afghanistan to a strength of about 4,00,000 at the earliest. Only then can a modicum of stability be expected to return to Afghanistan."[10] Also, since greater reach-out to the Pashtoons serves as the key, India's current $ 1.3 billion developmental assistance for infrastructure build-up and human resource investment requires to be stepped up, and remain focussed in Pashtoon-dominated areas where it has received widespread support. Furthermore, an effective engagement of Pakistan could help dilute its obsessions about India's reconstruction efforts not being a zero sum game. The interests of both India and Pakistan could be mutually reconciled, provided the merits of regional cooperative security that animates comprehensive security are at last sagaciously recognised by Pakistan's ruling elites. The five-fold growth in India-Pakistan trade from $400 million (2004) to $2.25 billion (2009) could serve as a significant pointer.

India also has vital stakes in other South Asian countries in its turbulent neighbourhood. Any socio-political instability would have inevitable spillover consequences on India's security. These states could do with abandoning counter-India policies and recognise that their growth, development and well-being are intrinsically linked to India as a growth hub. Sri Lanka is the first to derive value from such a realisation. In turn, India has moved away from its traditional reservations on the involvement of external powers like the US, EU, Japan, UN, etc which see in it the capability to promote democratic, pluralistic and secular values for greater regional stability. One could say that for India, the road to being a Great Power passes through Kabul, Islamabad, Kathmandu, Thimpu, Dhaka, Colombo and Male. India needs to (a) factor each country's sensitivities in the policy calculus; (b) leverage economic aid, and military and technological capabilities in pursuance of regional security; (c) induce greater attractiveness of economic integration for an eventual transformation to a common market; (d) encourage private sector investments; and (e) assertively enunciate India's security sensitivities and their non-negotiable status. In sum, India's neighbourhood must receive special attention in the decades ahead. The suggestion of a former Foreign Secretary to have a Minister of State in the Union mandated to "deal exclusively with neighborhood developments and relationships" appears a sound proposition.[11]

China's inexorable rise consequent to the dramatic geo-political and economic shifts unfolding in Asia would be India's most formidable challenge. As Jasjit Singh rightly observes in the opening chapter, the apprehension of a declining US aligning with a rising India has deeply riled China. Its multiple internal faultlines — political, economic and organisational — against the backdrop of a socio-economic pyramid said[12] to have a one per cent elite, 12 per cent middle class and 80 per cent peasantry/

working class (that reflects one per cent national income), generates deep insecurities in China's leadership. Perhaps the internal economic impact of global recession and increased restiveness in Tibet and Xinjiang compelled it to raise the level of tension with India during 2009.

To be sure, India and China have substantial convergence of interests as reflected in bilateral economic cooperation that has and must remain the *leitmotif* of cooperation. Equally, China could be a source of anxiety. Future competition over resources and influence would also animate this twin-dynamic. Some markers for a way forward could be (a) enhancement of trust and understanding and people-to-people contacts; (b) not allowing non-resolution of the territorial issue to cloud cooperation on other issues; and (c) accommodating each other's aspirations.[13] At the same time, by way of an insurance policy, India must remain committed to defence modernisation, including development of offensive trans-boundary capabilities and higher levels of nuclear and conventional deterrence. China has deep vulnerabilities in Xinjiang, Tibet and the Indian Ocean. The question is: what would be the circumstances under which India is compelled to leverage these vulnerabilities?

Defence Planning and Armed Forces

For defence planning to function as a key component of comprehensive security planning, the need for synthesised long-term assessments of the strategic environment needs no emphasis. These provide coherence to policies for meeting identified challenges and in evolving defence policy. In Chapter 5, N.S. Sisodia persuasively justifies defence reviews to assess threats and challenges, determine capabilities required and assist in a more rational resource allocation. Planning for worst case scenarios is clearly untenable. However, capabilities-based planning could assist in making cost-effective choices for building alternative capacities. Development of the defence human resource through a bold human resource policy that forms part of a national plan, hitherto neglected, is another key challenge. "This has a significant potential", as Vinod Misra cogently argues in Chapter 6, "in bringing down the substantial direct manpower and manpower-related costs, including defence pensions."

Given the competitive pressures on national resources to provide human developmental needs, defence budgets in the medium term are likely to hover around 2 percent of GDP or 12 percent of central government expenditure. Hence, a useful way forward would be utilisation of resources more effectively, faster decision-making on weapon system acquisitions and implementation of imaginative offsets regimes to leverage India's strengths as a major defence importer. The lesson, in true comprehensive security and defence, is to step up expenditure on R&D from 6 percent of the defence budget to 10-12 percent, revitalise defence Public Sector Undertakings (PSUs), and actively involve and encourage the burgeoning high-technology oriented private sector. Such a course would not only meet the rising modernisation needs of the armed forces but also boost industrial production and manufacturing. India can not hope to become a great power without a strong defence industry, notably in the high technology aerospace sector.

In an insightful and comprehensive piece (Chapter 8), V.P. Malik has justifiably drawn attention to the wide-ranging contribution of the Army in India's territorial integrity, internal security, nation building, disaster relief and many other spheres. These value additions to the nation's comprehensive security have achieved the success they have because of the Army's apolitical outlook, deep professionalism, its institutional strength and impeccable impartiality in communally sensitive internal security situations. Its meaningful bestowment to cooperative security (as a key ingredient of India's comprehensive security) through more than six decades of successful UN peace-keeping operations has received international recognition. In addition, situations could arise when vital national interests may compel India to undertake military expeditionary operations in its security arena either singly or as part of a coalition of the willing. The Army's "boots on the ground" would form a key component in such bi- or tri-Service force application scenarios.

Shifting gears to the maritime domain, Arun Prakash, in Chapter 10 has justifiably drawn attention to how India's continental mindset and landward orientation virtually invited invasion and domination by the sea-fairing European powers. He ominously warns that the country can ill afford such sea blindness ever again. Given India's utmost dependence on the Sea Lanes of Communication (SLOCs) in the ocean named after it – arteries that carry the lifeblood of its globally linked economy in trade, energy flows and raw materials, etc – their protection is vital to India's security. Nuclear and conventional deterrence, safeguarding and projecting national interests, protecting the SLOCs against interdiction by adversaries, non-state terrorist organisations and piracy, serve as drivers towards development and sustenance of a strong Navy. The strength would need to be in quality as well as quantity so that it is as good if not better than potential challengers like China are able to muster.

As to the third war-fighting dimension, the critical salience of air power – and with India's growing space profile, aerospace power – hardly merits emphasis. "Control and dominance of air and space places a nation in a situation of influence. The mandate is large; variables are many; when, where and how it is employed is important."[14] And an emerging India would be no exception, because air dominance, the *raison d' etre* of air power, could then allow the land and maritime forces to undertake their respective operations without what could prove mission threatening interference from the adversary's air power.

The key lies in integration of air and space with information and firepower. For years, aerospace power would remain the vanguard of coercion, and nuclear and conventional deterrence – an aspect that has also been so persuasively highlighted by T.M. Asthana in Chapter 9. On the other hand, technologies touch aerospace power the maximum. This special relationship requires constant leveraging because emerging technologies enhance every key attribute of aerospace power: surveillance and reconnaissance, reach, lethality, precision, survivability and strategic-cum-political effect.

India's challenges in the aerospace domain lie in a four-fold action plan. One, expand the strength of the Air Force to the sanctioned 45 combat squadrons. Two, exploit militarily the space and cyber domains and create dedicated organisations for the same, most of all dedicated military satellite systems. Three, identify technologies over which control is desirable, and develop the same indigenously. Four, strive to energise indigenous aerospace industry on the highest priority to build capacities in next generation combat aircraft, unmanned aerial systems, including combat variants, beyond visual range air-to-air missiles, cruise missiles, ballistic missile defences, hypersonic vehicles, scramjets and non-lethal directed energy/microwave weapons.

Nuclear Dimensions

In evaluating nuclear security, a fact that stands out is that nuclear weapons are progressively losing their Cold War salience. For India, as for other powers, these weapons would continue to serve as political tools of deterrence, and are neither usable nor can nuclear war be a feasible proposition. To be sure, India's armed forces would need to be constantly prepared for the unthinkable nuclear deterrence breakdown and the related criticality of survivability and appropriate retaliation. India's transformation of Pokhran II into a doctrinal posture of credible minimum deterrence, no first use and promise of 'assured' (later changed to 'massive') retaliation helped its "self-defence to ensure India is not subject to nuclear threats or coercion."[15] In Chapter 7, Manpreet Sethi has comprehensively analysed these major doctrinal attributes.

Evidently, the China-Pakistan nuclear-missile collaboration and their 'two front' conventional military modernisation would continue to remain India's key challenge. However, it is the impact of nuclear weapons on conventional war that merits careful analysis lest the adversary be compelled to breach the nuclear threshold. This imperative makes limited war — limited in political objectives, geographical scope, resources fielded, extent and duration — confine land forces' offensives to the vicinity of 'borders' or focussed punitive air strikes or limited outcome-centric naval engagements or selective special forces employment, the normative framework for conventional force application.

Recent US calls for freeing the world of all nuclear weapons are unarguably national security driven for fear of proliferation making the US more vulnerable to nuclear attacks by terrorists or rogue states. In conformity with its traditional stance, India must whole-heartedly support global, comprehensive and non-discriminatory nuclear disarmament. However, it is in the quicksands of non-proliferation and arms control that India would face major challenges. It cannot sign the NPT other than as a nuclear weapon state. As to the two subsidiary non-proliferation agreements, the Comprehensive Test Ban Treaty (CTBT) and Fissile Material Cut-off Treaty (FMCT) — that are unlikely to proceed beyond the "business as usual pace" — a more nuanced calibration is called for. If the US Senate ratifies the CTBT, India could reconsider its traditional opposition because this treaty has little impact on India's weapons status or nuclear arsenal. Likewise, if the US drops its opposition to the verification clauses of the FMCT, India could join, provided adequate fissile material both in the headstock and for weapons and reprocessing facilities is available.

Internal Security

Internally, at a broader conceptual level, India could be rightly termed a unique nation-building experiment of the 20th century in which hundreds of million people — deprived, discriminated and dispossessed over millennia — have been empowered through the press of a key on an electronic voting machine. They have found new hope in the socio-political churning of the democratic process that has thrown up huge opportunities. Such democratic and peaceful transitions have a reverse side: erosion in law and order, widespread corruption, criminalisation of polity and other societal ills that weaken the polity from within. Ironically, when the economic growth rate was hovering at 3 per cent, India's institutions were strong, but high growth rates over the last two decades have witnessed progressive weakening of institutions.

Continued lack of effective governance and inadequate enforcement of rule of law remain India's foremost challenges, for, as the epic *Mahabharata* suggests: "Governance is rooted in truth, people are rooted in governance; a society that neglects rule of law would soon have a law of the jungle." Political and bureaucratic corruption is rampant in India because of the role money and muscle power play in elections, thus, increasing the vulnerabilities of a corrupt and misgoverned polity. Tracing the evolution of left wing extremism and its causes, Prakash Singh, so perceptively avers in Chapter 12 that poor governance, endemic poverty, unemployment (increase from 7.32 per cent in 1999-2000 to 8.28 per cent in 2004-2005), near absence of land reforms and alienation of tribals have contributed substantially to this scourge. Ruling dispensations must not be tempted to believe that these systemic weaknesses can be left alone until hard security threats have been resolved.

There have been appalling failures on a number of other key socio-economic thrust areas — food security, nutritional levels, infant mortality, healthcare and primary education — in the development of comprehensive well-being and security. According to a Planning Commission document on the Tenth Five-Year Plan (2002-2007), 260 million people (about 24 per cent) did not have the income to access a consumption basket that defines the poverty line. More astonishingly, as a former Governor of the Reserve Bank of India, Bimal Jalan, stated in 2009, "The total value of the assets of the country's five billionaires equaled those of the bottom 300 million people". Later, in December 2009, an expert group led by Suresh Tendulkar, former head of the Prime Minister's Economic Advisory Council, estimated that roughly a quarter of the country's urban population lives on Rs 19 a day, while close to 42 per cent of the rural population consume goods and services worth Rs 15 a day. Thus, a sizeable proportion of India's population remains mired in abject and dehumanising poverty. The education system appears reasonably sound for the top 20 per cent, but it is abysmal for the rest. India cannot join the global knowledge economy as a leading player without a root and branch reform of its primary, secondary and higher education system.

Advances in Information Technology (IT) and reach of the media tend to aggravate the perceived expectation-actuality mismatch, leading to assertion of ethno-sectarianism, incipient sub-nationalism and religiosity, which, in turn, generate an environment of entrenched violence, militancy and terrorism. Jammu and Kashmir, the northeastern states and left wing extremism are instances in point. Unless we find ways to forge policies that address these serious shortcomings, effective internal coherence, a key comprehensive security objective, would remain a mirage in the midst of high economic growth. Kautilya's sagacious exhortation remains as valid now as in 350 BC: "Getting what has not been got, guarding it, developing it and then distributing it – these four constitute state policy." Government intermediated resource distribution would, thus, remain a key comprehensive national security imperative in the many years ahead.

As to having in place effective counter-terrorism and counter left-wing extremism strategies and policies that are a national level coordinated mix of long-term and a shorter/medium-term focus, India has not achieved much success. This combination would involve preventive socio-economic strategies, political grievance settlement and strengthening of national inclusive ideologies; integrated collection, assessment and dissemination of preventive intelligence on planned terrorist and Naxal attacks: high-quality physical security measures; and, readiness states for consequence/crisis management. In a penetrating and insightful analysis (Chapter 11), B. Raman justifiably places great value on "smart counter-terrorism." This would include the political class forging a consensus on eschewing use of terrorism as a political/electoral tool. Coordinated and well-synchronised actions by different elements of the counter-terrorism community are another aspect. Besides, completely revamped counter-terrorism strategies, policies and systems are among many other doable policy changes he recommends.

Deeper fundamental socio-economic stirrings that cause insurgencies, religious extremism and left wing militancy need objective analyses and actions. Lack of effective governance and tardy enforcement of rule of law against commercial groups that exploit the poor in far-flung isolated areas sharpen the challenge. There is need to address on priority the problem of major shortage of judges. India has 1.2 judges against the worldwide norm of 11.39. Adverse population-police ratio, 30-40 per cent deficiencies in the police establishment in states, poor training levels of one million Indian police personnel, their woeful infrastructure, accumulated capacity deficits and lack of Centre-state coordination exacerbate the problem. The spread of the Maoist-left wing uprising is emblematic of the deep malaise, compelling India's Prime Minister to term it as the most serious internal security threat that India faces.

To secure India against internal security threats, including mass casualty terrorist attacks, there is no alternative but to achieve higher levels of preparedness, both institutionally and at the level of government.[16] This would entail enhanced capacities in situational awareness and more optimal use of existing ones, including countering WMD and IT use for destructive purposes. Rebuilding covert capacities for offensive and deniable strikes against perpetrators of terrorism, as options short of military force application, could be another area. Strengthening national inclusive ideologies to de-radicalise homegrown terror groups and prevent radicalisation of others could be another thrust area.

Measures initiated post-Mumbai terror attacks, like the creation of the National Security Guards (NSG) hubs, greater control over intelligence coordination, assessment and dissemination; improved coastal/maritime security, setting up of a National Intelligence Agency and the planned creation of a National Counter-Terrorism Centre by end 2010 comprise a way forward. Likewise, the government could also gainfully set up the office of Director, National Intelligence, to control, task, interrogate, quality-audit and coordinate the functioning of all intelligence agencies.

Amidst other options, we could consider a dedicated internal security adviser to beef up the expertise in the National Security Council.

Key Non-Traditional Dimensions

Right to Information

With the challenge of removing socio-economic inequities being key work-in-progress for about two decades, India, meanwhile, could gainfully leverage information as an instrument to achieve people-centric comprehensive security. Wajahat Habibullah, reportedly played a key role in the conceptualisation and drafting of India's Right to Information Act (RTI) (2005). In Chapter 13, he very perceptively addresses the perennial question of ethno-national versus diversities-oriented inclusive state, liberty versus security and whether democracy negates security. He analytically brings out the huge security benefits the Act could provide to ordinary citizens.

The key to comprehensive internal security, in which people are the objective, lies in decentralisation of political and financial power for grassroots empowerment, development of local self-government (Panchayat Raj) and narrowing of the gulf between the people and the state. For instance, a diverse but culturally composite state like Jammu and Kashmir has witnessed religious, regional and ethnic radicalisation due to the two decades of conflict. Unhindered access to information would help address this challenge. The holding of political leadership and bureaucracy to sharp public scrutiny would serve to enhance accountability, eliminate alienation and foster in each citizen a sense of participation in governance.

The RTI Act will, in the end, as Wajahat Habibullah cogently asserts, "become a major tool in strengthening security, and restructure the debate on governance from what should be revealed to what must be kept secret". To meet "hard security" needs, Section 8 lists information that is exempt from disclosure. Significantly, the RTI Act takes precedence over the Official Secrets Act (1923), a colonial era relic, when the public was treated as an adversary. That an emerging India continues to be governed by this Act, as also the National Police Act (1862), speaks volumes about the glacial pace of work of India's political leadership and bureaucracy!

Energy

The key to India's energy security lies in assurance of continued availability of energy sources that are diverse, environment-friendly, available in sustainable quantity and at affordable prices, and freely accessible. Carbon-based fossil fuels (crude oil, natural gas and coal) account for 85 percent of the energy used by humankind. India is the fifth largest oil consumer in the world, and it currently imports 75 percent of its crude oil needs that are set to increase to 90 percent by 2020. The demand for energy is set to grow exponentially as India sustains high growth rates and modernises in the midst of a rural-urban movement that would create a further demand surge.

Alternatives like hydro, biomass, wind, tidal, solar and nuclear power, even in combination, are unlikely to replace oil for another 30-40 years. Besides, hydrocarbon resources are located in potentially unstable regions. Energy security, thus, intersects multiple security dimensions — geo-politics, foreign policy, environment and nuclear proliferation — that need integration. Moreover, linking sub-sectors like coal, petroleum, natural gas, renewable energy, transportation, power sector utilities and market driven energy pricing have to be part of the integration process. Some voices even suggest whether having a single Union Ministry of Energy, as in the US, would not serve India's long-term energy security interests better.

In Chapter 14, Shebonti Ray Dadwal has analysed in detail the supply side energy challenges in the oil and gas sector. In essence, these comprise five thrust areas. One is to step up indigenous exploration and production onshore, in offshore shallow waters and in the deep seas. Two, encourage and provide incentives to national oil companies and the private sector to acquire oil and gas blocks overseas on a large scale. Three, evolve import strategies focussed on diversification of sources of supply. Four, concentrate on energy pipelines as a means to reduce vulnerabilities of sea-based imports. (The launch of the 4,200-km Eastern Siberia-Pacific oil pipeline on January 1, 2010, that would enable Russia to ship oil to not only its traditional customers in Europe but also the ever-growing energy markets in East Asia, demonstrates the salience of energy pipeline geo-politics.)[17] Five, create and sustain adequate strategic crude reserves as a hedge against uncertainty and volatility. On the demand side, conservation in energy use, while highly desirable, has obvious outcome limitations. But as environmental concerns would serve to constrain India from exploiting its reserves of coal and hydropower, it may have no alternative to according nuclear power and renewable energy a high strategic priority now and in the future.

Eliminating shortages of power, especially electricity, to drive India's economic growth would be work-in-progress for decades – energy requirement doubling by 2020, doubling again until 2030 and being ten-fold of 2010 levels by 2050. In a comprehensive evaluation of the thermal-oil-nuclear energy landscape, Manpreet Sethi (Chapter 15) has cogently projected the cost-economics and environmental sustainability of adopting nuclear power as a key component of the future energy mix. This is in line with the global trend, more so due to concerns on global warming and climate change. The US and even the earlier Euro-sceptic countries are increasingly taking this path. In his 2010 State of the Union address, US President Barack Obama called for "nuclear renaissance." While French President Nicholas Sarkozy appealed for the developing world to seek "radioactive communion." China's nuclear power growth today is the highest and as Thomas Friedman avers, "It is expected to build some 50 new nuclear reactors by 2020; the rest of the world combined might build 15."[18]

India has the technological expertise in nuclear reactor design and development and related fuel cycles. Recent civil nuclear agreements with the US, Russia, France and the UK could further uplift the trajectory, facilitate export of smaller reactors, import of larger ones and help tide over delays in moving to the thorium cycle. In assessing the potential of the thorium cycle by 2050 there are vast differences in projections. The Prime Minister quoted 4,70,000MWe against the Kirit Parekh Committee's most optimistic figure of 2,75,00 Mwe.[19] However, unless the private sector is leveraged, even the modest installed capacity increase of MWe 20,000 (2020) and 63,000(2030) sought from the existing 4,120 MWe, would pose formidable challenges. Private sector participation would not be a choice but a vital necessity.

As to new and renewable energy, Ashok Parthasarathi, in Chapter 16, offers illuminating techno-strategic insights on current and emerging technologies. Key sub-sectors are wind power, solar photovoltaic systems, biomass, bio-fuels and fuel cells. Evidently, resource assessment, technology development and demonstration, and commercialisation would drive R&D. More importantly, political, legislative and infrastructural policies need frequent recast to stay abreast of the dizzy pace of technological transformation in renewable energy.

India would need to boost up R&D in hydrogen and fuel cell technologies as a source of primary power for transportation, apart from solar water heating and remote village applications currently at hand. The success story of India's electric car, Reva (targeted output 30,000 units in 2010-11 against 6,000 in 2009-10), mirrors the entrepreneurial vibrancy of India's private sector in dynamically leveraging emerging technologies. Trend lines suggest that a breakthrough in the employment of satellites to transmit perennial solar energy to the earth through microwaves may not be too far on the horizon.

Food

In the whole world, India has the largest number of malnourished men, women and children, the most malnourished being the landless labour and farmer consumers. Clearly, physical, economic and social access to food must remain the fundamental right of every Indian, even as the food security focus remains on indigenous production that is reliable and affordable." There cannot be import dependence for feeding over one billion people because of the vulnerability India would develop geo-politically".[20] By 2050, India requires to double its agricultural production. Further, unless the agricultural growth rate is stepped up and sustained at 4 per cent, the envisaged 8-9 percent GDP growth rate would remain a mirage. The hugely beneficial impact on 52 per cent of India's workforce that is still dependent on agriculture for livelihood and which would, in turn, drive demand in the manufacturing sector, is obvious. Significantly, the contribution of agriculture to GDP has fallen to 15.7 per cent.

To sustain food and nutrition security, Dr M.S. Swaminathan, the doyen of India's agricultural science community, in Chapter 17, unfolds a convincing multi-pronged strategy. This includes the critical need to enhance farm productivity in perpetuity without ecological harm, to serve as an "evergreen revolution", strengthening production systems under uncertain weather patterns, reducing wastage, providing credit support and giving a boost to post harvest technology and food processing. Food worth Rs 58,000 crore (40 per cent of 2010 defence budget) is wasted every year in the absence of grain storage and cold chain infrastructure for perishable farm produce. This calls for quantum increase in public investment in agriculture infrastructure and R&D, and measures to step up productivity and profitability in animal husbandry, horticulture, sericulture, etc. Such measures would serve to provide multiple livelihood opportunities to farmers, beef up their coping capacities during droughts and floods, and help in reducing their suicide rates.

Climate Change

It would be correct to say that climate change would pose for India non-sequential security challenges that are incremental, imperceptible and surrounded by uncertainty. "Changes in climate are not limited to an increase in temperature, but, in fact, involve several impacts such as increase in intensity and frequency of floods, droughts, heat waves and extreme precipitation events."[21] There are even obvious hard security implications as the armed forces' widespread infrastructure would be under threat even as dependence on them to provide humanitarian disaster relief, both nationally and internationally would increase. Reduced availability of water, arable land and food, and sea level rise, that could further exacerbate illegal migration, could generate societal turbulence and conflict.

Chandrasekhar Dasgupta, who has represented India in climate change negotiations since 2002, in Chapter 18, has provided an insightful perspective on its wide-ranging consequences on national and international security. Scripting a range of scenarios, he has sought to drive home how important it is to have international cooperation on this critical global security issue on all-are-equal basis without any country or group of countries claiming any exclusive right.

For the future, four points merit consideration. One, a large developing country like India cannot afford to de-link greenhouse gas emission cut commitments from financial and technological assistance from developed countries. Two, under the equity principle enshrined in the Kyoto Protocol, India must not abandon its known stance on per capita emissions as the developed world, the leading polluter, has already reaped the benefits of rapid and widespread industrialisation. Three, while it may be desirable to cut energy intensity of India's economy (2005) by 20-25 percent within the timeline of 2005 by 2020, this should not "involve needless pressure on itself" and be at the cost of developmental compulsions. Four, since India is still in the early stages of its development trajectory, it is "better equipped to demonstrate a low carbon lifestyle, which other societies could emulate." [22]

Water

Severe water stress (along with environmental degradation) could produce socio-political outcomes that threaten the survival of India's population and its effectiveness as a state. The 2005 World Bank Report "India's Water Economy: Bracing for a Turbulent Future" warned of a looming water crisis and declining per capita availability. No town or city in India gets uninterrupted 24-hour supply of water even as 70 per cent of irrigation water and 80 per cent of domestic water comes from ground water that is fast depleting. The expected 8 per cent GDP growth rate would impose further strains on available fresh water resources. In a perspicacious analysis of both demand and supply side projections, Ramaswamy R. Iyer, in Chapter 19, appears more sanguine and, quoting the Report of National Commission on Integrated Water Resources Development Plan, recommends a slew of water augmentation and conservation measures. These include higher water management efficiencies in agriculture and cropping patterns, enforcement of equities in domestic per capita distribution patterns, multiple recycling and reuse and a prudent combination of rain water harvesting, ground water drilling and storage projects.

Water would continue to be a source of contention between nations, more so when bilateral relations are under strain. India shares water systems with China, Nepal, Bhutan, Bangladesh and Pakistan. Excepting Bhutan (which has the highest per capita GDP in the region thanks to India's investments in its hydropower sector), water-irritants with others persist. As to China and Pakistan, water did not feature in bilateral talks with the former until a few years. However, since 2009, Pakistan has unjustifiably begun to perceive the Indus waters as the 'core' issue.

The reported Chinese plans to undertake major interventions on the river Tsang Po and divert its waters near the northern loop before it enters India (as the Brahmaputra) require constant monitoring. The issue also merits inclusion in the bilateral agenda. As to the India-Pakistan Indus Waters Treaty (1960) — that accorded 80 per cent of overall flows to Pakistan and 20 per cent to India — the difficulties persist. The rising population in Pakistan, wasteful water management in a situation of reduced flows and over-utilisation in West Punjab cause disgruntlement in the downstream provinces of Baluchistan and Sindh. To deflect this resentment, Pakistan, expectedly accuses India of violating the treaty, unmindful of the fact that it has a redressal mechanism that has stood the test of time. Water stress affects both India and Pakistan. Environmental degradation would exacerbate it further. For comprehensive security, cooperation should, thus, be a byword not only between India and Pakistan but with China as well.

Conclusion

Jawaharlal Nehru enunciated the concept of comprehensive security nearly six decades before Barry Buzan and others of the Copenhagen School in 1983 argued for the inclusion of non-military, societal, economic and environmental issues. Cooperative security has its own salience, for contemporary global crises can only be resolved through the significant cooperation of the international community. This community has not viewed India's emergence on the regional and global scene with disfavour. This has been quite unlike the earlier rise of powers in history, including the extant case of China.

To be sure, India would continue to face challenges from China and Pakistan, individually and in collusion, for which necessary political and military strategies require effective implementation. A paradigm shift in India's approach to Pakistan, at a grave juncture in its history, may be an idea whose time has come. This requires dexterous calibration, build-up of domestic public opinion and safeguards against the emergence of wild cards. India must also stay resolute in sustaining and perhaps stepping up developmental assistance to Afghanistan for safeguarding its traditional strategic interests there.

Key strands of India's security policy are to sustain a 8-9 per cent growth rate, stabilise the strategic neighbourhood, sustain higher levels of credible nuclear and conventional deterrence, induct and absorb technologies, and proactively seek to influence multilateral groupings. Economic integration of countries in India's turbulent neighbourhood, the security of West Asia, the vision of an East Asian Community, the stability of Central Asia, the heart of the Eurasian continent — all areas with various levels of vulnerabilities — must remain key priorities.

In the nuclear non-proliferation realm, while India can be a party to the NPT only as a nuclear weapon state (P5 or P5+1), it could sign the CTBT, provided the US and China ratify it. As for the FMCT, perhaps a more nuanced calibration may be necessary. For, should the US drop its opposition to the verification clause, and adequate fissile material, both in the headstock and for weapons and reprocessing, is available, India must not hesitate to join the treaty. In nuclear deterrence, regardless of probability, the system must be prepared for the 'unthinkable' nuclear deterrence breakdown scenario and criticality of the survival of retaliatory capacities. Further, the changing nature of conflict and impact of nuclear weapons on conventional force application must merit a review of doctrines, appropriate combat capacity build-up of land, maritime and aerospace forces, and the related technological intensities and inductions. Three imperatives of defence planning, that of strategic defence reviews, faster decision cycles for acquisitions and establishment of a strong defence industry, notably in the high technology aerospace sector, with a vibrant private industry and the entrepreneur as the key enabler, must receive the highest attention.

For internal security, India's critical faultlines lie in continued socio-economic inequities and near absence of governance that drive internally inspired terrorism and left wing extremism. It can ill-afford 30-35 per cent of geographical area in about 200 districts in 16 states being under Maoist ideology. Ineffective governance would constitute a colossal impediment in India's emergence, and needs fixing on a war-footing, even as left wing extremism and 'internal' militancy demand sensitively conceived force employment. Further, the people's grassroots empowerment, decentralisation of political and economic power, and more effective implementation of the Right to Information Act remain a way forward.

To sustain high GDP growth, continued access to energy sources that are diverse, environment-friendly and price-affordable is a national imperative. Nuclear power, along with new and renewable energy, must be a key priority for our clean and green energy strategy. As to food security, an emerging India, that has the world's largest number of malnourished men, women and children, is *non sequitur*. Perhaps strict implementation of the policy quartet of 4 per cent agricultural growth rate, greater public investment in agro-infrastructure and related R&D, enhancement of 'ever-green' farm productivity and setting up of a nation-wide post-harvest cold chain could enable 52 per cent of the agriculture-dependent rural populace to also 'emerge' with the rest of India.

Evidently, unmitigated climate change could reduce availability of water, arable land and food. Given the gravity of the problem, international cooperation on this critical global security issue has to be on all-are-equal basis. India needs to take the lead on adopting a low carbon lifestyle ,but emission cut commitments (on per capita basis) need to be inextricably linked to financial and technological assistance from the developed countries which have derived the benefits of rapid and widespread industrialisation. Further, India's voluntary cut-backs in energy intensity of the economy should not involve needless limitations on its developmental compulsions.

In a yet another dimension of comprehensive security, severe water stresses, should these occur, could produce unacceptable socio-political outcomes and make this natural resource an issue of confrontation in Southern Asia. Nationally, there is need to initiate multiple water augmentation and conservation measures. In bilateral terms, the problems of unimpeded flow of waters for growth and economic development in Southern Asia could serve as the ideal issue for resolution by India, China, Pakistan, Bangladesh and others through comprehensive and cooperative security. However, the key lynchpin must be strict adherence to time-tested international riparian laws, conventions and bilateral agreements.

Notes

1. S. Gopal, ed., *Jawaharlal Nehru: An Anthology* (New Delhi: Oxford University Press, 1980), p. 242.
2. *Report of the International Commission on Peace and Food: Uncommon Approaches* (London: Zed Books, 2005) for comprehensive security.

3. M.R. Srinivasan, "Technological Inequalities" in Jasjit Singh, ed., *Peace in the New Millennium* (New Delhi: Knowledge World, 2002) translates a few lines of the Rig Veda thus: "*Let us walk together, let us speak with one voice, let us think alike.*"
4. For a more detailed narrative on comprehensive and cooperative security in the external dimension, notably in Southeast Asia, see S.T. Devare, *Towards Security Convergence* (New Delhi: Central Publishing Company, 2003).
5. RAND Corporation, *Ugly Stability in South Asia* (Santa Monica: RAND Corporation, 1997), p.82.
6. Mohan Guruswamy and Zorawar Daulet Singh, *Chasing the Dragon: Will India Catch Up with China?* (New Delhi: Pearson, 2010), p.4.
7. K. Shankar Bajpai, "Our Grand Strategy," *The Indian Express*, January 1, 2010, p.10.
8. The strategic merits of early resumption of full engagement with Pakistan have been espoused by many Indian security commentators, of whom the principal ones are Shekhar Gupta, "No Silence Please," *The Indian Express*, January 9,2010, p. 12 ; C. Raja Mohan, "Learning to be Happy," *The Indian Express*, January 21, p.12; Pratap Bhanu Mehta, "Gamble on Pak, not Af,", *The Indian Express*, December 10, 2009, p. 12; Siddhartha Varadarajan, "Hard Line Diplomacy is not Homeland Security," *The Hindu*, December 15,2009, p. 10 ; Kanti Bajpai, "Look Nearer Home," *The Times of India*, January 21, 2010, p. 20.
9. Rajmohan Gandhi, "A Merger of Interests", *Hindustan Times*, January 2, 2010, p. 12.
10. S.K. Lambah, Indian Prime Minister's Special Envoy to Pakistan and Afghanistan proffered these views during his Inaugural Address at the International Seminar on *Peace and Stability in Afghanistan* organised by the United Service Institution at New Delhi on October 6-7, 2009.
11. Maharaja Krishna Rasgotra, "Hasina's Visit to India", *The Tribune*, January 9, 2010, p. 8.
12. Alka Acharya at a round-table discussion on "China's Rise to World Power Status" organised by the Centre for Land Warfare Studies at New Delhi, on September 24, 2009.
13. For a thoughtful expatiation of elements driving an India-China cooperative security construct, see the text of Indian Ambassador S. Jaishankar's speech at Sichuan University on December 14, 2009 excerpted in *The Indian Express*, December 15, 2009, p. 11.
14. Air Chief Marshal F.H. Major, former Chief of the Air Staff, during Valedictory Address at a seminar on "Future of Air Power" organised by the Centre of Air Power Studies at New Delhi on February 2-3, 2010.
15. Statement of Prime Minister Atal Bihar Vajpaye in Parliament on May 27, 1998. For details, see *Strategic Digest*, July 1998, p. 892.
16. For a comprehensive examination of the key issues of terrorism and Naxalism that impact India's internal security, see "The Big Threats", *India Today* (cover story), November 9, 2009, pp. 37-64.

17. For details of the ESPO pipeline, see Vladimir Radyuhin, "Changing the Rules of the Energy Game", *The Hindu*, January 12, 2010, p. 8.
18. Thomas L. Friedman, "The Green Leap Forward", *The New York Times*, reproduced in *The Indian Express*, January 11, 2009, p.10.
19. See Editorial, "Nuclear Power Still Locked In," *Hindustan Times*, October 2, 2009, p. 12.
20. Jayshree Sengupta, "The Big Challenge of 2010", *The Tribune*, January12, 2010, p. 8.
21. R.K. Pachauri, "Challenge of Climate Change, Post Copenhagen, *The Hindu*, February 1, 2010, p. 8.
22. For a few discerning and articulate perspectives on water issues, notably in the India-Pakistan context, see T.N. Krishnan, "Water and India's Constitution", *The Hindu*, January 25, 2010, p. 11; Ramaswamy R. Iyer, "Water, Aspirations and Nature", *The Hindu*, January 28, 2010, p. 9; Siddharth Varadarajan, "Water as a Carrier of Concord with Pakistan", *The Hindu*, February 25, 2010, p. 9; and, B.G. Verghese, "Do Pakistan's Claims over the Indus Hold Water? *The Indian Express*, March 12, 2010. p.15.

Index